The Emergence of Net-Centric Computing:

Network Computers, Internet Appliances, and Connected PCs

Bernard C. Cole

Prentice Hall PTR
Upper Saddle River, New Jersey 07458
http://www.phptr.com

ISBN 0-13-897869-7

Library of Congress Cataloging in Publication Data

Cole, Bernard Conrad.
    The emergence of net-centric computing   :   network computers,
    Internet appliances, and connected PCs   /   Bernard C. Cole
        p.   cm.
    Includes index.
    ISBN 0-13-897869-7
    1. Network computers.   I. Title.
    QA76.527.C65   1998
    004.6'186--dc21   98-34641
                                                              CIP

Acquisitions Editor: *Bernard M. Goodwin*
Editorial/Production Supervision: *James D. Gwyn*
Cover Design Director: *Jerry Votta*
Manufacturing Manager: *Alan Fischer*
Marketing Manager: *Kaylie Smith*
Editorial Assistant: *Diane Spina*

 © 1999 by Prentice-Hall PTR
Prentice-Hall, Inc.
A Simon & Schuster Company
Upper Saddle River, New Jersey 07458

Prentice Hall books are widely used by corporations and government agencies for training, marketing, and resale. The publisher offers discounts on this book when ordered in bulk quantities. For more information, contact: Corporate Sales Department, Phone: 800-382-3419; FAX: 201-236-7141; E-mail: corpsales@prenhall.com. Or write: Corp. Sales Dept., Prentice Hall PTR, 1 Lake Street, Upper Saddle River, NJ   07458

Printed in the United States of America
10   9   8   7   6   5   4   3   2   1

ISBN 0-13-897869-7

Prentice-Hall International (UK) Limited, *London*
Prentice-Hall of Australia Pty. Limited, *Sydney*
Prentice-Hall Canada Inc., *Toronto*
Prentice-Hall Hispanoamericana, S.A., *Mexico*
Prentice-Hall of India Private Limited, *New Delhi*
Prentice-Hall of Japan, Inc., *Tokyo*
Simon & Schuster Asia Pte. Ltd., *Singapore*
Editora Prentice-Hall do Brasil, Ltda., *Rio de Janeiro*

DEDICATION

To Toni, who has been a better friend to me
Than I have been to her, and who taught me the
Value of putting my head under my pillow
in times of crisis.

In Memoriam. As I was doing the final edits on this book I learned of the death of Dan Hildebrand. As senior software architect at QNX Software Systems Ltd. he played a significant role in the design of advanced microkernel operating systems. Dan had a unique ability to take technical knowledge and explain it in clear, understandable terms. A talented engineer and technical communicator, his articles have appeared in numerous technical and popular computing magazines. He was also an active user of the Internet and World Wide Web, particularly the various Usenet forums relating to his wide range of interests.

Conversations with him over the years provided some of the ideas for the genesis of this book. He was an enthusiastic believer in the transformation of the Internet and World Wide Web into a medium of intellectual interchange for the average citizen. I am sad that he will not be here to see that many of his prescient ideas regarding the emergence of net-centric computing are coming to fruition.

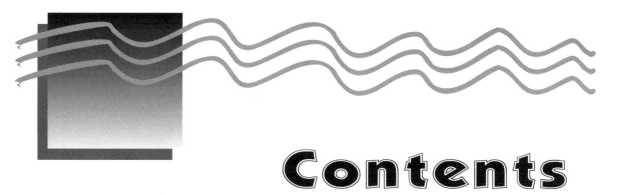

Contents

The response of the traditional computer industry to net-centric computing in the form of high-end personal servers, midrange Internet-connected multimedia PCs, and low-end NetPCs.

Part II. Building Blocks

The strengths and weaknesses of Java as a universal "lingua Internetica" and what is needed to improve its ability to handle data in real-time. The challenges the language presents to designers of both hardware and software.

Some possible alternatives to Java, such as Embedded C++, Lucent's Limbo, and AT&T's C@ + (CAT). Evaluation of some of the existing scripting languages such as tcl (tickle) and others, and how they could complement Java and other object-oriented languages on the Internet and World Wide Web.

How the emergence of network computers and Internet appliances independent of Wintel (Microsoft Windows/Intel) has created the need for a new generation of operating systems that meets the requirements of net-centric computing. The OS requirements of this environment and the OS alternatives that have emerged.

Alternative approaches to turning the network into a computer, from the use of remote procedural calls (RPC) to Java's Remote Method Invocation. The future of Java, ActiveX and a variety of proprietary distributed-object protocols in the context of an emerging standard called the Common Object Request Broker Architecture (CORBA).

Choices designers will have to make about the processors that they use to power their NetPCs, NCs, Internet Appliances and Web-enabled Set-top Boxes. Alternatives to the Intel x86 architecture, from the traditional RISC and CISC processors to the stack-based alternatives, such as Sun Microsystems' picoJava and Patriot's Shaboom CPU. Ways to keep hardware and software costs at a minimum, ranging from off-the-shelf standard integrated circuits to a variety of application-specific custom and semi-customized solutions.

Part III.   Listening for the Web Tone

Chapter 10: Building More Reliable Boxes   163

New methodologies that will be required in order to build hardware that is more reliable than the traditional desktop. The strategies of companies such as Intel and Microsoft to close up the desktop and some of the alternatives to current PCI bus architectures that are more reliable.

Chapter 11: Securing Network Connections   183

Strategies for protecting networks, clients and servers from viruses. A comparison of the solutions that Microsoft and Sun have come up with to a real-time, as well as a network-based virus hunting and destroying methodology developed by IBM that finds and destroys viruses just as fast as they are introduced into the network.

Chapter 12: Building Better I/O   201

The factors that have increased the load the "thin client" network computer model places on the servers that make up the computing backbone of the Internet, what is being done, and what still needs to be done to ensure that the Internet and net-centric model will continue to operate in the future. Strategies taken with the current generation of servers to improve the I/O capabilities of computers linked to the Internet.

Why it is necessary to move from 32-bit to 64-bit servers, routers and switches. A review of some of the 64-bit architectural alternatives available now and in the future.

What is necessary to ensure that reliable software is written, and, after it is in placed on the network, what can be done ensure its correct operation. Some of the ways that the problems that do occur can be identified, isolated, and corrected.

Part IV: The Future of the Internet

With the move to wider methods of transmission into the home and over the backbone, will rapidly evolve from the Internet to the "MultimediaNet," in which mixes of audio, video and 3D graphics as well as ordinary text and 2D graphics will be the norm. This chapter looks at how multimedia are being handled on the Internet and at such standards as MPEG-4, the next-generation standard for real-time networked interactive multimedia.

What is happening to the processors that will be used to build the connected PCs, NetPCs, and NCs, as well as the servers, as multimedia becomes more the norm on the Internet. First-generation multimedia processor architectures, such as Intel's Pentium II with MMX, and a comparison of this desktop-oriented technology with some of the more network-oriented multimedia-capable solutions.

The limitations of the multimedia processors and why some other alternatives may be necessary to achieve the 1,000-fold improvement in processor performance that higher-bandwidth Internet connections and multimedia data will require. The capabilities of some of the more representative multimedia coprocessors and their abilities to perform networked interactive multimedia in real-time.

Chapter 18: Moving from GUI to NUI...To XUI? <span></span> 309

The limitations of the present graphical user interface with the emergence of the Internet and World Wide Web, especially as far as "getting lost in hyper-space" is concerned. What new kind of user interface is needed to allow the average user to find out where to go; how to get from here to there; and, once at a desired location, how to find out what else is in the neighborhood. Some new alternatives to viewing masses of data on the Internet, and the impact of new standards such as the Extensible Markup Language (XML) on simplifying the way in which we view and navigate the World Wide Web.

Appendix A <span></span> 329

References and further reading.

Appendix B <span></span> 341

Web sites for Further Information.

Index <span></span> 349

Preface

Connecting the Dots: Engineers and the Net-centric Paradigm

The idea for this book grew out of a process that I do periodically to track developments in the computer industry. I call it "connecting the dots." But it also describes, I imagine, the process by which every engineering manager, vice president of technology, and strategic marketing executive assesses the nature of a market or a technology.

What I do is look for the hot spots, the areas where development is occurring, and then try to see how each one relates to the other. It is in those inter-relationships where I find the "hot stories" in technology, and where the vice presidents of engineering find the opportunities and where engineering managers who do similar analyses find information on what new products to develop and how.

But in the mid-90s applying this methodology to the computer market made me aware that a fundamental change was taking place — a paradigm shift — from traditional desktop computing to net-centric computing.

This change can be seen in the non-linear way in which the "dots" are connected. In the relatively linear days of the original PC revolution the process of connecting the dots was straightforward. There was one reasonably well-defined market and that was developing linearly: faster processors, bigger monitors, faster modems, bigger disk drives, bigger and faster DRAMs, applications that required more display, more memory, etc. So, when connecting the dots was done, it was basically a matter of straight line extrapolations.

The imposition of the World Wide Web and Internet, which grew up in parallel with and somewhat independent of the PC revolution, has changed all that. Now we have a relatively mature personal computing technology that is being superimposed on the WWW and Internet where there are various developments occurring in isolation — the dots — which can change the way in which computing will evolve in a number of different ways.

Rather than straight line extrapolations, what we have now is something like a children's drawing book in which images are created by connecting the dots outlining a somewhat amorphous, but barely discernible, form behind the not-so- random placement of the dots.

When we look at computing against this background what appears to be an essentially random placement of dots — technical development of various sorts that may or may not be connected, but which are also somewhat dynamic, some fading in and some fading out — is coalescing into a discernible form.

The challenge for me in this book, and for the engineering manager trying to assess the future direction his company must take, is to pin down as much as possible some of the dots, the technical hotspots that seem certain to remain in the foreground, and then determine how they are connected to one another.

One thing is certain: things have changed fundamentally. As one of the many who has participated in the rapid growth of the personal computer industry over the last decade or so, until recently I was convinced that nothing would ever again compare with the excitement and technical challenge of those days.

I was wrong. For we are all on hand for the birth of a new computing paradigm, one as different from the traditional desktop PC and workstation as the PC was from the minicomputers and mainframes that it replaced in most corporations.

This new paradigm goes by several names: the information or Internet appliance, the connected PC, the NetPC, and the network or net-centric computer. What they all have in common is the idea of the network as computer, not just as a repository of information that can be accessed via a modem or local-area network, but as an environment for collaborative computing. In this new environment of the network as computer, the chores that the user will perform will not necessarily be done by the system on his desk or on top of his TV set. Instead, they will be done at the appropriate place on the network, on the processors — servers, in the lingo of the Internet — that were designed for the job.

The net-centric computer has not come into existence in a vacuum. The desktop and workstation computer came into existence because of a demand from society for a computer that was easier to use and less expensive than the mainframes and minicomputers that it replaced. So, too, the net-centric computer has emerged to fulfill a promise that the personal computer did not quite fulfill: a computer that is easy to use and no more complicated than the telephone, television, or VCR; that gives the average user access to services that are actually useful and entertaining; and lastly, is in the price range of the average consumer.

To make all of this possible requires that engineers and systems designers rethink the way they have been designing computer hardware and software. While they have done a reasonably good job of keeping the costs down, the personal computer has, if anything, gotten harder to use.

In the same way that occurred in the television and telecommunications industries, designers of net-centric computers are going to have to find ways to hide the complexity of the system from the user without reducing its power and functionality. Just as the telephone or television hides a complex and sophisticated infrastructure from the end user, the wealth of functionality that users have gotten used to on workstations and PCs will have to be hidden, but be accessible when needed by a push of a button or the click of a mouse. As in the television or the telephone, most of the complexity will be in the network, where the main job of computing will be done.

The purpose of this book is to analyze the significant technical challenges and opportunities ahead for engineers and systems designers if this new computing paradigm is to make it through its childhood and advance toward adulthood. For the purposes of this analysis I have divided the book into four parts: The Varieties of Net-centric Computing, Building Blocks, Listening for the web tone, and The Future of Net-centric computing.

I. The Varieties of Net-centric Computing

In this first part of the book, an examination is made of the first generation of net-centric devices which depend on their links to more powerful servers on the Internet and intranet for full functionality. At one end of the NC spectrum are designs essentially targeting the office and corporate workplace. At the other end of the spectrum are a host of information or Internet appliances.

An attempt is made to analyze this first generation of net-centric information appliances, how they are similar, how they are different, and the problems faced in bringing this new computer into existence. I investigate the ways in which net-centricity is being achieved, how some of these computers achieve their ability to parcel the computing chores out over the network, and the new forms these systems are taking.

II. Building Blocks

Part II of the book is an examination of the hardware and software building blocks with which the engineer will have to deal, the alternatives available, and the technical choices he or she will have to make.

On the software side, a major focus of this section is on newly emerging languages such as Java from Sun Microsystems, Inc., and Inferno from Lucent Technologies. Right now, the bets are on Java as the way to build a network software infrastructure that allows developers to create applications that execute independently of the underlying hardware or OS architecture. An important consideration that engineers will have to keep in mind is that there are fundamental differences between the way Java distributes applications and the way we do things now on the desktop. In the Java approach, an application might consist of numerous applets working cooperatively, some on the server and some on the client, to accomplish a particular task. By comparison, the ICA and X-windows approaches are more monolithic, with the entire application running on the server, with only the results displayed on the client terminal or network computer.

This section will also look at the competing interoperability standards, the advantages and disadvantages of each, the cost and consequences to the industry if they are adopted, and the decisions that engineers and systems designers will have to make.

In this section I also discuss some of the processor alternatives and implementation choices facing the engineer. These chapters will look at not only existing desktop stand-bys such as the x86 and the PowerPC, but a variety of reduced instruction set processors as well, analyzing them for their suitability in processing programs and applications written in Java and other object-oriented languages.

III. Listening for the Web Tone

In Part III Three, the aim is to look at the critical need for increased reliability — on the desktop, in the network computer and Internet appliance, and on the servers, routers and switches. While the Internet and World Wide Web are on their way be becoming a ubiquitous part of our lives, the increasing number of clients — PCs, NCs, Web-connected set-top boxes and Internet appliances — are creating the conditions, both directly and indirectly, for increased unreliability. These conditions include too many users causing traffic jams which slow down server and router response, and PCs that are being worked harder to act not only as standalone workstations but as communications terminals, and that show the overload in degraded performance and lockups.

An effort will be undertaken in this section to look at what the industry is doing to improve the hardware reliability of the net-centric devices, be they network computers, Web-enabled set-top boxes, or NetPCs. We will look at present initiatives, mostly aimed at closing up the systems and less amenable to user "fixing," as well as what needs to be done in the future to make the average computer as reliable as the average TV set.

Also addressed in this section are the reliability and throughput of the servers, routers, and switches on the Internet and how they can be improved without pulling out all the hardware and starting from scratch. Among the alternatives that will be discussed are such things as improving the I/O processing capabilities of computers, an ignored aspect of their design until the emergence of the Internet as a mass medium. It will look at the impact of going to 64-bit architectures to relieve the bottlenecks and redesign the computing infrastructure to match the increased bandwidths available to end users as higher-bit-rate technologies such as ADSL and cable modems become more common.

This section will also deal with the issues of network reliability and how it can be improved through the use not only of hardware but of software redundancy, better methods of monitoring and testing network servers, routers, and switches. Also addressed is the greater need for software testing to eliminate as many unknowns as possible in the software that is provided to run not only the net-centric clients, but the servers, routers, and switches.

An important aspect of reliability of net-centric computing is the increased sensitivity to viruses in the form of malicious ActiveX and Java applets. This section will look at the alternatives to controlling the penetration of viruses and suggest solutions that better fit the real-time dynamics of net-centric computing.

IV. The Future of Net-Centric Computing

This section will deal with the emergence of multimedia on the Internet and World Wide Web, a trend that will increase as available bandwidths available get larger. In addition to assessing what this new "MultimediaNet" will look like, this section deals with the new architectural choices that engineers will have to make on the desktop and in the servers that are employed. Some of these new multimedia-enabled processors are already making their way to the market, but the requirements of the Internet will accelerate this trend, as well as influence the directions in which these architectures go.

Finally, we will take a look at another needed paradigm shift, from a graphical user interface (GUI) to a network user interface (NUI) to deal with the enormous amount of information available and the larger number of Web sites and the diverse types of data. Traditional GUIs and browsers, no matter how sophisticated, are not adequate to the job of helping the average user avoid "getting lost in hyperspace."

Engineers and systems designers who participated in the growth and evolution of the desktop computer over the last 20 or so years may have thought that they were lucky to have been able to participate in a revolution that not only fundamentally changed computing but changed many aspects of modern society as well.

Luck may strike twice. For we are now in the beginning changes of another revolution. Just as the desktop computer was a radical break with the previous mainframe and minicomputer approach, so too the net-centric computer is a fundamental break with the past twenty or so years of computing. As with the earlier personal computer, the net-centric computing represents not only challenges, but opportunities, to the engineers and designers who can think far enough ahead to take advantage of them.

One thing engineers and managers of engineers will have to make up their minds about is whether they are willing to take a fundamentally new direction. For it will not be without its costs. For one thing, the entire software industry is going to have to shift from a monolithic of-a-piece strategy to a more modular one, in which an application can be pared down, reassembled, and repartitioned among the client and server participants. What this means is that virtually every familiar piece of application software will have to be rewritten and modified to fit this new paradigm.

For the net-centric computing paradigm to fulfill its promise requires universal interoperability, and this will require that engineers and developers learn new tools and new ways of thinking. One thing that will have to change is the tendency to create the best and the fastest to the detriment of interoperability.

It is also clear that hardware and software designers will have to become much more rigorous about how they define and deal with reliability issues. The broader consumer market, which will be the main consumers of the net-centric computer in all its forms, is much less forgiving than the average desktop PC, who has accepted as normal that the system will freeze up occasionally and that it will be necessary to reboot. Rebooting of the desktop, server, or switches and routers is not an option.

It will not be clear for many years which of the competing alternatives will win out. Maybe "none of the above." What will be the impact of this? Maybe none, for the desktop computing industry has survived for many years with three or four alternative hardware and software platforms. But so much of the idea of the network computer depends on interoperability that the lack of such agreement could have serious implications. Engineers will have to understand the impact of such developments and make their technical decisions accordingly.

List of Trademarks

Postscript is a registered trademark of Adobe Systems, Inc.

emWeb is the registered trademark of Agranat Systems, Inc.

Rompager is the registered trademark of Allegro Software Development Corp.

AM29000, AMK5 and AM K6 are registered trademarks of Advanced Micro Devices, Inc.

Apple I, Apple II, Hypercard, Macintosh, Quicktime and Quickdraw are the registered trademark of Apple Computer, Inc.

ARM and StrongArm are the registered trademarks of Advanced RISC Machines, Ltd.

PocketNet is a trademark of AT&T Corp.

mPact is the registered trademark of Chromatics Research, Inc.

ICA, Thinwire and WinFrame are the registered trademarks of Citrix Corp.

FlexSMP is a registered trademark of Compaq Computer Corp.

MediaGX and 6X86MX are the registered trademarks of Cyrix Computer Corp.

emObjects, emManager, emMicro, emNet, emServer and Microtags are trademarks of emWare, Inc.

@workstation is the registered trademark of HDS Network Systems, Inc.

PA-RISC is a registered trademark of Hewlett Packard Co.

SH and SH3 are trademarks of Hitachi Electronics America, Ltd.

pSOS and pSOS+ are trademarks of Integrated Systems, Inc.

AGP, i486, i960, MMX, Pentium, Pentium II and Pentium Pro are trademarks of Intel Corp.

I2O and I2O are the registered trademarks of the Intelligent I/O Special Interest Group

IBM, OS/2, OS/2 Warp, and PowerPC are the registered trademarks of International Business Machines Corp.

Dis, Inferno, and Limbo are the registered trademarks of Lucent Technologies, Inc.

ActiveX, Authenticode, Direct3D, Microsoft Windows, NetShow, and WebTV are the registered trademarks of Microsoft Corp.

VRTX, VRTXsa and VRTXmc are trademarks of the Microtec Division of Mentor Graphics Corp.

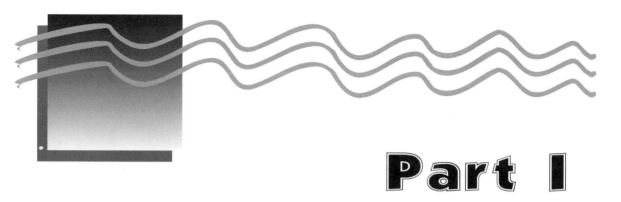

Part I

The Varieties of
Net-centric Computing

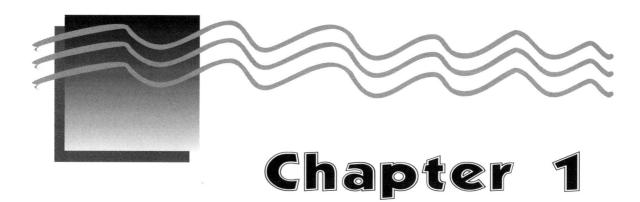

Chapter 1

The Network Computer:
Reinventing the Wheel?

With net-centric computing a reality on the Internet and on the many corporate intranets, the need to control costs and ensure reliable computing has led to the invention — or more accurately, the reinvention — of the concept of the network computer.

The main motivation behind the push by Larry Ellison , chairman and chief executive officer of Oracle beginning in late 1995, was to create the "Network Computer" that was not only easier to use, but lower in cost. His target: $500 or less for the basic system and no more than $2,500 for the total cost of ownership to a large organization.

In many of its essential elements the network computer is similar, if not identical, to the "dumb terminal" used in pre-PC days, with a centralized mainframe computer containing all the computing and data storage resources and the terminal used only for data entry and display.

This scheme is still used in many large corporate environments, except that the mainframe was replaced first by the minicomputer and now by the "super server." And in the era of the Internet, intranets and the World Wide Web, the dumb terminals have been replaced by Windows-based smart terminals and, eventually, network computers.

The network computer has a wide variety of generic meanings, both explicit and implicit — from network-connected PCs and Web-enabled set-top boxes to mobile computers with communications capabilities and dumb terminals. However, a very precise network computer platform specification has been defined by an industry organization lead by Oracle, Sun Microsystems, IBM, Apple and Netscape. In fact, the terms Network Computer and NC have been trademarked by Oracle.

But this attempt at coming up with an all encompassing industry-wide standard for a new kind of desktop computing system that would be competitive with the defacto standard established in the marketplace by Intel Corp. and Microsoft Corp. has given rise to at least six other platforms, These include ones from National Semiconductor, which has acquired x86 vendor Cyrix; Advanced RISC Machines Ltd., which defined the ARM architecture and licensed it to numerous companies around the world; IBM; and Sun Microsystems Inc., which in the latter part of 1997 introduced a more tightly defined specification for the NC.

One reason for this proliferation has been the fact that the specification is broadly written, allowing virtually any computing system to meet the requirements. The NC Reference Profile as defined by Oracle has two parts: one for hardware and the other for software. Of the two, the hardware reference is the most general, while the second is somewhat more specific. The aim: production of a network-connected computer whose bill of materials, if not the end user cost, was below $500.

Like the NetPC specification, the NC requirements include at least a 640 by 480 pixel display; a pointing device; provision for some form of text input; and audio output. Also required is a minimum set of networking features capable of carrying Internet-based packet messages, such as a standard analog modem, a cable modem, a wireless modem, ISDN, Ethernet LAN, or Asynchronous Digital Subscriber Loop (ADSL) protocol.

The most interesting difference, from a hardware point of view, is that, similar to dumb terminals but unlike most desktop computers, the NC specification in its original form has no provision for local storage: no floppy disk, hard disk, or CD-ROM. All such devices are defined under the specification as optional. While a network-centric computer built with such capabilities would be NC compatible, it would not be an NC according to the definition.

Under the hardware portion of the specification no provision is made for the use of a specific CPU or even a specific type of CPU. Unlike the PC platform, which depends on the use of an X86 processor, virtually any processor can be and has been used as the central processor of an NC. These include pre-Pentium X86s, Sun Sparc processors, and a number of different variants of the ARM architecture, including the blazingly fast StrongARM from Digital Equipment Corp. Acquired by Intel Corp .when DEC was bought by Compaq Computer Corp., the StrongARM has, ironically, has become a strong contender in NC designs, because Intel has committed major resources to the architecture as a part of its strategy to dominate the non-X86, nonPC computing market.

The NC Reference Profile(see Table 1.1) is much more specific about the software requirements. NCs are required to communicate over a network using standard IP protocols such as TCP (Transmission Communications Protocol), UDP (user diagram protocol), DHCP (dynamic host configuration protocol), Bootp (bootstrap protocol) and SNMP (simple network management protocol). UDP allows Sun's Network File System to be used to set up end-to-end application-specific communications, while DHCP allows an NC to automatically acquire an IP address, as well as send configuration data over a network when it is booted up. SNMP is designed to ensure that an NC will perform as a client on managed, internal intranets. Bootp enables an NC to boot over a network.

Specifications	Network Computer	NetPC	ConnectedPC
Processor	Any RISC/CISC	X86, Pentium	Pentium II
Main Memory (MB)	4 - 64 (8 typical)	4 - 32 (16 typical)	32 - 128 (64 typical)
Mass Storage	Flash (8-16) or optional disk	Hard disk	Hard disk
Network Interface	10/100BaseT, 33.4 kbps ATM, ISDN	Any	Any
Protocol	TCP/IP, PPP	Any	Any
Boot options	Server (local optional)	Local (server optional)	Local (server optional)
Boot server requirements	Host required	None	None
OS requirements	Optional (any)	Windows	Windows
Terminal emulation	Optional	Required	Required
Graphical user interface	Any (X-windows, MS Windows)	MS Windows	MS Windows
Java	Required (remote)	Optional (local)	Optional (local)
Remote Windows emulation	Optional	Not required	Not required
Remote operation	RMI, RPC	RPC	RPC
Price	Under $500 target	Under $1000	$1500 and up
CD-ROM	Optional	Optional	DVD

Table 1.1 **NC versus PC versus NetPC**

Because the base-line specification has no provisions for local storage, if a user wants to store data it must be done remotely. Sun's Network File System has been designated as the standard method for mounting remote drives. Alternatively, as an option, an NC can support remote connections to other systems via FTP or Telnet and can establish secure communications using the SSL (secure sockets layer) protocol.

The most basic requirement of the Profile is support for the Java application environment, including the Java Virtual Machine, the Java Run Time Interpreter, and the standard Java Class Libraries. Interestingly, unlike the PC platform, there is nothing in the NC Profile that specifies a particular operating system layer. This is a bit misleading, however, because in its support for Java, it does require that the operating system support preemptive multithreading. While this is common to most embedded system real-time operating systems, it does rule out many of the desktop operating systems, where support for multithreading is uneven at best. For example, Windows 3.1 is not very good at preemptive switching. Windows 95 is better, as is Windows NT. Mac OS, however, is not very good.

Also stacking the deck in favor of embedded Real Time OSes and away from OSes such as Windows NT/95/CE is the requirement of the Profile that the primary job of the OS is to maintain the TCP/IP stack so that the NC can communicate with other devices on a network. Unlike traditional desktop OSes, all other functions are either secondary or nonexistent, even file management and the user interface.

The most noticeable difference between an OS for an NC and one for the desktop is that there are no previsions within the OS for supporting a graphical user interface. A primary assumption of the Profile is that the NC's primary function is operate in a network-centric environment in which a Web browser will in all likelihood be present. It is assumed that any GUI functions that are required by the NC will be handled by the browser. While the NC Reference Profile does specify the ability to read and interpret Hypertext Markup Language (HTML) — in other words, run a browser — it is not specific as to which version or tags will be needed. As far as sending and receiving email, the profile is pretty flexible: SNMP, the Internet Message Access Protocol, or POP, the Post Office Protocol. The profile also requires that the NC be able to recognize and deal with some of the more common multimedia formats on the Internet: JPEG and GIF graphics and WAV and AU audio files.

Supporting the NC Profile were a wide range of companies (see Table 1.2) that fall into three broad categories: systems manufacturers such as Acorn Computer Group, Boundless Technology, Noika Group, Olivetti Group, and Wyse Technology; technology partners such as ARM, Cirrus Logic, Corel, Digital Equipment, Mitsubishi Electric, Motorola and VLSI Technology; and systems integrators such as Hitachi, Japan Telecom, NEC, and NTT Data, amongst others.

	Sun JavaStation	IBM Explora/Pro	NCD Explora	HDS @workstation	Boundless TCXL Series	Wyse Technology 2000/4000 Series
Processor	microSparc (microJava)	PowerPC 403	PowerPC 402/403	i960	I960	X86/StrongARM
DRAM	8 - 128	4 - 64	4 - 36	4 - 128	4 - 64	8 - 16
Mass Storage(MB)	Flash 4 or 8	4 (PC Card)	4(PC Card)	1000 (PC Card)	1 - 4	0.5
Network Interface	10/100BaseT	10/100BaseT Token Ring	10BaseT	10BaseT. ATM	10BaseT 10Base2	10BaseT 10Base2
Protocols	TCP/IP, PPP	TCP/IP, 3270 SNA, DecNet	TCP/IP, 3270 DECnet	TCP/IP, PPP, tokenring	TCP/IP, PPP, SLIP	TCP/IP (PPP optional)
Boot options	Server, local	Server (BootP, MOP), local	Server (BootP, MOP), local	Server, local	Multiple server options	Server, local
Boot server requirements	Solaris	NT, OS2, UNIX	NT, UNIX	NT, UNIX	WinFrame For Windows	NT, UNIX
OS requirements	Java OS	NCDWare Java OS	NCDWare	netOS	XEasy, ntEasy	JavaOS, WinFrame
Terminal emulation	3270, VT320 Windows, x-windows	3270, VT320 Windows, x-windows	3270, VT320 Windows, x-windows	PC, UNIX, Internet, mainframe	PC, UNIX, Internet, mainframe	WyseWorks (VT320, x-terminal)
Graphical user interface	WinFrame, WebTop, Views	Citrix, WinFrame	WinCenter X11R5	Ngrigue	Citrix, Winframe	Citrix Winfrare WebTop Views
Java	Local	Local	Local	Local	JVM, Java applets, local	Local
Price	Starts at $800	Starts at $700	Starts at $700	Starts at $750	Less than $1000	Less than $1000

Table 1.2 **Typical Network Computer Configurations**

Beyond these initial participants, the builders and proponents of the NC had not grown substantially by the end of 1997. Moreover, unlike the PC and NetPC market, which has some standardization, one implementation does not share much in common with another.

HDS Network Systems (King of Prussia, Pennsylvania.), in its initial implementation built a system configured around an 66MH i960 RISC processor, a good choice considering the fact that this CPU has had wide use in network routers, bridges and switches. Boundless Technologies (Austin, Texas) has also chosen to use the i960 in its XL and XLC series of NCs.

IBM's initial offering, the IBM network station, as well as its follow-on network computers, was built around a 32-bit PowerPC RISC chip, the same CPU family used in the Apple Macintosh. Network Computing Devices Inc. (Mountain View, Calif.) has also entered the fray, but with a design built around a 64-bit MIPS R4000 RISC.

For its part, long-time terminal manufacturer Wyse Technology Inc. (San Jose, California) has chosen to base its Winterm 4000 series of NCs on the 200-MHz StrongArm 110, a RISC processor built and designed by Digital Equipment Corp. using the ARM architecture. Finally, Sun Microsystems has produced its own version of an NC called the Javastation (see Photo 1.1), initially built around its own 100-MHz microSparc II RISC processor.

Photo 1.1 **Javastation.** Sun's Sparc-based network computer uses its Java OS to distribute computing chores between client and servers on a network. © 1998 Sun Microsystems Inc. Used by permission.

The fact that a number of different processors have been used to implement these designs should not be troubling. That is in fact one of the purposes of the Reference Profile: to allow designers enough flexibility to achieve their design goals and still meet the spec. What is troubling is that, even among designs using the same processors, the implementation for each processor is different, as is the choice of operating system.

To provide companies who want to enter this market with a common framework within which to design their systems, a number of reference platforms has begun to appear. One of the first was a reference design from Oracle's Network Computer Inc.(see Figure 1.1), developed in cooperation with Advanced RISC Machines Ltd., with a bill of goods estimated at about $300, allowing an end system to be built for about $500. Based on ARM's 7500FE processor, the reference design includes 1 Mbyte of boot ROM, no more than 4 Mbytes of ROM for storage of the operating system and applications code; and no more than 4 Mbytes of DRAM. Beyond this bare minimum, a number of other features of the reference design are optional. While the NC must have a text input device, it does not have to be a keyboard. It could also be a pen input device or a touch sensitive screen. The pointer device in this reference platform can be a mouse, joy stick or pen device.

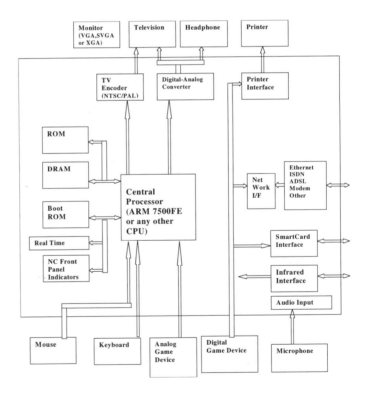

Figure 1.1 **The NC platform.** Initial implementation of the NC specification as proposed by Oracle Corp. and its partners is based on the ARM RISC microprocessor designed by Advanced RISC Machines Ltd. and manufactured by about a dozen companies worldwide.

The ambitious cost constraints of this initial reference platform have not been met by many of the participants so far, with initial end-user pricing ranging from about $750 to $1,000. To bring costs back under control, further hardware platforms have emerged, each built around a specific processor and specifying very precise hardware and software elements.

Motorola has worked with NCD on a reference platform built around an MPC8xx embedded PowerPC RISC and using the OS-9 modular operating system from Microware Systems Corp. ARM has also come up with a reference design using the ARM7500FE, as has Digital Semiconductor Corp., built around the StrongARM processor.

By late 1997, Sun Microsystems had attempted to jump-start the NC effort with its own reference platform, a flexible design that uses the Java OS and initially the microSPARC II, upgradable to a microJava CPU that implements the Java "virtual machine" in silicon.

The success of the network computer, in the corporate environment in particular, will depend not only on the ability to reduce the total cost of the system, but the total cost of ownership. According to one study (by the Gartner Group) the total cost of ownership of a desktop PC is made up of four elements: the actual cost of the unit, the cost of support, the cost of administration, and the cost of end-user operations.

The PC's actual cost, given the continuing drop in the cost of high-performance processing power, makes up about 21% of the total cost, according to the Gartner study, while support expenditures account for 27%, and administrative overhead about 9%.

The majority of the cost of a PC, about 43% according to the study, is what corporate managers of information services refer to as end-user operations. This is a politic and indirect way of describing the costs incurred when the end user is doing other things than job-related activities, such as adding special software and hardware and using it for tasks not related to the job. This cost also includes the cost to the corporation when a specialist is required to come in and repair the system after the user has "fixed" it and the cost in downtime incurred while the user is waiting for his or her computer to be repaired.

Ironically, for all the focus on hardware costs, the long-term success of the NC as a cost-effective alternative to the PC, or even the NetPC, lies in the software infrastructure that is needed, sufficient performance, and its ability to supply the one thing that has been missing on the traditional desktop: reliability.

In terms of both the hardware and software used to build the network computer, two basic trends are emerging. At one end of the spectrum is the pure Java machine, typified by Sun's Javastation and the software it uses, in which applications and applets are downloaded to the thin client.

At the other end is an extension of the traditional Windows terminal concept, which uses the remote procedural call mechanism to provide a connection between the terminal and the server, on which the applications reside and are executed. The new alternative is a more independent version of the Windows terminal, with its own minimalist operating system. The addition of the OS allows the extended Windows terminal to operate in a number of different modes. These modes include as a pure remote terminal, executing and storing on the server; as a minimalist local computer with local storage and execution; and, depending on the operating system used, as a Windows system, a UNIX system, or a purely Java-based machine.

The Javastation Alternative

The Sun Microsystems Javastation is a thin-client desktop that employs modern technologies such as the Java language/runtime environment, and Sun's microSparc and microJava-based designs. In some respects, the Javastation itself represents a compromise with the original view of Java-based remote computing in which Java applets and applications are downloaded to the thin client when necessary. It does download Java applications on demand and executes them locally; but all applications and data reside on centralized servers for ease of administration, using mechanisms built into the Java specification such as Java Method Invocation (JMI), a remote-procedure-call look-alike.

The Software Infrastructure

At the heart of the Javastation is the JavaOS, a small and efficient operating environment that executes Java applications directly, without requiring a host operating system. JavaOS consists of a small Java kernel, an embedded Java Virtual Machine, Java windowing and graphics primitives, and a suite of networking protocols and device drivers written in Java. Its streamlined design enables the entire JavaOS environment (including HotJava Browser) to run in as little as 2.5 MB of memory. JavaOS is not a traditional operating system because it does not have (or need) a file system, a virtual memory subsystem, multiple address spaces, or support for multiple language programs.

Similar to what Microsoft did with Windows, the Javastation is available with two easy-to-use graphical environments: HotJava Browser or the HotJava Views environment. Just as Microsoft is trying to integrate Internet and Web capability, with its traditional desktop management systems, into a single graphical user interface that eliminates the border between data outside and data inside the computer, HotJava Views is an attempt at a similar integration.

Based on HotJava Browser, HotJava Views(See Figure 1.2) features lightweight communication and information access tools in an intuitive user environment. Similar to the variety of utilities supplied with Windows, HotJava Views include a number of standardized "views," such as Selector, MailView, NameView, CalendarView, and InfoView.

Selector is a Java application run by JavaOS once a user login on the JavaStation is verified. It is a tool for selecting applets with a push-button metaphor. MailView is a simple email tool for editing, sending, and saving electronic mail. It features a native Java implementation of the IMAP4 protocol that communicates with a Solaris SMP server. Tight integration between MailView and the other tools in HotJava Views allows users to attach CalendarView appointments to messages; to send messages to names looked up in NameView; and to view live URLs contained in messages. NameView is a database giving users access to a name directory. CalendarView is a personal calendar that allows simultaneous viewing of multiple calendars and group scheduling. InfoView is the Web-browser component of HotJava Views. It supports exactly the same version of HTML that HotJava Browser supports. In addition, it allows system administrators to determine browsing access rights for users.

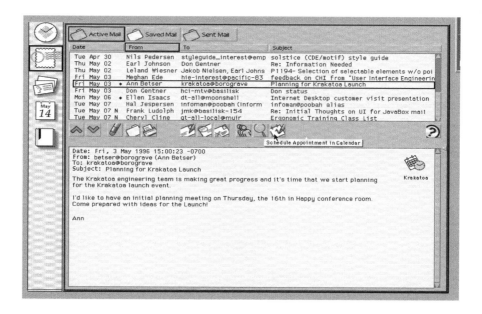

Figure 1.2 **WebTop.** As an alternative to the Windows-based desktop graphical user interface, Sun uses its HotJava-based Views GUI to access the World Wide Web. © 1998 Sun Microsystems, Inc. Used by permission.

Even Sun, with its purist approach to both Java and network computing, has had to make compromises with the fact that there is a lot of legacy software out there that will not go away for a long time. In the Javastation, this has been dealt with through the use of emulators, which give access to legacy data and applications without costly and disruptive reimplementation or redesign.

One category of emulators is terminal emulators, which give Javastation users access to mainframe data. For example, these emulators translate 3270, 5250, VT220, and NVT datastreams into Java datastreams, and vice versa. Another type of emulator gives Javastation users access to legacy Microsoft Windows applications, similar to what Citrix has done in the Windows environment. It allows display of a Microsoft Windows application that is running on a Microsoft Windows NT server. However, much like an X-terminal, the client handles only window drawing commands.

Windows is more than just another pretty user interface, incorporating a wide range of application programming interfaces to simplify integration of third-party software into the Windows environment. A new desktop metaphor standard, developed by Sun and Oracle, tries to do the same thing for the pure Java environment. Their API solution is called WebTop. Sun's WebTop defines a complete set of APIs for developers to create services, protocols and components. In the absence of any other running applications, the WebTop is simply a Java Runtime Environment (JRE) with a graphical container. Developers can interchange this definition with the simpler term "Java-enabled browser." The WebTop provides developers with APIs that fall into three major categories: Platform APIs, Connectivity APIs, and Service APIs.

The Platform APIs contain the fundamental programming interfaces that all applications will need. Examples include input-output, thread management, internationalization, graphical user interface, and foundation classes. The Connectivity APIs include class libraries sufficient to build WebTop clients that can participate in multitier client-server infrastructures.

Sun's WebTop supports open network protocols, including any IP-based protocols that ride on top of TCP/IP. The Service APIs abstract common and useful services, protecting them from the underlying implementation. Some specific services include the Java Naming and Directory Interface, the Java Mail Interface, the Java Print Interface, and the Java Schedule Interface.

Just as Microsoft has adopted ActiveX as its component-building mechanism, Sun has developed JavaBeans for the WebTop. By building software libraries as Beans, developers can produce cross-platform, reusable components.

The Hardware Underpinnings

Originally, the Javastation was designed using a 100-MHz microSparc II, which is, for a general-purpose microprocessor architecture, reasonably good at fast execution of stack-oriented Java commands. However, better performance is possible with a Java-specific microprocessor. So the plan is to upgrade the Javastation to one of the first devices in Sun's JavaChip family, the microJava 701. With the use of a just-in-time byte code compiler, this will allow the execution of Java code on roughly the same par as a 200-MHz Pentium.

The initial implementation of the Java Virtual Machine in actual silicon — the picoJava core — accomplished what previously was not possible. It simultaneously maximized the efficiency (minimized the overhead) of executing a Java program, while at the same time preserving all the benefits of platform independence and run-time security associated with the byte-code program format.

But as originally implemented, the picoJava had a very important limitation arising from the fact that it is a new CPU architecture. While the overall JavaChip architecture is not limited to executing Java code, this is not what it is best at performing.

For applications such as network computing, where there is non-Java code written in any other high level language, for best performance it is necessary to use a compiler to convert these programs into Java code recompiled specifically for their target JavaChip. This is an effort that becomes more burdensome in direct proportion to the amount of legacy (non-Java) code that the system must run. And as we have seen earlier in this chapter, even on a purely Java-based network computer, there is a significant amount of legacy code that it is not realistic to convert to Java byte-code.

To address this issue, the microJava 701 was designed to run non-Java code about as efficiently as comparable RISC CPU architectures. While it offers no tangible price-performance advantage when executing non-Java code, as a JavaChip Sun claims that it executes Java byte code faster than the alternative: using a general-purpose chip optimized for other high-level languages.

The 701 is designed to be able to facilitate the creation of inexpensive thin-client machines of all types. This means it enables complete systems that can be both designed in minimal time and built for minimal cost. The 701 attempts to minimize system-level cost by building the core logic chipset into the CPU itself, and at the same time tries to preserve enough design flexibility to fit readily into a wide range of possible applications falling within its price-performance budget.

Samples have been scheduled for availability in mid-1998 at an initial target frequency of 200 MHz, roughly equivalent to the performance on Java code on a 200-MHz Pentium, supplemented with the latest JIT compiler from Microsoft. C code performance has been estimated at about 200 Dhrystone MIPS.

From Windows Terminal To Multipurpose NC

Sun has made compromises to the existing environment in which desktop paradigms such as Windows are still dominant. In the same way, many of the builders of network computers, including Oracle, are making compromises to what they anticipate will be the Java-enabled future. Many of the builders of network computers are not new to the business, having provided less sophisticated systems in the form of dumb terminals and Windows Terminals. These systems are based on protocols such as X-windows and RPC in the UNIX environment and Citrix Systems' Intelligent Console Architecture (ICA) protocol in the MS Windows environment. They are now moving to compromise designs based on small-footprint operating systems that provide local storage and I/O capabilities and are adopting mechanisms that allow them to operate in a Java environment.

One of the first to make the move was Network Computing Devices, Inc., with its next generation of Windows terminals, codenamed "Thumper." In addition to ICA and X-terminals emulation, it incorporates Windows CE 2.0 as a local operating system to satisfy organizations whose Information Systems executives want some degree of local storage and I/O capabilities.

Who's on First Now? And Who Will Be on First Tomorrow?

What will determine the future of net-centric computing on the desktop? Certainly cost will be a factor. But just as important will be the impact of interactive networked multimedia on the underlying architectural requirements, the ability to perform Java-centric, stack-based code at clock rates that allow a net-centric computer to operate sufficiently fast, and the development of an appropriate OS underpinning.

Costs

The network computer in all of its guises seems to have all of the advantages in the cost of materials arena. Ironically, in their battle to compete with the challenge of the NC, Intel and Microsoft and their system vendor allies are adopting many of the features of the challenger. The end result may be that it will be harder to differentiate between an NC and NetPC. But the bottom line will remain the same: the traditional concept of the desktop computer has changed fundamentally. Initially, however, the network computer has had all of the advantages. For example, a RISC processor with a respectable 200-MHz clock rate is priced in the sub-$50 range, compared to an equivalent Pentium processor, which when introduced cost about $750 each and is just now pushing through the $150 to $200 range. This is due to a number of factors. Its stripped down architecture is one: a reduced-instruction-set-computer requires fewer gates to be implemented in silicon than an equivalent complex-instruction-set-computer architecture such as that used in the Pentium and its x86 progenitors. And as the fabrication technology used moves to sub-quarter-micrometer geometries, the silicon area required to implement even a fat RISC core of even 400,000 gates would be less than 10% to 20% the die area or less than a comparable Pentium design.

This difference in die area is of special importance when, to gain an edge in both costs and reliability, builders of net-centric computers take advantage of the fact that their processors require less silicon to add more functions for the same or lower cost of fabrication. For example, in an IC with roughly the same number of gates as a Pentium for just the CPU alone, a RISC vendor can put not only a RISC core, but many other peripheral functions that were previously off chip. It also allows the integration of additional on-chip SRAM and cache to boost performance.

Executing Java Code

An issue facing designers of connected PCs, NetPCs and network computers is how best to execute Java code without giving up its "write once, run everywhere" advantages, but at the same time retain the ability to run non-Java code and applications as well. Sun's microJava is one alternative, but there are others. A number of general-purpose processors, such as the x86 and Sun's own Sparc architecture are relatively good at executing stack-oriented Java code. Among the other non-x86 alternatives are the CPUs available from Advanced RISC Machines and its partners, Silicon Graphics and its partners, SGS Thomson, and Patriot Scientific.

If one assumes that Java will become the lingua Internetica of the future, then a purely Java-specific CPU might be an appropriate choice. However, if CORBA becomes the defacto way to do distributed-object-based network computing, then Java becomes just another network programming language, albeit an important one. Then, a CPU that is capable of handling both Java and non-Java code would be the right choice. In its first implementation of the picoJava core in a full-function microprocessor, the microJava 701, Sun has realized this too, and integrated hardware features that allow it to execute both Java and non-Java code expeditiously.

The Impact of the MediaNet

The speed at which the Internet and the World Wide Web become real-time multimedia-capable could also have a profound effect on the choices that a designer makes. The multimedia-enabled x86 architectures would be the best alternative if the designer wants to pick an architecture that has "legs." However, one should not rule out many of the non-x86 architectures. Multimedia extensions to the MIPS and Sparc family are already being implemented in high-end systems and will proliferate down fairly quickly if the market demands it. A number of media coprocessors might also be suitable.

OS Issues

One final factor that could make or break the net-centric computer, be it Oracle's NC or Intel-Microsoft's NetPC, is how quickly the underlying operating system is adapted to reflect the requirements of operating in a network-connected environment.

As far as the connected PC, the NetPC and the NC are concerned, to date most of the efforts have been oriented toward decreasing memory footprint and improving real-time and deterministic response. Important as these are to the future of net-centric computing, especially as the Internet and World Wide Web become more multimedia-intensive, there are a number of features that a network-capable operating system should have, and which desktop OSes have yet to incorporate.

Tight integration between the real-time operating system and network software is a must for network-centric systems where a premium is placed on small code size, high through-put, and maximum flexibility. Beyond these features, however, some of the facilities that a network-enabled OS will need are real-time multitasking, memory management, file system design, and network software layering. Real-time multitasking improves system reliability and flexibility by enabling the efficient scheduling and prioritizing of network tasks.

Memory management facilities improve performance by minimizing OS service calls. Prudent file system design and network software layering enhance portability. They also reduce cost by making possible distributed designs that leverage the use of intelligent, low-cost peripheral devices.

Scheduling network, application, and system-level tasks predictably is particularly important in many of the emerging real-time networked computing applications. For example, in the next generation of network computers and connected PCs that must operate in a much more multimedia-rich environmen, the OS residing in the desktop device must manage MPEG data delivery, audio-video de-multiplexing and synchronization, and the user interface. At the same time, the OS must provide the networking facilities needed to communicate with service providers via a variety of public and private phone and cable communications networks.

Though most OSes provide multitasking, only a few offer it in real time, and even fewer at the application level. As a result, user-state application tasks can preempt each other, but not system-state tasks and OS services. The problem with this is that it is impractical to run network software at the system-state level, where it can be done more efficiently. Without support for system-state preemption, there is no way for an application-level task to preempt network tasks, regardless of the application task's priority.

While the layered approach has numerous cost and efficiency advantages, it must be done very carefully with an eye to its impact of network response time. A good example of how not to do it is in layered approach used in OSes such as Windows 95 and CE, which rely on the network software to perform tasks at an application level that could be more efficiently executed at the system level by a network-aware OS. This not only increases code size, but results in additional system calls that degrade performance.

There are other problems vis-à-vis memory management. For example, most network software must ask the OS to allocate user memory in order to pass packets up and down the protocol stack. A more efficient strategy is to preallocate memory for the stack at the OS level. In this way, the OS need only manipulate a bitmap, effectively eliminating the need for memory allocation service calls.

A well-designed network-capable OS can boost protocol performance and improve efficiency if it provides automated facilities for allocating and initializing data storage for a protocol's state machines. Every protocol needs initialized storage for its state machines, and this storage must be provided each time a network path is opened and a protocol module is invoked. The way in which this gets allocated and initialized affects not only protocol performance, but the degree to which memory is fragmented as well.

What must also be taken into account is that most network modules have to make service calls in order to perform initialization and allocation functions. Having the OS do it automatically eliminates the service calls. It also minimizes memory fragmentation by taking advantage of the OS's ability to efficiently allocate memory for path and process descriptors.

Something that should also be considered is that in order to effectively support the distributed processing that net-centric computing will need, the network software must be efficiently partitioned between the CPU and peripherals. Indeed, what is needed is a unified I/O system, one in which device specifics are decoupled from application programs. In such a system, rather than referencing specific devices or networks, applications would use generic I/O commands such as read, write, initialize, and terminate. The OS would invoke a driver for the appropriate device based on the system configuration.

Of the OSes that are in common use on network computers, JavaOS certainly comes close to providing all of the requirements, except for the necessary hard real-time capabilities. Another that ranks toward the top is OS-9000 from Microware. Other good candidates are RTOSes from Wind River and Microtec. All three have focused on telecomm and networking applications and have many of the necessary features and real-time determinism as well. Of the various Microsoft alternatives, Windows NT comes closest to providing many of the needed networking enhancements. However, it is neither real time nor deterministic. Moreover, it is also more of a memory hog than Windows 95. Win95/98 is at the bottom of the list as far as many of these requirements are concerned, as is Win CE, which in its initial implementations was aimed at standalone, hand-held PCs and relatively low bandwidth Internet appliance applications.

Chapter 2

The Web-enabled Set-top Box

One of the most ubiquitous forms that the network computer will take in the home will be the familiar set-top box, but vastly modified and upgraded. Indeed, the Web-enabled set-top box will become the main conduit for the Internet into the home. In combination with the emergence of digital television and the availability of much wider bandwidth on the Internet, the Web-enabled set-top box will also be the main mechanism for the delivery of interactive digital services into the home. With the cable companies, regional Bell operating companies (RBOCs), and video entertainment content providers all scrambling for position in this potentially lucrative market, the stakes are extremely high.

There are currently more than 200 million televisions installed in 96 million U.S. homes. Over 85 million of these homes can be accessed by CATV, and there are already 64 million active subscribers to CATV services, forming an industry worth over $20 billion. Virtually every home is also served by the access lines of the $300 billion telephone industry. In addition, the demand for home video can be gauged by the rapid growth during the past decade of the $10 billion video rental industry. These powerful market players are already realigning themselves to deliver digital in-home services, such as video-on-demand, interactive news and information services, interactive games, and in-home banking, insurance and other consumer oriented offerings, all of which require a set-top box to decode them.

Set-top's Internet Challenge

While the interactive TV set-top box presents some exciting marketing opportunities, it also presents the systems designer with some very challenging and fast-changing, technical problems. And the introduction into the equation of Internet access and the World Wide Web along with the growing availability of megabit-per-second cable modem and satellite access, makes for an even more complex matrix of opportunities and problems.

Even when considered separately, both net-centric computers and set-top boxes are very cost-sensitive devices and this restricts the amount of memory in the unit. But together in a Web-enabled set-top box the constraints are even more severe. Typical configurations (see Figure 2.1) provide 1-megabyte (MB) of read-only-memory (ROM) for the operating system; 1-MB FLASH (electrically erasable ROM) for the resident application, system patches, and additional system software; 1-MB DRAM (dynamic read-only-memory) for downloaded applications; and 2-MB RAM for MPEG decompression and graphics. Such memory constraints force the developers of system software and applications to implement a variety of unique and imaginative architectures to enable compelling services. Silicon integration plays a large role in making the boxes cost effective, but this integration can also require the CPU to handle a large set of general housekeeping activities that can potentially affect performance.

In almost every important aspect, the Web-enabled set-top box can correctly be designated the prototype of the future multimedia-enabled net-centric computer. In its ultimate all-digital form, it is connected to a high-speed broadband network capable of delivering data at up to 38 megabits per second (Mb/s). For two-way installations, a return path equivalent to a fractional-T1 is provided. High-end multimedia capabilities are standard, including MPEG-2 video, Dolby AC-3 surround sound, high-quality graphics display, video manipulation and compositing, and locally generated PCM audio.

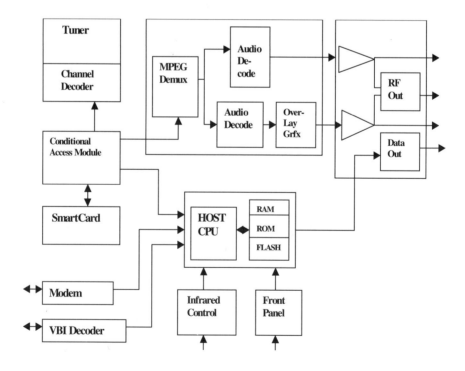

Figure 2.1 **Web-enabled Set-top.** In a typical all-digital set-top box linked to a cable modem for Web access, the basic processor is aided by a complex mix of functions to enable seamless operation at high bandwidth.

But set-tops are also consumer devices and as such must function reliably. There is no reset button. And when the unit does not work that is a direct loss of revenue for the operator. Reliability must be addressed in terms of both the set-top's internal functionality as well as its operation when the network delivering the content becomes impaired. In such a complicated network environment there are many potential sources of failure to deal with while minimizing the impact on the end user.

Another Web-enabled set-top requirement is the ability to support multiple applications executing at the same time in a memory-constrained device. These include the resident application, a downloaded application, and perhaps a background task, such as an SNMP (simple network management protocol) agent. All of this affects how the operating system is designed, because it must be able to support multiple applications and to handle resource management for access resources such as the tuner, network interface, display, etc.

This environment has many challenges for a system developer. For example, the low-memory environment requires very careful management of resources to avoid fragmentation and performance degradation. To support multiple interfaces and housekeeping tasks means optimizing low-latency threads and real-time interrupt management.

As mentioned earlier, the Web-enabled set-top box provides multiple interfaces to both the network and locally attached devices. A forward application transport (FAT) channel delivers MPEG-2 transport streams to the set-top. Depending on the sophistication of the system, up to 38.8 megabits per second (Mbit/s) must be delivered in the bandwidth previously occupied by a standard six megahertz (MHz) analog television channel.. There are multiple FAT channels in cable systems where typically over 150 MHz of their RF spectrum is being set aside for digital transmission. The MPEG-2 transport stream includes system specific information used to control the set-top, such as tuning. In a Web-enabled configuration, conditional access information. Internet protocol (IP) packets may also be delivered in the FAT, in addition to MPEG-2 video and AC-3 audio streams.

A forward data channel (FDC) transmits IP messages to the set-top over a 1.5Mbit/s link that is sent on a separate carrier to the FAT. This guarantees that no matter what channel the set-top is tuned to, messages may be sent to it and received "out-of-band."

For the return path, a reverse data channel (RDC) transmits IP messages from the set-top to the network using a programmable fractional-T1 line, using among other options, a dynamic reservation protocol that enables multiple set-tops to share the same return path bandwidth efficiently.

A limited-bandwidth channel is available through the use of the vertical blanking interval (VBI) in analog channels. The use of the National Association of Broadcasters (NAB) standard for carrying data permits applications to receive data while tuned to an analog channel. This is particularly useful for augmenting analog broadcast video with program synchronous information, such as enhanced advertisement, sports, or weather information.

In addition to the above mentioned video and audio output ports, the new generation of Web enabled set-tops includes two additional output ports: a universal serial bus (USB) port and a 10Base-T Ethernet port. These ports are used for connecting the set-top to a variety of peripheral devices. The types of devices that can be connected to the set-top via the USB port include an infrared (IR) transmitter for VCR control, keyboard, joysticks, and a printer. The types of devices that can be connected to the set-top via the Ethernet port include personal computers and, over time, video game players. Since the set-top uses IP as its data transport for the out-of-band communications channel, the Ethernet interface to a computer is an obvious extension to the set-top. Thus configured, the set-top can act both as a symmetric cable modem and a normal set-top terminal at the same time, giving users the ability to have a faster connection to the Internet for their home PCs.

The Web-enabled set-top box in many ways represents the future of net-centric computing as bandwidths get wider and standards such as MPEG-4 drive the Internet and World Wide Web further in the direction of networked multimedia. And one very obvious requirement is the need for real-time and deterministic response. The Web-enabled set-top box needs to process a large amount of data at very high data rates in a constrained memory environment. The only way to do this is to implement a deterministic system using a decoder model that ensures that the data transmitted to the set-top are received, filtered, processed, and passed on to applications, where necessary, using a buffering scheme that fits into the constrained environment.

The rendering of graphics objects also requires tight real-time control so that objects are rendered on the display without introducing tearing artifacts. Desktop systems solve this by using double buffering. But in the set-top environment the use of memory for extra buffering is a luxury that cannot be afforded.

In designs that take advantage of advances in silicon process technology to integrate virtually the entire set-top system on a single chip, the CPU has to process requests from multiple functional blocks within the set-top and respond in a deterministic manner, making it important that both the processor and the OS have very low interrupt latency and context switching.

Classifying Set-top Box Performance

Many of these requirements can be ameliorated by the system designer, depending on where in the spectrum he or she decides to set up shop with a Web-enabled set-top box. The Interactive Multimedia Association has developed a classification system that divides the various systems into six board categories.

At the lowest level are Class 1 systems, with all analog channels with no return channel, which covers most existing cable channels. Typically, most of the control and housekeeping functions in such a system can be handled with a 16-bit CPU, such as the 68000 or a 386 type processor with no more than a few thousand kilobytes of memory. Class 2 STBs are also analog- based, with a return back channel of limited capacity for use in simple set-top device identification and simple play per view requests, requiring the use of a slightly more powerful processor, such as the 68020/030. Class 3 and 4 STBs retain the same back channel bandwidth as Class 2 systems, but split the download channel between analog and encoded video such as MPEG2. The main difference between the two is that Class 4 allows for a greater degree of interactivity in terms of services. In addition to the processor, an MPEG decode chip is required as is some form of video controller to give the system a graphical rather than a text based look. Class 5 and 6 STBs are all- digital systems. The former is asymmetric with a wider downstream path than an upstream path and the latter provides the same wide bandwidth back and forth, allowing a full range of interactivity. In the digital realm, processing requirements increase substantially. Not only is a high end RISC machine such as a PowerPC needed, but at least 4-Mbyttes of main and graphics memory is also required, not to mention the MPEG chip and its associated memory as well as the network interface modules, error correction circuitry, and modulation-demodulation ICs.

Major silicon vendors (Hitachi, Hyundai, IBM Microelectronics, LSI Logic, Motorola, National Semiconductor, SGS-Thomson, and VLSI Technology, among others) are hard at work developing strategies to reduce the component count and complexity from about 6 to 10 ICs to 1 or 2. Some, such as Motorola, VLSI, and LSI are proposing an exclusively hardware solution, using leading-edge process and ASIC technologies to come up with solutions that integrate all of the elements into as few chips as possible. Others, such as Hitachi, Hyundai, SGS-Thomson and National are proposing mixed hardware-software solutions that depend on the high MIPS (million instructions per second) rates of relatively cost effective RISC processors. With the higher clock rates, the CPUs can handle many of the provider/service specific protocols in software and using dedicated logic to provide only those functions that are specific to particular protocols. Many of the network interface protocols have many commonalties. There are various modulation-demodulation schemes, such as 64 and 256 QUAM and QPSK that have a lot of common features and could be done in software with a sufficiently powerful processor.

While Class 1 to 4 Web-enabled STB designs allow developers to keep a clear boundary between the controller-processor subsystem necessary to process Web and Internet data and the video-graphics subsystem used for TV access, more advanced digital STBs will begin to merge these into the same common set of ICs. Maintaining a separation will be even more difficult as next generation processors such as the PA-RISC and SPARC with multimedia extensions begin their migration into net-centric computing. This means that traditional compilers, debuggers, and emulators will have to be complemented with multimedia-aware tools and design environments. For example, the video function libraries will need to be written in C++ or Java, allowing the exploitation of the object-oriented advantages to simplify data flow between objects, with formats encapsulated in the library elements to reduce the tedium of redefinition at each stage of the simulation.

As of this writing in mid 1998, it is still unclear which route to producing a cost effective, but powerful Web-enabled STB will win out or if any of the will result in the under $300 per box the market demands. In the interim, many cable vendors are scaling back their efforts at Class 5 or 6 STBs and putting more effort into the mid-scale Class 3 and 4 systems, combining largely analog systems with a quasi-digital based back channel that allows a wide range of nonvideo services to be provided.

One of the ironies of all of the effort going into the high end interactive systems is that there have not been any content providers and services out there in the market that justify going to such an expensive solution. However, there are services on the Internet for which the relatively modest bandwidths available on the existing cable plants are more than enough. Even 1 or 2 Mbits/second, which would eat up only a small portion of the bandwidth of existing cable systems, would be a bonanza for many Internet service providers and users.

Already, a number of companies are jumping into the market to take advantage of this convergence. For example, General Instrument (Norcross, Ga.), one of the leading manufacturers of analog set-top boxes for the cable industry is not only remaining active in the trials and initial high volume implementations of advanced digital STBs from the likes of Pacific Telesys, NYNEX and Time/Warner, but has also upgraded its existing analog STB product line, replacing it with its new "advanced analog" 8600X family. Containing the same 68000-based controller used in its earlier analog boxes, it has added a custom graphics chip that generates a simple 16-bit color 320 by 200 pixel graphical user interface (GUI) as well as special logic that turns the vertical blanking interval used for transmission of text translations for the deaf into an upload/download channel for nonvideo services. Each of the 100 or so analog video channels in a typical cable system has eight VBI lines per channel transmitting at 9,600 bits/sec.. This adds up to between 7 and 8 Mbits/sec of bandwidth, more than enough to add a number of "virtual channels" for the delivery of a wide range of nonvideo information services.

Such services might include menu and program information, shopping, banking, and news. Originally the thought was that many of these services could be provided by the cable provider, but with the availability of the Internet and World Wide Web a wide range of services are now available provided the cable vendor installs the appropriate T1 or primary rate ISDN. With the number of news, informational shows, and dramatic series already advertising Web sites, the tie-in and interest in the Internet is already there.

Taking advantage of this conjunction of interests, Scientific Atlanta, General Instrument and a number of cable equipment providers have been working with WorldGate Communications Inc. to provide a new Internet access service called TV On-Line (TVOL). It enables cable systems to offer subscribers low-cost access to the Internet via a simple wireless keyboard-channel changer using a cable-ready TV or advanced analog set-top box. Via a TVOL server located at the cable head end, proprietary software enables the set-top converter to function as a remote Internet terminal and provides two-way communications to the World Wide Web at two to three times the rate of typical analog telephone line modems.

Building an STB's Network Infrastructure

Another set-top box provider -Welcome to the Future, Inc., of Columbia, Maryland - has developed an "information highway access" scheme called ConnectTV (see Figure 2.2). It uses a 386-based set-top box that operates in parallel with existing cable systems, using

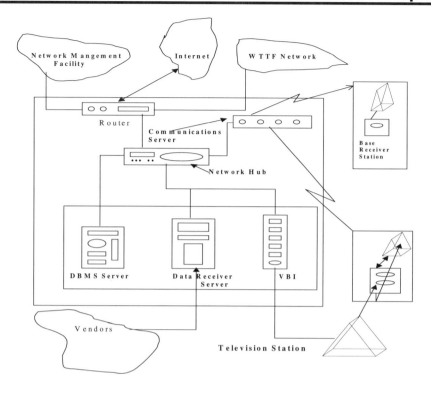

Figure 2.2 **Set-top Box Infrastructure**. For a set-top box to operate both as a TV receptor and as a Web-access device, Welcome to the Future has built a complex infrastructure of servers to provide necessary information.

the vertical blanking interval (VBI) for downloading of information services and a 220 MHz synthesized RF modem for uploading of data from the set-top unit.

While it is neither the largest or the most well known of the companies providing Web-enabled set-top box technology and services, the box and the infrastructure it has developed to support it are illustrative of the challenges facing the system developer. The Welcome approach is a good example of the significant modifications to the underlying infrastructure that must be made to make net-centric computing in the home a reality.

Services provided in addition to the basic TV channel offerings include a variety of text and simple VGA graphics-based services, packaged in hypertext markup language (HTML) form and delivered to the user via a generic Web browser over the vertical blanking interval (VBI). Present offerings that a cable vendor can offer over the VBI link include distance learning, on-line news and information, home shopping, on-the-fly purchasing, and a variety of text and graphics- based games and entertainment. Enhancements include linkage of the system to external telecommunications connections for delivery of Internet Web-pages to the TV viewer that he can select and manipulate via a simple infrared-based hand-held unit or standard keyboard via a wireless link.

In addition to the set-top box, the system architecture for this Internet-enabled interactive system consists of a Base Receiver Station (BRS), a Data Server and a VBI Server. In addition to the above capabilities, the set-top also monitors TV usage for Nielsen type statistics and demographic data. The set-top software also provides the option to globally or individually download new software into the set-top. Software program files can be transferred down from the Data Server to the set-top which can in turn replace the existing version with the new one. The file transfer takes place over the same transmission paths used for data services. When the user has chosen an interactive application and has selected a menu choice, the set-top transmits the user selection-response to the BRS via UHF at 218 to 219 MHz. The BRS forwards data transmission from the set-top to the Data Server. A single BRS can serve thousands of customers..

The Data Server retrieves service data from the various locations, including Internet data sources, modem data services such as news, and direct-connect data services such as airline reservation systems. The Data Server categorizes the service data, formats it for viewing on the set-top, and stores it for interactive use. This Data Server is also responsible for satisfying inbound set-top requests for data. It retrieves the requested data from the database, remote Data Servers, Internet applications, or vendors as appropriate. It then sends this information to the VBI Server for transmission to the set-top.

The interactive data are then returned to the home via the VBI of the television signal. The VBI Server is responsible for collection of the interactive data from the Data Server and formatting it to NAB standards before sending it to a VBI encoder for embedding into the TV signal. The entire process takes 5 seconds to fulfill a customer transaction.

The Data Server is a sophisticated 32-bit server based on Hewlett-Packard's PA-RISC running a UNIX operating system. In addition providing data storage and retrieval for the various interactive applications available, it also acts as an agent in satisfying data requests from the set-top units. It interacts with the vendor server and Internet applications to complete purchase transactions and data requests. The Data Server for a given region maintains a virtual view of each set-top box, each interactive data subscriber, each interactive data provider, and all the vendors within that region. In addition, it also maintains a view of the entire network of Data Servers. Thus, Data Servers may access data associated with other regional Data Servers. Access between Data Servers is provided via connectivity over dedicated, high capacity land lines.

UNIX was chosen as an operating system because of its fault-tolerant capabilities. It quickly and easily performs fail-overs and clustering and manages RAID hardware. A RAID (redundant array of interchangeable disks) used to keep a system on-line in the event of disk failure. More importantly, UNIX systems provide the scalability required to support a system that will need to expand rapidly.

Another critical element in the server infrastructure is the HP LM90, a WindowsNT-based platform that is physically separate from the Data Server, but may optionally reside in the same platform as the VBI Server. It provides connections to the various vendors within a single market. The vendor services are unlimited and include such service capabilities as pizza ordering, grocery ordering, airline flight reservation and ticketing, home shopping, and credit-debit card transactions and confirmations, to name a few. Windows NT provides advanced capabilities for interprocess communications and multiprocessing. It was primarily chosen for the ease with which access to remote systems can be achieved, using the well-known remote procedure call (RPC) protocol..

The VBI Server is also a WindowsNT-based platform that is physically separate from the Data Server but may optionally reside in the same platform as the Vendor Server. It provides data communications protocol and scheduling for data being sent from the Data Server to the various set-top boxes in a given market. The VBI Server controls the transfer of data to the VBI injector which encodes the data and then inserts it into the VBI of the standard NTSC Television broadcast signal.

The Packet Router is a process that runs on either a WindowsNT- or UNIX-based platform. This process may be housed on the same physical platform as the VBI Server, the Vendor Server, or the Data Server. It receives the incoming set-top data transmissions from the BRS and routes them to the appropriate destination as follows: Data requests are forwarded to the Data Server which acts as an agent in satisfying the request and then passes the requested data to the VBI Server for transmission to the set-top. Data retransmit requests are forwarded to the VBI Server, which pulls the requested data packet from a circular buffer of transmitted packets and schedules the packet for retransmission. File Transfer data are forwarded to the Data Server which processes the inbound file. These files are used for statistics, accounting, and troubleshooting. The Packet Router process code is written in POSIX Compliant C++ and can executed with only slight modifications on either the WindowsNT or UNIX OS.

Critical to keeping the cost of the entire interactive client-server architecture within the range of the average cable vendor and viewer was the cost of the Web-enabled set-top box. The key here was the use of mature technologies that are well understood and that have been on the market long enough to achieve significant economies of scale. To this end, in the initial designs the set-top unit used a 25 MHz 386EX embedded microprocessor, complemented with a number of necessary peripherals, all of which have been widely used in other applications. These peripherals include a VBI data downlink with separate tuner, similar to those already used in many TV sets to provide text services to the deaf; an RF modem uplink; a selectable VGA or TV output with separate tuner; an internal 14.4 or 28.8 kilobit per second modem; and circuitry to provide an infrared link to a remote control unit or infrared-based PC keyboard. Rather than use the cable hookup for delivery of the data services, these are provided through an radio frequency (RF) modem that takes a small asynchronous serial data stream and transmits via an antenna located near the set-top unit to a receiving antenna located within 3 to 5 miles of each set-top unit.

In the set-top box, software is provided to strip data from the incoming TV signal and process it. In addition to handling all of the functions involved in cable signal conversion, the set-top box software also handles the display of VGA computer graphics, and composite video capabilities. This allows the display of video in three different formats: standard, full-screen television programs; interactive services menus and applications; and interactive services menus and applications with standard television programs visible in a resizable Picture In Picture (PIP) with computer-generated graphics overlaid on the television image.

Interactive data service displays are based on standard Hyper-Text Markup Language (HTML) display definitions, the same as used in defining Internet Web pages. This provides several benefits, including a standardized display definition, compact file size for display definition, ease of portability between services and Internet services and incorporation of text and graphics specifications into a single display. These HTML files are displayed and manipulated via a "browser" that translates the HTML files into visual displays and translates mouse activity into system events. The browser is similar to many of the existing World Wide Web browsers that are publicly available. All the various text- and graphics-based services are written using the Internet HTML Web protocol and presented to the set-top box user via an X-Windows version of the Spyglass Web browser that has been ported to the QNX RTOS.

In addition to the above capabilities, the set-top software also provides the option to download new software into the set-top. Software program files can be transferred down from the Data Server to the set-top which can in turn replace the existing versions with new ones. Since this file transfer takes place over the same transmission paths used for data services, it allows the set-top capabilities to be modified, fixed, extended, or even replaced without the need to replace the hardware or return of the set-top to a central location. Furthermore, these software upgrades can be made without the user's knowledge during non-use periods or at the user's option.

Other Configurations, Other Choices

Depending on where the systems designer decides to enter the Web-enabled set-top box fray - at the low end with a mixed analog-digital combination or at the high end with an all digital- approach - different choices as to hardware and software will have to be made.

For example, at the low end of the spectrum, indeed, barely registering on the Interactive Multimedia Association's scale is a system from WebTV, now owned by Microsoft Corp.. Neither a set-top box in the accepted sense nor a fully functioning computer with a browser, it essentially bypasses many of the problems outlined earlier. It does this by using the TV set as a display device, with the modem and Internet access performed independently of how the TV set is receiving its video, be it an antenna or a cable hookup.

Manufactured by Sony Corp. and Philips, the WebTV box acquires and displays Web pages provided by a special WebTV Internet service, using ordinary telephone lines with an integrated 33.6-kbit/sec modem. Both boxes closely follow the WebTV specifications, differing only in the choice of processor. The Sony box uses a 112-MHz, 64-bit MIPS R4640 processor while the Philips design uses a 32-bit customized MIPS R3000 core that it has used in a variety of Internet appliances. Both use a special graphics chip designed by WebTV that generates a standard NTSC-compatible TV display, rather than the standard RGB(red-green-blue) output for computer monitors.

During its evolution, the WebTV has used a number of different operating systems, first an in-house kernel and then several off-the-shelf RTOSes. With its acquisition by Microsoft, a move was made WinCE 2.0. Since the current requirements of the system depend on a relatively low speed 33.6 kbits/sec modem connection to an analog tele-

ITEM CHARGED

Patron: Ashish A Desai
Patron Bar 60179600188808033
Patron Gro Undergraduate
Student

Due Date: 3/8/99 23:59

Title: Emergence of
net-centric
computing :
network computers,
Internet appliances,
and connected PCs
/ Bernard C. Cole.
Author: Cole, Bernard
Conrad.
Item Barco 3934600467691

phone line, the RTOSes were used not for their real-time determinism but for the small memory footprint. Win CE 2.0 should be more than adequate as far as its interrupt response is concerned at this data rate, but requires much more RAM and ROM than its RTOS competitors. However, in higher performance asynchronous transfer mode, ISDN and cable modem applications, the data rate may be such as to stretch WinCE's ability to handle them.

Trying to fit into both worlds, France's Minitel system, the government-sponsored interactive email system that was introduced to the general populace in the mid-1970, has been moved into the world of interactive TV and Web-access with the introduction of an upgrade to the original test-oriented terminal that every French citizen had installed in his or her home. Built by Com 1 Communications of Bordeaux, France, around the 32-bit MediaGX, a low cost multimedia-capable x86 clone from Cyrix Corp., now a part of National Semiconductor Corp,, the system is called the SurfTV (see Photo 2.1) and can operate in either of two Web-enabled set-top modes, as either an IMA Class 1 analog system or as am all-digital Class 4 or 5 mode. In the first mode, video and video and audio is received via a standard television connection, and Web-access is provided via a 33.6 or 56 kbps analog line or a 128 to 256 kbps ISDN connection. In the second all-digital mode, it accesses the cable system via an 10 megabit/second Ethernet interface that connects to a high speed cable modem. Because of the much higher data rates and the need to deal with both video and digital data in real time, it uses the QNX RTOS to provide the necessary deterministic interrupt and response time performance. For the user interface it uses the QNX's own small footprint Photon GUI and a proprietary browser it calls Voyager.

Photo 2.1 **SurfTV's Up.** Built around 32-bit MediaGX, French Minitel's SurfTV Web-enabled set-top box replaces older text-based home terminals with multimedia-based unit that operates with analog TV and 33.6 kbps modem or as all digital cable modem system. (*Source*: Com1).

At the other end of the spectrum is a system implementation by Hong Kong Telecom IMS Ltd., (see Figure 2.3) which is designed to deliver Java- based applications and interactive services such as video-on-demand (VOD), home banking and home shopping to homes. It uses a Web-enabled set-top box built by NEC Ltd. that uses a PowerPC 602 CPU from

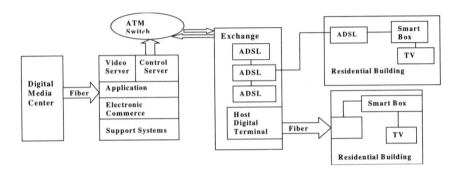

Figure 2.3 **ATM-based Web TV**. Using Microware's OS-9 operating system, Hong Kong Telecom has built a multi-megabit/sec. digital set-top system that allows both TV and Internet connections to operate in real time without compromising quality.

IBM, a MPEG-2 chipset, 12-megabytes of DRAM, 2-MB ROM, and 2-MB flash, all of which communicate with each other over a standard PCI bus. It also has an asynchronous transfer mode (ATM) interface which allows delivery of digital video and Internet/Web functions at very high data rates, on the order of 1,000 to 10,000 times what is now possible into the average home. The system uses Microware's OS-9000 real-time operating system with Java extensions and DAVID (Digital Audio-Video Interactive Decoder) system software platform for digital television.

The challenges faced by designers were in two areas: presentation and infrastructure. Internet applications are designed to expect a traditional PC or workstation as the presentation vehicle, not set-top boxes that in the past have made use of slower CPUs, less RAM, no permanent storage, low-resolution TV displays and lower bandwidth. Since the set-top box designed by Hong Kong Telecom has a powerful 32 PowerPC CPU and as much dynamic RAM as many desktop systems, as well as the high bandwidth ATM-based network architecture, many of the traditional PC versus set-top box design issues did not arise. The box is powerful even when compared to some of the U.S. all-digital set-top box systems and the ATM architecture allowed Hong Kong Telecom to experiment with different infrastructure bandwidth schemes to improve network performance.

Over the long term, Hong Kong Telecom expects to implement a complete net-centric computing infrastructure based on the Common Object Request Broker Architecture (CORBA). The distributed object software architecture, in combination with Java allows a great deal of flexibility in application development since the objects requested by Java can be dynamically de-referenced by the Iona's Orbix Web CORBA object request broker. This gives Hong Kong Telecom substantial control over the applications themselves and lets them ensure that Web-oriented content is presented effectively on a standard TV display, in this case, using the European PAL rather than U.S.'s NTSC format.

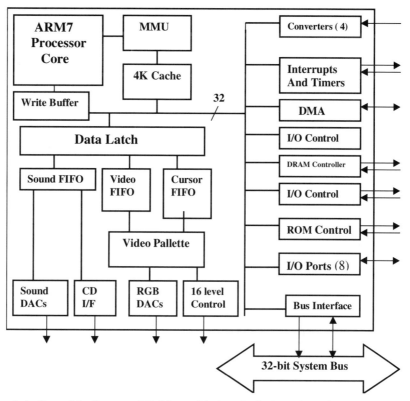

Figure 2.4 **One-chip Set-top.** Working with Arm Ltd., U.K.-based cable and Internet provider OnLine Acorn Media has integrated almost all of the basic elements of a Web-enabled set-top box onto a single chip, the ARM7500.

OnLine Acorn Ltd.'s Integration Effort

Illustrative of the high levels of integration that will be needed in both today's Web-enabled set-top box, and tomorrow's multimedia-enabled net-centric computers is a joint effort underway by Advanced RISC Machines Ltd. and OnLine Acorn Ltd. They developed a highly integrated microprocessor chip - the ARM7500 - with sufficient processing power and the right mix of functions to satisfy a high bandwidth Web-enabled set-top box design (see Figure 2.4). In addition to the core 32-bit RISC processor, the chip also includes all of the necessary user I/O, a separate dedicated graphics/text processor, sound DACs, serial I/O, DRAM/ROM controller, and some basic glue logic. It uses unified memory for both the video buffer and main program to reduce memory costs. Performance is maintained by using direct memory access and other hardware to transfer the video, cursor (mouse), and sound data. When clocked at 33Mhz, the chip delivers 30 MIPS of processor performance. It is designed for use with MPEG 1 or MPEG 2 decoders. The next stage of integration will see the ARM processor combined on chip with both an MPEG 2 video coprocessor and an AC3/MPEG audio coprocessor. This will eliminate the need for separate MPEG memory space.

Using the ARM7500, Online Media is looking at several approaches to delivering high-bandwidth digital video and on-line Internet and Web services over the local loop and into the home. Currently, Online Media , and several regional Bell operating companies (RBOCs), have opted for Asynchronous Transfer Mode (ATM) to pipe services into the home. Another approach is to use a combination of traditional coaxial and digital Integrated Services Digital Network (ISDN) telephone lines. The asymmetrical digital subscriber lines (ADSL) , an upgrade of the traditional telephone line, is also being considered as a Web-enabled set-top box interconnection.

In the United Kingdom, user trials of video-on-demand (VOD) and other interactive services such as home shopping, are being carried out in Cambridge by a consortium of companies including Online Media. The aim is to assess both the technology and the services. The trial system uses ATM. In the Cambridge set-up, broadband services are delivered to the home via ATM switching modules, each serving a cluster of houses. Multi-channel services are delivered over fiber-optic links running at 155 Mbits/sec. to the curbside unit and then distributed to individual houses at rates of up to 2 Mbits/sec.over coaxial cable and twisted pair. The curbside switch uses Asynchronous Transfer Mode switching technology developed by Advanced Telecommunications Modules Ltd., Cambridge.

Learning By Doing

As these examples illustrate, if a designer who wants to prepare himself or herself for the future of net-centric computing, the one place to get the right training - and on-the-job training at that - is to get involved in a Web-enabled set-top box design. The issues with which the designer must deal - real-time versus non-real-time performance, integration of digital video and on-line network access, standard processors versus highly integrated application-specific designs — are the same ones that designers will have to face as the Internet in general moves to higher data rates and as standards such as MPEG-4 establish a beachhead.

Chapter 3

The Evolution of the Net-centric Appliance

The confluence of increased consumer interest in the Internet and low-cost processor technology of sufficient performance has made possible a wide range of Web-enabled consumer computing devices that directly appeal to the average nontechnical user. Most, if not all, are taking take advantage of the built-in capabilities, services, and information resources now available on the World Wide Web via the Internet. And as the Web and Internet evolve, the capabilities of these net-centric appliances will continue to grow.

These net-centric appliances are evolving in two directions: as Web-enabled information appliances and as ordinary appliances and consumer electronic devices that can be connected to the Internet. The first are essentially devices that take advantage of the information available on the Internet and World Wide Web and present it to the consumer in a compact form on an Internet-enabled telephone or as a handheld PC with wireless connection to the Internet. The other type of net-centric appliance is farther in the future and will typically be the existing consumer electronics device, such as a VCR or stereo system, but with an Internet access built in, using the TCP/IP protocol. The initial use of this second type of net-centric connectivity will be by the providers of the variety of consumer electronics and home appliances of all sorts. What this kind of connectivity will give them is access to the devices in which a variety of microcontrollers and microprocessors have been embedded. Using this connectivity, they can now make such devices even more reliable by allowing the designers to correct, modify and upgrade the capabilities of their products by accessing the innards of the electronics to change the programming. Further in the future is the promise of the "electronic home" in which the consumer has access to and control of a variety of consumer electronic devices and appliances via his or her desktop computer.

Web-Enabled Telephones as Information Appliances

One type of Internet appliance that has hit the ground running is the Internet-enabled telephone, a new product category that combines telephony features with computing and Internet access capabilities. Specifically designed for consumers who are not familiar with PCs, the Internet smart phone integrates hardware and software to provide a low-cost, easy-to-use tool. Not coincidentally, this is a design philosophy that is much the same as that taken in the construction of the original telephone.

Although the computer industry has assumed that Internet access would become available in the home to the average consumer via Web-enabled set-top boxes, folding a Internet access into a wired or wireless connection to the telephone network has many advantages. While TVs are primarily for entertainment, often for an entire family, a screen phone is a natural information-gathering tool for individuals. Financial programs (including home banking and stock trading) and email are good applications for the Internet-connected smart phone.

Many home financial applications have already migrated to both the telephone Touch-Tone pad and the computer, requiring two devices for home access. The Internet smart phone brings all the necessary capabilities together in one device. Moreover, combining voice and email messaging with smart-phone features such as message forwarding and caller ID creates a powerful communications package. From a list of voice messages on the screen, for example, users could select any message and have the phone automatically call back the person who left the message based on caller ID information.

Maximizing the ease of use and functionality of such a product, while minimizing cost, is possible only with a high level of hardware integration and support from a capable operating system. In both the standard telephone connection and in the wireless form as well, systems designers are addressing the issues of what services to provide and the technologies needed to provide them.

Both wireless and tabletop versions of Internet-enabled telephones are now entering the market. Some of the first generation of wireless units included AT&T's PocketNet, built by Mitsubishi and Samsung, and the Nokia 9000i. The PocketNet is a 9.6 oz. unit, about the same size as a cell phone, with a 1 by 2 in. screen and a calculatorlike keypad for single finger entry. The Nokia unit is weighs 13.9 oz., measures 3 by 5 in., and has a 4.75 by 1.5 in. screen and a 65-key QWERTY keypad. Because of the limited wireless bandwidth, currently 9,600 bits/sec. and projected to reach 28.8 kbits/sec., real time response is not required, so both use a Windows look-alike, the Geoworks OS.

As far as tabletop units are concerned virtually very major long-distance and local provider of telephone services (AT&T, Southwest Bell, Pacific Bell, U.S. West, Bell Northern and Nokia, among others) has come up with designs that combine Internet access with ordinary telephone services. Illustrative of some of the trade-offs that must be made to achieve Internet and Web access in such small-footprint devices is the SmartPhone from Navitel, which has been acquired by Microsoft Corp.

Designing an Internet Smart Phone

To achieve the mix of capabilities envisioned for the Internet smart phone, it was necessary to combine several elements. First, to enable users to access any page on the World Wide Web with a standard browser, for example, the phone had to incorporate a full VGA screen (640 by 480 pixels). Almost all Web pages as currently defined display correctly at that resolution. A lower resolution would force users to scroll repeatedly to read many text displays, and some graphics would not display correctly.

For the most versatile Internet access, the Internet-enabled smart phone must use standard Internet access protocols such as TCP and PPP. These protocols give the phone users the same broad choice of Internet service providers available to PC users. As we shall see later, by contrast, some consumer-oriented Internet access solutions work through a specialized server, which becomes a bottleneck when large numbers of users are involved. This latter type of service can also lock users into high-priced, low-featured arrangements that could disappear at any time and leave users with a unit that is useless for any other means of access to the Web. The use of standard protocols in the Internet smart phone avoids these problems and permits use of Internet functions such as email and Web browsing. A full QWERTY keyboard is also necessary to allow users to enter alphanumeric information. The design that the Navitel developers came up with, in addition to a VGA-sized LCD screen, includes a keyboard that slides out of the front of the phone when needed. Instead of a mouse, Navitel designers assumed that most users, especially in the home environment, will rely on the phone's touch-screen capability. For business users, a version with a mouse was designed. Navitel developers found that pointing devices such as the mouse are surprisingly awkward for most novice users, but that a touch screen is fairly intuitive. In addition to providing a pointing method for Web interaction, the touch screen is used to control telephony and other communications applications (voice mail, caller ID, etc.). The Navitel unit is also a full function telephone, with a speaker phone, digital answering machine, and support for the Analog Display Services Interface (ADSI) and functions such as caller ID. As mentioned earlier, these telephony features are integrated with the phone's computing resources.

Thus, users access both voice messages and email from one mailbox. To make this possible, a Navitel application program extracts data from the system's off-the-shelf email manager, combines these data with voice-mail information, and displays listings for both types of mail in one integrated view. The phone can also make intelligent responses based on caller ID. When a specific caller ID is detected, for example, the phone could provide a distinctive ring, page the user, play a personalized voice message, send an email message to a specific address, or furnish name and address information available from services on the Web. The ability to use a popular Web browser is important because HTML and other Web-related technologies are evolving rapidly. To keep up with the latest technology, users must be able to take advantage of the leading browsers.

Enabling Technologies

Delivering the functions needed for the Internet smart phone would be prohibitively expensive if a traditional PC architecture were used. Navitel's goal was to offer benefits similar to those of the PC (including versatility and upgradability) without the PC's drawbacks (mainly cost and difficulty of use for novices). Fortunately, many of the resources and most of the expense of the PC architecture are unnecessary. The product needs no floppy or hard disk drives, no mouse, and no x86-compatible processor. The phone does require a fairly large amount of DRAM (though less than some low-end PCs), but the costs of this memory and the phone's VGA LCD panel have become quite reasonable. Beyond these items, the phone requires several crucial hardware and software technologies to meet its ambitious performance and versatility goals.

Most of the phone's processing and peripheral functions are provided by a two-chip set from Philips Semiconductor (see Figure 3.1). The TwoChip PIC was originally designed primarily for use in hand-held PCs (HPCs), and it supports the Windows CE platform by Microsoft. The TwoChip PIC includes the PR31500 all-digital IC and the UCB1100 mixed-signal device. The PR31500 is a customized version of the MIP R3000 RISC CPU for which Philips is a licensee. In addition to on-board cach memory and a controller for several types of external memory, the Windows CE-compatible device includes a video controller, dual UARTs, an infrared interface, a real-time clock, counter-timers, a power-management subsystem, and a number of different I/O ports. The video controller in the PR31500 provides all the circuitry, except the LCD drivers, necessary to drive the Internet smart phone's VGA screen. Because a portion of main memory serves as a video buffer, the PR31500 needs only one memory subsystem, thus reducing costs, integration effort, and

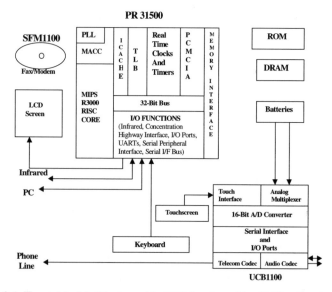

Figure 3.1 **Two-chip IA**. The two chip PIC developed by Philips Semiconductor is based on the 32-bit MIPS CPU and the Windows CE 2.0 operating system and allows telephones to be turned into Internet information appliances.

system size. Tightly integrated with the PR31500, the UCB1100 furnishes audio and telecom functions along with other mixed-signal capabilities. Features include audio and telecom codecs, a touch-screen interface, four general-purpose analog-to-digital converter inputs, and 10 programmable digital I/O lines.

In keeping with the goal of minimizing external components, the UCB1100's four-wire touch-screen interface connects directly to the phone's resistive touch screen. The UCB1100 includes the necessary switch matrix; voltage- and current-generating circuits; and sensing and control logic, including a 10-bit analog-to-digital converter. The interface measures x/y position, pressure, and plate resistance to detect valid inputs and their positions. The UCB1100 can generate an interrupt when a touch is sensed. Similarly, the UCB1100's 12-bit sigma-delta audio codec interfaces directly to a microphone and speakers. No other components are necessary because the UCB1100 includes amplifiers and digital filters, plus digital-to-analog and analog-to-digital converters. To support a variety of different audio streams, such as those needed for telephone, speakerphone, and voice-messaging services, the codec allows digital settings of sampling frequency, sound volume, noise shaping, anti-aliasing, mute, and loop-back.

Conspicuously absent from the Navitel unit is a hardware modem. The omission might seem strange for a product that connects to the Internet via analog telephone lines, but, in fact, the Navitel Internet smart phone uses a V.32bis software-based fax/data modem (soon to be upgraded to V.34bis). Part of a suite of DSP software functions licensed from Philips, the modem software runs on the PR31500's processor core and takes advantage of a multiply-accumulate unit in the PR31500 and the telecom codec in the UCB1100. The use of the software modem saves a significant expense. An external fax-data modem chipset would cost about $30, about as much as the PR31500, UCB1100, and software modem combined. In addition to the modem, the DSP software suite includes routines for voice compression (used for the digital voice messaging service) and acoustic echo cancellation (useful for the speaker phone). Also important for any smart phone today is support for Bellcore's Analog Display Services Interface (ADSI). This interface provides services such as caller ID and caller identity delivery on call waiting. The signal processing software module available from Philips takes advantage of the TwoChip PIC's vector-processing capabilities to meet ADSI signaling requirements. In addition to simplifying development of features such as selective call blocking and automatic call back, the ADSI software enable the Internet smart phone to interact with home banking services.

Developing A Standardized Platform Architecture

More than any other implementation of the net-centric computer, Internet appliances will over the years exhibit a wide range of forms and functions and use a wide range of CPU and operating system alternatives. This evolution and proliferation of devices will continue until consumers determine which form appeals to their tastes and needs or until some form of de facto standardization occurs: either imposed by one company or group of companies or via a commonly agreed to specification.

In other areas of net-centric computing such standards have emerged. On the desktop, Intel and the variety of computer makers have agreed to a common NetPC platform. In the corporate-oriented networking computing arena, Oracle Corp., Network Computing Inc., and Sun Microsystems have done a good job proselytyzing a common NC specification. In the set-top box market, cable providers have started to come to a common agreement on a set of technical requirements that a Web-enabled set-top box must provide, as well as a common cable modem standard. .

Some of the larger consumer electronics companies, such as Thomson, Philips Electronics, NEC and Sony, through their sheer size and the fact that they play in so many of the key consumer markets, may be able to achieve the first kind of standardization in Internet appliances. And Microsoft and Intel (Wintel in the industry parlance) are betting that their dominance in the desktop PC market will give them an edge in the newly emerging net-centric computing market, even in consumer-oriented products such as Internet appliances.

More likely, though, standardization will occur only with a commonly agreed to set of hardware and software specifications. In the consumer-oriented, Internet-connected information appliance area, one effort at "rationalizing" the design of such devices is a hardware/ software architectural specification (see Figure 3.2) developed by by Diba Inc., which has been acquired by Sun Microsystems, which has gained favor amongst a number of integrated circuit companies and consumer product makers. Unlike present point source solutions which address one specific issue (hardware, software, applications, etc.), Diba's information appliance architecture is all encompassing. It deals with both hardware and software requirements, from both the client and the server points of

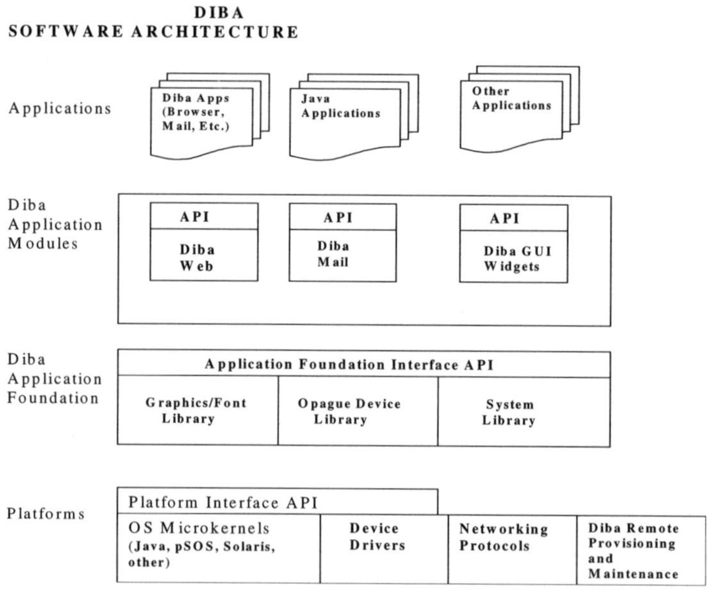

DIBA
SOFTWARE ARCHITECTURE

Applications	Diba Apps (Browser, Mail, Etc.)	Java Applications	Other Applications

Diba Application Modules	API / Diba Web	API / Diba Mail	API / Diba GUI Widgets

Diba Application Foundation	Application Foundation Interface API		
	Graphics/Font Library	Opague Device Library	System Library

Platforms	Platform Interface API			
	OS Microkernels (Java, pSOS, Solaris, other)	Device Drivers	Networking Protocols	Diba Remote Provisioning and Maintenance

Figure 3.2 **A solid** software architecture developed by Diba, Inc., a variety of consumer electronics companies are building Internet-connected information appliances using a set of standard interfaces and building blocks.

view. As defined by Diba, the architecture of such an information appliance consists of several layers. First there is the platform, consisting of a microprocessor, a real-time microkernel, peripherals, and memory. Next are applications, which provide and even define the bulk of the user interface. Interaction with the hardware and software is not done necessarily via a mouse or any other sort of computer-oriented input device, but most commonly via function-specific functions, either on the screen or in physical form, as buttons on the actual device. This is in startling contrast to connected PCs, network computers, and NetPCs, which are by and large aimed at existing users of computers already familiar with graphical user interfaces and the other sophisticated accoutrements of that environment. A primary function of the information appliance is to keep the appliances as easy to use as possible and as "thin" as realistically possible relative to microprocessor intelligence, memory space, and user interaction. Most of the applications in this conception are located out in the network, on the servers, with only results distributed on demand to the specific Internet-attached information appliance.

There are five basic elements in the Diba software platform: device drivers, networking protocols, microkernel, remote provisioning and maintenance software, and the operating system specific layer. Diba device drivers support hardware components designed for a particular appliance platform: network connections, I/O devices, graphics chips and touch screens. Only device drivers that correspond to the specific hardware for a particular appliances are included in any particular implementation, considerably reducing size and cost. Developers can license the Diba device drivers as is, modify them, or add their own, provided they include all the "hooks" that link to the rest of the architecture. Diba has designed a portable TCP/IP stack for network access so that it can reside on any microkernel. It can support either TCP/IP or UDP/IP, the two most common Internet protocols, and at the transport layer can communicate with both SLIP- and PPP-based hosts.

In terms of operating system support, the Diba software architecture gives the systems designer two ways to go. First, it is possible to use essentially any real-time kernel that meets system requirements as far as interrupt response, program and application memory size are concerned. The only proviso is that it offer certain basic functions, including multitasking and multithreading support. As of this writing, operating system kernels that have the features that meet these basic minimum requirements are RTOSes such as pSOS from ISI and the ITRON kernel developed by Japanese consumer, home, and industrial electronics firms. Also satisfying these requirements, but certainly not meeting the small-footprint requirements of most Internet appliance applications are Windows NT, Linux and Sun's Solaris operating systems. Others are being added constantly. Diba, in addition, offers its own native mode real-time microkernel, optimized specifically for use in a wide variety of information and Internet appliances. Highly portable and easily configurable, the Diba OS is based on the use of multithreading, message queues, semaphores, mutexes, and timers. In the Diba OS, interrupt latency is deterministic and does not increase as the number of tasks increases. Important in information appliances where memory space is at a premium, the Diba OS requires no more than 100 kbytes at most.

Constructed on this foundation of hardware and software is the Diba Application Foundation. It consists of a graphics library, a font library, an opaque device library, and a system library. Developers can write code directly to the Foundation Application Interface to create customized applications or use a number of high-level application modules and applications already created by Diba. The graphics library includes a powerful set of software functions for drawing and displaying graphics, fonts, and images and is designed to support a range of displays including VGA, LCD and ordinary TV. The font library supports most industry standard fonts, font styles and display types. It also includes antialiased characters for TV display to achieve clear and crisp displays on lower-resolution TV sets, as well as for a variety of LCD and computer monitor types. Two extremely important components in the Diba software architecture are the opaque device library and the system library. The first is a means by which developers can create a range of applications that will support most peripherals without getting into the details, through the use of a single interface for multiple types of devices. So far, the device drivers that are supported include network connections such as standard analog telephones, RS-232 serial links, and Ethernet; I/O devices such as keyboards, mouse, and infrared remote controls; parallel printers; touch screens and smart cards; flash memory and a variety of display devices.

Even more critical to the operation of the Internet appliance is the system library consisting of five components: memory management, error handling, thread and state management, messaging and real time clock-timer interfaces. Cognizant of the fact that the consumer market requires a much more hardy system than even traditional industrial systems and certainly much more than the traditional desktop, a lot of attention was paid to making the various elements in the library as robust and fault tolerant as possible. In the Diba memory management scheme, for example, a robust mechanism for detection of memory leaks was built in, as was a more efficient means of handling caching. When the application downloads content from the Web, the memory manager caches the content for use later. If the application requests content that is already cached, it is displayed immediately. When full, cache contents are disposed of on a least recently used basis. Also incorporated is a memory compaction algorithm that corrects for the memory fragmentation that always occurs over time in devices used for extended periods. For error handling, a "catch and throw" algorithm was used. With this model, errors are automatically directed to the code segment that can take appropriate action, replacing it, or redirecting them to a general error handler.

To speed application development time and move more quickly to market with the finished device, Diba has an ongoing effort underway to create not only complete systems applications, but also application modules for Information Appliances that can be used as is or customized for a specific application. For example, rather than use existing browsers, Diba has built its own from the ground up, optimizing it for Internet Appliances such as smart phones. For example, the browser interface does not employ scroll bars or pull-down menus. Diba's mail application has been developed in a similar way. It includes only five different screens, none of which requires a pull down menu or a scroll bar. The various modules (Web browser, mail, GUI and telephony) are provided with a tool kit that includes the C-code.

A Mark-up Language for Information Appliances

Because of memory and display limitations, in addition to the design of the basic unit, the systems developer will have to tailor the infrastructure to fit the capabilities of the hand-held or telephone-set-sized Internet-enabled information appliance. For example, just as Sun has come up with a subset of its full Java specification, called personalJava, for use in such devices, so, too, will modifications have to be made in the very foundation of what makes the World Wide Web so ubiquitous: the Hypertext Markup Language (HTML). The most widespread alternative to HTML is a specialized adaptation proposed by Unwired Planet, Inc., which it calls HDML, for Hand-held Device Markup Language. Submitted to the World Wide Web Consortium as a candidate for standardization HDML is part of a package that Unwired calls UpLink, which also includes a minibrowser and a gateway server. Similar to one of the first successful hypertext programs, Apple's Hypercard, HDML uses a card metaphor. Each card represents a task to the Internet-connected wireless or smart phone to carry out. There are cards, for example, for displaying data, for displaying menu selections, and for prompts to capture text that the user enters. All the cards in a deck constitute a single transaction and decks are allowed to exchange data and processes with other decks.

HDML is not a true markup language, as such. Rather, it is more a set of commands or statements that specify how a hand-held device, such as a smart phone, interacts with a user. HDML statements display information on a phone and specify how the phone responds to user input. For example, an HDML statement can instruct a phone to display text and, based on the user's input, either display additional text or send a request to the UpLink gateway server.Because HDML is designed to be used on a variety of hand-held devices, with significantly different capabilities, it is defined with respect to a reference, or abstract, phone. This abstract device has a minimum feature set, including a fixed width character display, vertical scrolling, ASCII printable character set, a message-waiting indicator, numeric and alphabetic character entry, choice selection using arrow or numeric keys, accept or previous card keys, up to two programmable keys. It also has commands to direct the device to store the current state, including the current deck, card, choice, activity stack, variables, and destination card. The abstract phone that all physical imple-mentations must support includes a number of optional features as well: variable-width characters, Latin and Unicode character sets; bold and italic characters; and bitmap graphic display.

At a minimum, HDML supports four types of cards: entry cards, which display a message and allow the user to enter a string of text; choice cards, which display a list of options from which the user can choose a single option; display cards, which simply display information; and no-display cards, which execute an action but do not appear on the phone display. Under HDML when an Internet-enabled phone receives a deck, it normally goes to the first card in the deck, displays the information on the card, and allows the user to respond to it. Depending on the card type, the user can respond by entering text or choosing an option and then pressing a function key. The phone maintains a history stack of the cards that a user visits, enabling the user to navigate backward through the cards by pressing a previous key. And each time the user navigates forward to a card, the phone pushes the card onto the history stack. Each time the user presses the previous key to navigate backward, the phone pops the current card off the stack, leaving the previous card at the top of the stack.

As must be clear by now, this approach differs significantly from the model used by most Web browsers. Where most browsers provide Back and Forward commands that allow a user to navigate backward and forward through the history, in HDML the current card is always at the top of the history. HDML applications are structured around activities, which are tasks that he user wants to carry out.

As part of the Up.Link package, Unwired also supplies Up.browser, a small footprint browser and messaging "engine" that flashes messages that email is waiting or that the user should call the office. This is similar to protocols in many of the pagers that are now in common use. Unlike the traditional browser, where all the capabilities are resident on the client system, in the Unwired approach, many of the capabilities are resident on a specialized gateway server, called Up.Link Gateway. It is the gateway server that makes the HDML approach work. It is the mechanism by which the smart phone or wireless unit is connected to traditional Internet and Intranet servers. In addition to handling many browser functions, it also acts as the translation service taking user commands in HDML and converting them into standard HTTP code, and vice versa.

The Future Of Information Appliances

Where is the Internet-connected information appliance going? Certainly Java will have an impact as will real-time video, especially as the data pipelines get larger. Basically, as originally conceived, Internet-enabled appliances such as smart phones and wireless personal digital assistants (PDAs) are designed simply as information-access devices. The introduction of Java and features such as dynamic loading and the ability to run programs both remotely and shared with the appliance expands greatly the possible features that could be added. Among the many things that developers of such devices will have to take into account are the needs and desires of the average consumer. Unlike many PC users, who are somewhat more open to spending an hour or so learning to operate a system or to run a program, no more than 5 or 10 minutes will be all the user of an Internet appliance will allow.

There is also the issue of memory. Unlike a network computer or a Web-enabled set-top box, where anything less than 4 Mbytes is small, memory is more constrained in an Internet appliance, where 1 Mbyte of user memory is about the limit. As we shall learn later, a lot can be done to pare down the memory requirements that Java imposes, but it requires discipline and careful attention to including only those features that are absolutely required. What this means is that the content of every memory location needs to be justified. For example, features such as double precision floating-point numbers, a feature of Java, makes no sense in most Internet appliances. More than likely all that will be needed are Boolean, byte, and short integer types. Not required are things such as 16-bit Unicode characters, 32- or 64 bit integers or 32- or 64 bit floating point operations.

What will be required, however, are the full suite of Java programming constructs, including all operators, flow of control statements, and object-oriented programming features. As far as system services, many Internet appliances will not be supporting more than one operation at once, therefore, the ability to run multiple processes and threads will be not necessary. And depending on how the application is designed, exceptions and garbage collection may not be needed either. With dynamic memory space at a premium, there had better not be any garbage to collect. On the other hand, identity and security features such as PIN handling and file access control and basic terminal functions would be needed. As far as memory space requirements are concerned, WinCE and personalJava should be more than enough for some types of information appliances, such as Web-enabled screen phones, but not for all wireless devices, which no doubt would be better served by the use of a much more compact RTOS.

Megabit/Second Wireless?

Systems designers will have to pay particular attention to their choice of operating system in such an environment. Real-time performance in such devices is not a major factor, as long as data rates to not go much beyond 28.8 Kbytes/second. In wireless Internet-connected information appliances, in particular, the bandwidth is likely to lag the rest of the Internet by many years. Current implementations are doing well to achieve data rates of 9,600 bits/sec. So in this kind of environment alternatives such as Win CE 2.0 and personal Java should provide more than enough performance.

That may change as initiatives such as IMT 2000 gain momentum. Proposed by a consortium of most of the world's major manufacturers of wireless telephones, this standard will push data rates over wireless phones to as high as 2 to 4 Mbits/sec., enough bandwidth to start sending multimedia data over the ether. In this environment, efforts by Sun with Java and Microsoft with WinCE will likely find it hard to keep up with the real-time requirements of this new medium.

This possibility has occurred to many of the major wireless telecom companies such as Ericcson, Motorola, and Nokia, who together have backed development of a new real time operating system targeted specifically at this new megabit/sec.wireless environment. The real time operating system is called EPOC32 and was developed by Psion plc specifically for wireless systems is now being developed further by Symbian Ltd., to extend the OS into the real time IMT 2000 environment as a part of the next-generation Universal Mobile Telecommunications Standard (UMTS). Among the advantages EPOC32 has to wireless phone and Internet appliance makers is its scalability from a very small kilobyte sized kernel to several megabytes, which would allow it to be used in all sorts of wireless units from smart phones up to multimedia-capable units. It supports Internet connectivity and allows customizable user interfaces, important to wireless vendors because gives licensees to differentiate their products. By contrast, Windows CE has a common look and feel across all platforms, which makes it hard to builders to establish a separate identity in the market. A real-time, multithreaded OS that currently only runs on the ARM architecture, EPOC32 is written in C++

Connecting Ordinary Appliances To the Internet

There is an even more fundamental change to the Internet and the way it is implemented in progress. It is the use of the Internet and its HTML, HDML and HTTP protocols as a mechanism for providing engineers, and ultimately the average user, with access to ordinary appliances and consumer electronics devices, such as microwave ovens, stereo players, and VCRs.

The mechanism for providing such capabilities has a number of different names -daemons, servlets, and embedded Web servers, micro-servers, micro-webservers, and picoservers, among others. Virtually every embedded software tool vendor has entered the market with such products, targeted at numerous applications that in the past were deemed impractical. This was because there was no low cost and code-preserving or memory-conserving way to achieve such functionality. Such servlets can be used to allow an engineer to remotely diagnose an appliance or consumer electronics device with an embedded processor or controller by simply dialing up the application and accessing it with a standard browser.

TABLE 3.1
Selected Web Servelets

Company	Agranat	Allegro	Magma	Quiotix	SpyGlass	Web Device
Product	EmWeb	RomPager	Lava	QEWS	MicroServer	Pico Server
Version	3.0	2.0	2.0	1.0/2.0	2.0	5.0
HTTP Version	1.1	1.1	1.0	1.0	1.1	1.1
Security						
— Standard Auth.	Yes	Yes	Yes	Yes	Yes	Yes
— Digest Auth.	Yes	No	No	No	Yes	Yes
— SSL	No	No	Yes	No	Custom	Yes
Compression	Proprietary	Dictionary	Proprietary	GZIP	None	ZIP-like
Code Footprint						
— Min	10K (CISC)	10K (CISC)	15K (CISC)	45K (x86)	35K (x86)	15K (RISC)
— Max	30K (CISC)	40K (CISC)	40K (CISC)	80K (x86)	110K (x86)	35K (RISC)
— Typical	20K (CISC)	20 - 25K (CISC)	18 - 25K (CISC)	70K (x86)	35 - 50K (x86)	30K (RISC)
Connection Manager	Single thread or OS bindable	Single thread	Single thread	Single thread	Single thread	Multiple threads
Licensing						
— Source	Included	Included	Included	Included	Included	Optional
— Scheme	buyout	buyout	buyout	buyout or royalty	royalty	buyout or royalty
Source Language	ANSIC	ANSIC	ANSIC		ANSIC	ANSIC, JAVA
Web page URL	ww.agranat.com/	www.allegrosoft.com	www.magmainfo.com	www.quiotix.com/	www.spyglass.com/	www.webdevice.com/

Table 3.1 **Some Web Servelet Providers**

Via such servelets an engineer or end user can monitor the operation of an embedded application, make changes, and determine not only the state of the system and when the system needs to be serviced again. Embedded software vendors are also using these embedded servelets as a way to more easily access the inner workings of their operating systems, using the browser GUI as the means for modifying the features and capabilities of an operating system to meet specific user requirements.

The list of vendors providing such "embedded servelets" is large and still growing — Accelerated Technology, Arganat, Tasking, CniT, emWare, ISI, Microtec, Pacific Softworks, Pharlap, and Wind River, to name a few (see Table 3.1). Even Sun has introduced a Java-based servelet API. But the basic technology and methods used are similar. Depending on the degree of sophistication, the additional memory space required for such servelets in an embedded application can range from as little as 5 to as much as 100 kbytes.

Essentially, these servelets are no more than a combination of C and HTML commands that access data collected by the embedded processor, configuring it for later access by a browser. Some also incorporate a TCP/IP stack for easy connection to the Internet or Intranet, as well as the HTTP protocols. In most POSIX-compatible RTOSes, the servelet creates a TCP socket, binds the socket to a local address, and starts listening or collecting data on that interface. Once the client browser makes a connection to the servelet, it services the connection as well, possibly by spawning a new process. With a servelet, the developer can insert specialized template tags in an generated HTML page at locations where dynamic information, such as current information about the working status of a given home appliance, can be inserted. Each tag can be associated with a C-language call-back function so that when an HTML page is requested, the output of the call-back function -which contains dynamic information such as variable values - replaces the tag in the page. Some servelets also contain a symbol table that effectively controls the users' interactions with the target application. Since the symbol table links all tags to C functions, only those functions available in the symbol table are accessible from HTML pages, ensuring a secure environment. With the addition of Java capabilities to these servelets, designers now have the ability to not only monitor and debug the network connected appliance they have built, but to also update, modify, or correct the embedded code by downloading the appropriate code by means of a Java applet.

A number of problems face developers who want to build such capabilities into their traditional appliances and consumer electronics devices. The most obvious is that the inclusion of servelets eats up memory that might be devoted to the actual application. Second is the amount of system resources allocated to each servelet. While a servelet is usually blocked until a client attempts a connection, it does consume a task control block, even in its blocked state, and the operating system must monitor it periodically, taking away time and processor resources that might be devoted to the application. With the introduction of Java into the servelet equation other problems surface. For one thing, although the interactions with the browser are, in a sense, offline and not in real time, the servelet takes away resources from the embedded processor and OS. So, in effect, the servelet for all intents and purposes must also operate in a rea- time and deterministic manner.

Although Java, as it is presently constituted is neither real time or deterministic, Sun is making noises indicating that it is moving in this direction. Also, there are alternatives to Java, such as pERC from NewMonics, Inc. that add real-time extensions to Java. Although this makes the implementation nonstandard it is being used in a relatively closed environment. So the need for a standard version of Java is moot, since the "write once, run everywhere" capabilities of Java are not required. Also, the memory-space problem in the construction of such servelets is further exacerbated, since Java in its standard form requires anywhere from 1 to 2 Mbytes of DRAM for application code and at least 500 kbytes for program space, far beyond the requirements of home appliances and consumer electronics units in the home. However, Sun's Javasoft is working on versions of Java, called personalJava and embeddedJava, that considerably reduce the memory footprint. And real-time versions of Java, such as pERC, are available that push the memory footprint down to a more reasonable 20 to 100 kbytes.

Despite these problems, the use of embedded Web servelets to provide engineers and eventually average consumers with access to the innards of their microwave ovens, TVs or stereos will continue to accelerate. Why? Because they provide a range of capabilities previously not open to designers and at only a nominal cost. Other than the cost of the actual servelet software and development of the C/C++ or Java code virtually everything else is free. Nor is there any expense at the user interface end of the connection. Unlike other control-oriented protocols, which required the use of specialized GUIs, the browser acts as a universal GUI, one that comes free, since most operating systems for the desktop come supplied with a free browser.

Chapter 4

Connected PCs and NetPCs

In the emerging era of net-centric computing, the traditional desktop computer will never be the same. Even before the emergence of the network computer and the Internet appliance, the impact of the World Wide Web and the Internet were having an impact, first giving rise to the connected PC and more recently the NetPC. And further transformations are in the works. As the idea of having your own Web page or Web site becomes more popular and as the communications pipelines become wider, it is likely that a new kind of desktop system with capabilities beyond those of the NC, NetPC or connected PC will emerge: the Personal Server.

The Connected PC

The concept of the connected PC and more recently the "advanced connected PC" was first articulated by engineers and executives of Intel Corp. in late 1995 and early 1996, just as Internet and Web access was catching on with the general public. What computer industry executives such as Intel Corp.'s chairman of the board Andrew Grove saw was the emergence of an entirely new digital medium and a computing platform capable of delivering and running compelling content with the real-time immediacy of television, but with the interactivity of desktop computers.

What is driving the emergence of the connected PC are two forces at once both complementary and in opposition to one another, at least in the near term. One is the transformation of the PC from an exclusively text based business machine into a multimedia-enhanced entertainment and educational device for the home. Second is the continuing growth of the Internet into a ubiquitous medium accessible by millions and its transformation from a primarily text and graphics environment into what some are calling the "MultimediaNet." This is an environment where the Internet provides multimedia in the form of 2D and 3D graphics, animation, video and audio, as well as traditional text.

Intel's grand plan for the Advanced Connected PC initially had six basic hardware building blocks: a high performance 150 to 200 MHz processor, such as a Pentium or Pentium II, with multimedia instructions added; an enhanced PCI bus; an advanced graphics port, a new generation of 3-D graphics controllers, DVD-ROM, and AC-97 quality stereo. As processor technology has advanced and as communications become faster, this specification is evolving as well. But at the core of specification are at least two key technologies: multimedia extensions to allow the basic CPU to handle a wide range of multimedia data types and higher speed internal busses to handle the increased data throughput load that multimedia requires. At the other end of the spectrum, Intel is also reaching out to the average non-PC, non-Internet consumer with configurations that have features and a cost range that are comfortable, the NetPC.

Intel's MMX Solution

While the addition of multimedia instructions and extensions to the existing x86 architecture and instruction set was seen by the industry as a way to make its traditional desktops more attractive to the home and entertainment user, the other half of the momentum behind them is their usefulness in furthering the concept of the connected PC.

With multimedia applications moving into the mainstream of desktop computing and the Internet, microprocessor vendors such as Intel Corp. are faced with at least two problems they must solve: (1) how to push the multimedia and communications performance curve beyond levels expected with process technology; and, (2) how to do it without increasing component cost or harming compatibility with existing software.

To solve these problems, Intel Corp. has developed a set of architectural and instruction set enhancements to be added to all its future microprocessors - collectively called MMX technology - that offer the promise of significant performance improvements when handling audio, video, 3D and communications. The improvements to the basic architectures are completely compatible with existing operating systems, applications, drivers, and BIOS. To take advantage of MMX technology and achieve significantly higher overall performance, the software developer need only rewrite a small segment of code using MMX Instructions.

In a typical example of corporate-think and lawyer-speak, Intel executives insist that MMX does not stand for "multimedia extensions," but is a trademark symbol that does not stand for any particular technology, but rather describes a collection of methodologies that address issues such as multimedia.

To keep the system price of the connected PC in the mainstream, the CPU must take on more and more of the processing burden associated with multimedia and communication applications. This problem has been solved in the past by adding sound cards, video cards, modem cards, and other hardware. However, the key is to get these features at the lowest possible PC price point without compromising performance, and to do this will require that the CPU accomplish a lot of the work in software.

Assume an MPEG1-encoded video clip is running on a 100-MHz Pentium Processor. The video clip is being decoded in software and runs just fine in a window in the corner of the screen. However, if this video clip provides entertainment while the computer works on another task, the other task will be running quite slowly as the CPU tries to maintain the clip at 30 frames/second. Clearly, the most direct solution to this problem is to simply use a faster processor, or buy more hardware that can decode the MPEG clip and reduce the CPU burden. If the CPU could handle the task alone, MPEG decoding would become a feature available to the base PC. But the performance needed to do this will not be available until later when a higher-performance processor is available. Thus, the PC user waits for architectural improvements (generations) and speed (MHz) improvements to increase the capabilities of the base PC.

Most other multimedia and communications applications have the same set of characteristics: small native data types (8-bit pixels, 16-bit audio); regular and recurring memory access patterns; localized, recurring operations performed on data; and compute-intensive calculations that can be done in parallel. Rather than select a set of highly complex instructions to speed up specific multimedia and communications functions, Intel instead used a more general instruction set with the Intel architectural look and feel. This required that the new MMX-enhanced processors use an architectural approach first used in high-end scientific computers: a single-instruction, multiple-data approach (SIMD) that would allow developers to take advantage of additional parallelism without significantly modifying the existing design.

The MMX technology uses four new data types, 57 new instructions, and eight new MMX registers. It requires no new modes or states. The instructions are accessible at every privilege level. Thus any application or operating system developer can use them. Designed to enable SIMD operations using the MMX instructions, the four new data types are packed bytes (8 bytes), packed words (four 16-bit quantities), packed double-words (two 32-bit quantities) and quadwords (64-bits). The instructions fall into four general categories - arithmetic, logical, conversions, and transfers - and operate on these data types using eight new MMX registers. By operating on a packed data type, the CPU can perform multiple, parallel operations all in one instruction, and usually in one clock cycle. In general, all the instructions behave this way. All the MMX instructions use the same addressing modes as integer instructions and use the same two-operand encoding scheme, but allow the results of an operation to be either saturated or unsaturated (wraparound) and operate on both signed and unsigned operands. The arithmetic and logical instructions operate on MMX operands the same way as arithmetic and logical instructions do with integer operands. To retain computability with existing software, MMX is implemented using the floating-point register set. The MMX registers are mapped into the 64-bit mantissa of the 80-bit floating-point registers (see Figure 4.1). The FPU tags are used to indicate valid MMX register data. Since all major operating systems are aware of the FPU unit, the MMX state is automatically comprehended and addressed.

In the first Pentium processor implementations, all of the MMX instructions execute in one clock cycle except multiply and multiply-add. Added to the chip are a multiplier, two arithmetic logic units (ALUs), and one shifter. The multiply unit is a new, three-clock latency, single-clock sustained multiplier. It is capable of 16- x 16-bit multiplies, or two 16- x 16-bit multiplies followed by an add (multiply-add). The ALUs can each perform adds, subtracts, and compares of SIMD data types in one clock cycle. The shifter performs SIMD shifts, packing, and unpacking also in a single clock cycle.

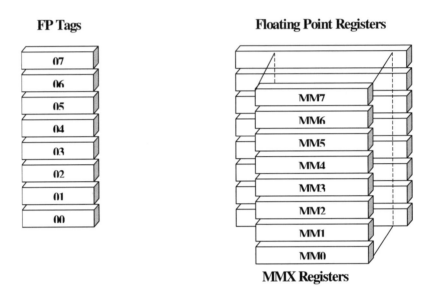

Figure 4.1 **MMX Registers.** To conserve silicon area and avoid redesign of the basic Pentium architecture, Intel stores instructions for multimedia operations in the same registers as those used for floating point operations.

Even though a specific implementation of the MMX architecture on an Intel CPU could vary slightly, the data types, instructions, and operations were designed to execute, for the most part, in a single cycle, in parallel and pipelined. The new technology is capable of improving application performance in three different ways. The most broadly applicable is the use of a 64-bit data path (see Figure 4.2). Second, and more specific to multimedia, is the use of SIMD, which offers the most dramatic performance improvements. Third is the use of saturating arithmetic operations to extend the architecture's ability to perform signal and media processing applications. The 64-bit data path has, to a great extent, gone unnoticed in most discussions of Intel's implementation of multimedia extensions. It is true that since the introduction of the Pentium, Intel processors have had 64 bits worth of data path. However, it has always required two 32-bit instructions to move 64 bits worth of data. With the introduction of a new MMX instruction, MOVQ, it is now possible to load or store 64 bits of data with a single instruction. This improvement has its greatest impact when the increase in data transfer bandwidth is taken into consideration. Although several other x86 instructions have been modified to exploit this capability, in many cases the fastest method for a copy routine on an MMX will be hand-coded loops using the MOVQ instruction and the developer's knowledge of the data to be moved.

The SIMD capabilities of the MMX technology are by far the most publicized but in some cases are not well understood by software developers. The theory behind SIMD capabilities is that a single instruction is issued and the operation it performs is repeated over more than one pair of input data. In the case of MMX technology, the number of times that an operation is repeated is dictated by the size of the input data. Since the registers are 64 bits wide, they are capable of storing two double words (32-bit elements), four words (16-bit elements), or 8 bytes at a time. So, an MMX technology ADD instruction that operates on a packed byte data type is defined to add eight pairs of bytes together in parallel - eight times more work than the regular (X86) single-byte add. SIMD is a technique that exploits the parallelism that exists within algorithms. The burden will fall on software developers to identify the parallelism that exists within a given algorithm and develop the MMX code to exploit that parallelism. In some cases the algorithm needs to be changed to expose more of its inherent parallelism. For example, many multimedia and net-centric communications applications spend a large amount of execution time in a small amount of code, usually well-defined subroutines or code loops that have a high degree of internal parallelism. Recoding these subroutines or loops using MMX technology will substantially boost the loop's performance. However, not all algorithms have internal parallelism. So, those that do not, or that only have a small amount, will not necessarily realize significant improvements from the new technology.

Mrw: MMX read and write
Mex: MMX execution
WM: Write to memory
WMul: Word multiply
PF: Instruction prefetch

IF: Instruction fetch
D1: Instruction decode
D2: Instruction paring and dispatch
E: Execution and memory access
WB: MMX Write Back

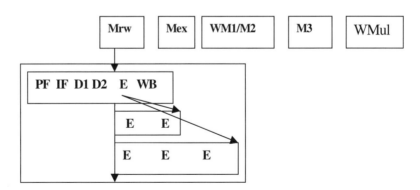

Figure 4.2 **MMX Pipeline.** Intel uses the Pentium architecture's internal high speed pipelined bus structure to execute multimedia instructions in parallel.

The current mix of multimedia instruction extensions employed in MMX are a good first step. But as we shall see in Chapter 15, while they are appropriate for a desktop computer with only occasional access to the Internet and for playing video games and interactive multimedia via CDROM, they will not be enough in the new environment of the Multimedia Net. There, driven by the interactive networked requirements of MPEG 4, a net-centric PC will have to have a wider range. Initially there will have to be better support for 3D, given that MPEG 4 incorporates almost completely the specifications for the newest implementations of VRML (Virtual Mark Up Language).

Intel still has much work to do as far as making their multimedia extensions appropriate for servers and the next-generation of Connected PCs. Because it was aiming at the current generation of desktops, the focus was on accelerating playback of multimedia through efficient decompression of video images from CD-ROMs or off-line, in non-real time after downloading from the Internet. Little thought seems to have been given to the efficient decompression of video of the Internet in real time, certainly not in the interactive networked multimedia environment that MPEG 4 will make possible. And in the server environment much more attention will have to be given to video compression as well, an area where CPU vendors to the server market such as Compaq's Digital Equipment subsidiary Hewlett Packard, Motorola, Silicon Graphics, and Sun have been very aggressive.

PCI and AGP

Where Intel has done its homework is in the improvement of the internal bus structure of the PC to the level that will be necessary when MPEG 4-based interactive networked multimedia becomes common on the Internet. Calling it the Advanced Graphics Bus, Intel's short-term goal was to increase the capability of the desktop PC to run highly graphical, 3D and video base games and to appeal to professional users of 3D. However the ultimate result is a bus architecture that will serve the needs of the Multimedia Net well into the future. If Intel's engineers had not invented AGP for 3D games and professional graphics, they would have had to invent it to deal with the stringent requirements of the Internet and World Wide Web in future years.

The most obvious response to the increasing requirements of multimedia and communications-oriented data types on the PC architecture is to improve the ability of the internal bus to handle greater amounts of data. Compared to the original 16-bit wide ISA bus, which at most had a bandwidth of no more than 10 to 20 Mbytes/second, the original 32-bit wide Peripheral Component Interconnect (PCI) bus proposed by Intel boosted that to about 80 to 133 Mbytes/sec. The original PCI specification has proven to be inadequate for the needs of multimedia computing, connected or not, with typical bandwidth requirements across the bus for video and high quality 2D graphics pushing 300 Mbytes/sec, and 3D in the future pushing the requirement to about 500 Mbytes/sec.

The PCI Special Interest Group, the industry organization that took over the PCI specification from Intel, has extended the original definition to boost performance. First it has widened the bus size from 32 to 64 bits resulting in a bandwidth of about 250 to 275 Mbytes/sec. Second, changes were made to the way in which signals on the PCI bus carry information. In the original specification, data is carried, or clocked on the leading edge of the sine or square wave-like signals, allowing all components in a system to synchronize their operations to the operate on the data in a coherent manner. New versions of the PCI specification now allow data to be clocked not only on the leading edge, but on the trailing edge of the signals. This effectively doubles to about 500 Mbytes/sec the rate at which multimedia data is transferred between the main processor and memory and the graphics-video subsystem. But even this is not enough for some of the sophisticated images required for 3D, high resolution animation and broadcast quality video, where bus bandwidths need to be at least 2X higher. One major reason is that these high bandwidth data types had to share the PCI bus with other functions that needed access, so sustained bandwidth was far below 500 Mbyte/sec peak potential

This situation lead Intel Corp. in late 1996 to define a dedicated graphics port, the Accelerated Graphics Port (AGP), to speed up graphics processing on the PC. Since then every major graphics vendor has introduced a version of their product that supports the new specification. Through the AGP, the graphics subsystem connects to the system core logic controller, providing a more direct access to the CPU and main memory. The AGP interface uses as its starting point the PCI bus 2.1 specification, and manages to achieve four to eight times higher data transfer rate, but without the radical redesign a move to a more complex PCI bus specification would have required. It includes three major extensions. First, AGP allows deeply pipelined Read and Write requests to memory in order to hide memory access latency. Second, new sideband signals allow demultiplexing of address and data to overlap processing of memory requests. Third, the clocking on the bus increases bandwidth to 533 MBytes/s peak for 2x mode and 1,066 MBytes/s for 4x mode.

The specific changes from the PCI specification involve both physical changes and architectural changes. AGP is implemented as a physically separate bus from the PCI or ISA buses, which may also be part of the PC. In a move away from 5.0 V bus operation, AGP requires 3.3 V I/O only (for 1x and 2x mode), which is consistent with the PCI 2.1 specification. The 4x mode also introduces new signaling. But the most significant architectural change is the disconnection between address requests and data, effectively allowing split transactions. Unlike PCI, AGP allows a large number of memory requests to be queued in the core logic chip set. There is an option to allow the address to be transmitted while data are being transferred, yielding higher bus utilization. All AGP transactions must include information on the size of the data block to be transferred, which is unlike PCI. AGP removes almost all graphics traffic from the PCI bus (except incoming video streams), allowing much more headroom for other PCI traffic, such as disk or network data transfers. In this way, AGP allows both higher-performance graphics processing and more balanced PC architecture, as graphics were, by far, the highest consumer of PCI bandwidth.

One of the main benefits of the AGP is to reduce the amount of memory that a system must maintain in the graphics frame buffer subsystem. However, this places much higher bandwidth requirements on main memory. Since the AGP port has as much bandwidth as today's Pentium processor, main-memory bandwidth needs to increase. Whereas the Pentium processor cannot saturate its memory bus, future Intel processors with MMX extensions will greatly increase processor bus traffic. Higher speed processor busses are planned for next-generation processors and Intel has plans for a 4x AGP that will need even more bandwidth. These factors will further exacerbate the need for more main-memory bandwidth.

In some ways AGP simplifies working with main memory for systems designers, since it allows controllers to be constructed with a simple pointer that can point anywhere in memory, internally to the graphics memory or externally to main memory. However, the design of these graphics controllers must make use of two pools of memory with very different latencies and bandwidths. This presents some serious design tradeoffs. Just as the emergence of the PCI bus did not let systems designers do away with the arcane and vestigial ISA bus, the AGP has been forced to deal with it as well. Also, as of this writing, the problems of adding PCI-like bus to the system are unresolved, raising questions as to the best ways to support bus-to-bus transfers. The AGP specification does not require the core logic to implement a full PCI-to-PCI/AGP bridge function. A device on the PCI bus can write to a device on the AGP bus without full bridge functionality. But the specification does not allow for writes to go from AGP to PCI.

The degree to which the multimedia-enhanced connected PC architecture moves into the mainstream depends not only on how quickly Intel can lower the costs of such "improvements" but on whether the consumers perceive a need for such capabilities. Until relatively recently, there has been no compelling "killer app" appealing to the broad range of consumers. As the Internet and World Wide Web move to connections with higher bandwidth and as more multimedia content is available there, multimedia on the desktop will become more of a bread-and-butter requirement for the average consumer, rather than just cake, as it is now.

Improving Connected PC I/O

Another area where Intel Corp. will have to do a great deal of work is improving the input/output capability of the connected PC. The only environment in which the company is directly dealing with this issue is in servers where it has had a major role in defining the Intelligent I/O (I_2O) specification. It has also redirected its embedded 32-bit processor family, the i960 series, away from general embedded applications to specific support for I_2O with a new family directed at the requirements of its Pentium Pro-based servers for better I/O response to network transactions (see Chapter 12).

In the multi-transactional environment in which servers operate, the need for better I/O performance is becoming critical as network bandwidths increase and as the number of users and transactions increases. In the past server manufacturers have dealt with the problem on an individual company-by-company basis. And while a number of intelligent I/O subsystems did emerge, the cost was such that only high end systems benefited, because it was not possible to prorate development costs over a large enough customer base. With this new I/O standard, low cost, high performance I/O processing is now possible in the server environment.

The big questions facing Intel and PC builders in transitioning this technology to the connected desktop PC are (1) Is such a solution necessary for the desktop connected PC now? (2) If not necessary now, when will it be? And (3) When it is necessary, what form will this enhanced I/O solution take?

Certainly the desktop connected PC does not necessarily need such a solution right now in most of the present environments in which it working. But there are specialized environments, such as PCI workstations based on the Pentium/Pentium Pro, the Sparc and the Alpha, where this venerable bus is running into bottlenecks as the PCI interfaces with the most common networking environment, 10/100 Mbit/sec.Ethernet. Most home users of PCs will not face this problem until 1 to 50 Mbit/sec. asynchronous digital subscriber loop and cable modem access become a common way to access the Internet and World Wide Web. But it is useful to see what some of the bottlenecks that PCI faces currently on networks where multiple PCs are linked together and to analyze what kind of improvements will be necessary.

In the standard high volume (SHV) server that Intel Corp. has defined, each I/O device interfaces to network environments that place differing load types on the PCI bus. In particular, 10 Mbit/sec. Ethernet networks operate with TCP/IP protocols where multiple small transfers are created, with a maximum size of 1518 bytes and a minimum of 64 bytes. The TCP/IP protocol creates a request and acknowledgment system that sends requests for data, blocks of data, and then sends acknowledgments of that data. The data transfers are generally large, usually very close to the 1518 byte limit. The acknowledgments and requests are generally small, and hover around the 64-byte minimum. On average, a TCP/IP Ethernet packet is approximately 256 bytes, corresponding to 64 data transfers on a 32-bit PCI bus and 32 data transfers on a 64-bit PCI bus.

To handle this kind of data, Ethernet network interface cards (NICs) employ a scatter-gather technique for moving data from the network to the operating system and back. In this environment the CPU sets up a command structure in memory, on the order of 16 bytes. It contains the location of packet data and the length of the segments of the packet data. The data for a single packet is not usually within a contiguous memory space, a result of multiple CPU processes setting up the packet data. The combination of the scatter-gather structure and TCP/IP packet segmentation decreases an Ethernet NIC's capability to burst on the PCI bus. Each fetch of a single scatter-gather structure is a data burst on the PCI bus that requires four clocks. The 16-byte MAC header requires another 4-clock data burst, while the TCP/IP header is 40 bytes without TCP header options and requires another 10 clock data burst. The result is that even when there are large packets to transfer, the nature of the network structure segmentation and the scatter-gather command structure forces bus masters to burst for short periods on the PCI bus.

Using the average size of a TCP Ethernet frame of 256Bytes, the scatter-gather structure will cause burst cycles of only 25 data transfers which reduces the realized PCI throughput to 60%. Aside from the packet data movement, Ethernet traffic creates other system performance limitations when it comes face to face with the PCI bus. For example, when an Ethernet NIC receives a new packet from the network, the it fetches an available scatter-gather structure and transfers the packet data to memory. It then informs the CPU of the new packet. It does this by asserting an interrupt to the CPU. For a 256 byte average packet size this begins to add up to some serious numbers. For example, a 100 Mbps Ethernet link would generate more than 97,000 interrupts per second, about one interrupt every 350 PCI clocks at 33MHz PCI. Extrapolating into the future, a 1000 Mbps (1 Gbps) Ethernet link would create one interrupt every 35 PCI clocks. Oh boy!.

The impact of such a high interrupt rate on the PCI bus is disastrous, since the effect of interrupts on the CPU is to cancel the CPU's pipelining and parallel execution nature of advanced CPUs. So, if the pipeline or cache of the processor is stalled or flushed, the CPU must wait for the pipeline or cache to refill before it can operate at full capacity. Each interrupt also generates several I/O reads and writes on the PCI bus by the CPU. What makes the situation worse is that these I/O reads and writes are costly in terms of PCI bus bandwidth because they are single data phase transactions. They are also expensive in terms of CPU processing capability, since the I/O reads and writes cancel some of the internal pipelining of the CPU.

The NetPC Alternative

Another challenging technical and marketing problem facing Intel Corp. is how to deal with the emergence of lower cost alternatives to the desktop PC, not only from its X86 competitors, but from builders of the many network computers now entering the market. It is clear to everyone, including Intel Corp., that with the emergence of the Internet and the Web, there is no longer just one personal computer market. Whereas in the past one basic platform, one set of components, and one set of applications programs could be customized for use in either the home or the office, the requirements of each environment are diverging rapidly.

While a multimedia-enabled connected PC, as envisioned by Intel and others, may be appropriate for the home and the power user, it is too rich, both in terms of dollars and features, for most corporate environments.

What the information systems managers at major corporations are looking for is a high-tech equivalent of an alternative that almost disappeared with the mainframe: local information stations that were low cost, easy to upgrade, easy to maintain, and immune from the well-meaning machinations of the local user who wants to "improve" the performance and capabilities of his or her local unit by adding hardware or software.

One solution proposed by Intel, Microsoft, and a group of PC vendors is the NetPC, which was introduced as an effort to counter the possible impact of the so-called Network Computer proposed by Oracle, Sun Microsystems and IBM among others. In many respects the NetPC is an extension of the low-maintenance "sealed PC" specification targeted at large corporations where the cost and complexity of the newest generation of x86 PCs was rapidly increasing their cost of ownership, especially as it relates to down time and repairs. In an effort to reduce sources of "error," Intel, Microsoft, and the PC vendors who have joined in the NetPC initiative have retained the functionality of the original PC - floppy and hard disk drives, DRAM, CD-ROMs, mouse and keyboard input - but have tried to take any modification of the system out of the hands of the user and put it into the hands of the systems administrators and technicians (see Chapter 10).

Under the NetPC specification, Intel and its hardware partners have defined the specification so that users are not allowed to have access to the internal workings of the system and are not allowed to replace or upgrade any of the add-in cards. In terms of software,

Microsoft has developed server-based administrative software that does not allow the user of the desktop to install or remove software. All installation or modification is done either via the server-based administrative software or by company approved technicians who have received extensive training and have been certified by either Intel or Microsoft. Many aspects of this specification have been rationalized by PC hardware and software vendors as part of a self-protective delusion that many of the reliability problems associated with the desktop problems are due mostly to user inexperience. Any long-term user of desktop PCs who has gained any degree of technical knowledge knows this to be far from the case.

Windows 3.1 and Windows 95, as well as the Microsoft applications software, have gone through numerous revisions over their lifetimes to fix problems ranging from simple coding mistakes to unrecognized interactions between system software components. And while problems with the desktop occur when a user inadvertently accesses a program in the wrong way, this is not the user's problem, it is the software designer's. While Windows 95 is a significant improvement in some respects over the earlier Windows 3.1, neither is particularly intuitive and no attempt has been made to prevent users from accessing programs or hardware resources in a manner that would create problems. Recognizing that its user interface still has a lot to be desired as far as ease of use is concerned, Microsoft as of late 1997 has announced its intention to further "simplify" its Windows user interface.

As far as software reliability is concerned, it is only with Windows NT, and more recently with Windows 98, that a range of features have been built in to prevent catastrophic failures of both the operating system and the applications programs when a malfunction occurs. Ironically, one of the capabilities Microsoft has made use of in these latter two implementations is memory segmentation and protection, a feature that has been available on the X86 since the early 1990s. Virtually every operating system vendor, except Microsoft, has made use of this feature to build a more stable platform: Digital Research in its competitor to Windows in the mid-1980s; IBM with its competitor to Windows on the desktop in the late 1980s: OS/2; virtually every vendor of UNIX; and almost every supplier of real-time operating systems that support the X86.

The Future of Connected PCs and NetPCs

The success of either the multimedia-rich connected PC or the NetPC depends on several key factors. First, there is cost. For the home-entertainment PC, the cost of the basic platform certainly cannot be more than the $1000 to $2000 for the baseline desktop PC. In the corporate environment, the targeted cost is in the $500 range. Second, there is the issue of maintainability and reliability. The traditional desktop is notoriously unstable, and if these systems are to become at least as reliable as the television in the home, the laser printer, or the old electric typewriter in the office, significant changes in the way both the hardware and the software are manufactured and tested will be required.

Cost will be influenced by two important factors: volume and competition. One reason Intel Corp. has been able to dominate the market has been the fact that, with an almost

exclusive ownership of the X86 architecture, it has a potential market for any new generation of its processors that ranges from 10 million to 30 million units, at a minimum. This
allowed it to lower the cost and learning curve as yields on its integrated circuits improved, so that a basic desktop PC at one point costs $2000 to $3000, but within a year or
so is in the $1000 to $1500 range. This is because much of the cost of the system was tied
up in the main CPU, which is initially introduced in the $300 to $800 range, dropping to no
more than $80 to $150 for more mature versions of the desktop device. Chip costs for more
mature versions of the X86 are now down to under $50.

The problem that Intel Corp. is facing is that the personal computer market is now fractur-
ing into at least two or three different market segments, each with its own optimized CPU:
the Pentium, the MMX-enabled Pentium, the Pentium Pro, the Pentium II, and the 64-bit
version of the architecture that may be available by the time you read this. As of the end
of 1997, it is clear that Intel Corp. will cannibalize the market for its existing 32-bit Pentium
with MMX-enabled versions. This means, however, that the company will have to start
over on the learning and price curve and will have to quickly drive down the cost of the
MMX-enabled versions to allow it to compete both in the connected PC and the NetPC
segments, each of which has vastly different price elasticity.

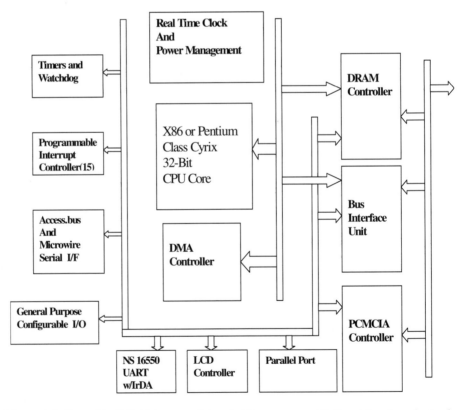

Figure 4.3 **One-chip PC.** Using a variety of x86 processors, National Semiconductor is
looking to lower the cost of NetPCs by integrating all of the basic functions onto a single
chip.

No matter how fractionated the market, even if volumes do not justify further lowering the price of such devices, another factor will: competition. Intel Corp. already has competition in other segments of the net-centric computing market. Using licensed derivatives of the original 386 and 486, companies such as Advanced Micro Devices, SGS Thomson, and National Semiconductor have already been able to penetrate the less graphics/video-intensive segments with highly integrated core chips that significantly lower the cost of materials of such designs. Now, AMD and National have set their sights on the network computer and NetPC market and the low end of the connected PC market. Locked out of the Pentium-PentiumPro market because they have been unable to gain licenses to these chip designs, both companies acquired start-ups who have developed "clean-room" equivalents, as well as their MMX-enhanced successors.

AMD acquired NexGen with its Pentium clone and come to market and went head to head with Intel in the connected-PC and NetPC market. National Semiconductor has acquired Cyrix Corp. with its Pentium, Pentium II, and Pentium Pro equivalents. With the 80486 architecture, National has already created a NetPC reference platform (see Figure 4.3) that would allow developers to quickly and easily create a system that goes a long way toward achieving the magical $500 price, which is perceived as the price point at which both corporate IS managers and the average consumer will buy into the concept.

Personal Servers?

 Further differentiation in Intel's desktop market is in the offing as the companies building the Internet and intranet infrastructure shift from the use of 32-bit CPUs and operating systems to 64-bit architectures. As we shall learn from later chapters in this book, some of the Internet traffic issues can be solved by simply widening the processor word size to 64 bits, effectively doubling the capabilities of the servers now used in intranets and in the Internet backbone. Using them in the necessary and ubiquitous routers and switches alone will go a long way toward solving many Internet traffic problems. The shift is already underway in the Internet backbone and is likely to begin on the many intranets as the number of users and the time they spend on the intranets continues to increase.

Does this mean that the day of the existing 32-bit Pentium Pro and its competitors is passing? Not likely. First of all, while sales of systems based on 32-bit designs such as the Pentium Pro to the server market have far exceeded those of many competing designs, it has certainly not reached the volumes we traditionally associate with the desktop. Second, there will emerge in the not too distant future a market for "personal servers" that will appeal to the power users in the corporations, in the home, and in small businesses who now buy beefed-up desktops for multimedia content and desktop publishing. Third, there are the continuing delays Intel has announced on its next generation 64-bit CPU, which is not likely to start shipping to high end server applications much before the end of 1999, much less be available at a cost structure that would be attractive for the desktop or as a personal server.

As the bandwidth available to the average, noncorporate user increases from 56 kbits/sec to millions of bits per second, the appeal of setting up not just personal Web pages, but personal Web servers will become more attractive. Making this even more likely will be the availability of 24-hour access to the Internet by average home and business users, in the same way that the telephone is available, at a reasonable cost.

But for the equivalent of the existing telephone dial tone - the "Web tone" - to become an everyday part of our lives will require more than just availability. It will require rigorous attention to reliability: on the desktop unit, on the server, and on the network.

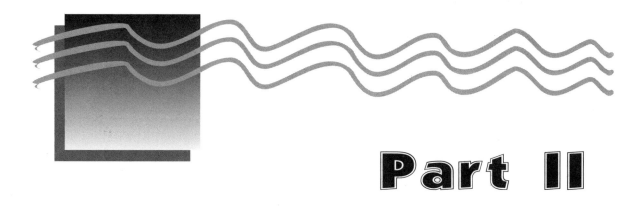

Part II

Building Blocks

Chapter 5

Java and the Internet:
Do We Need a "Lingua Internetica?"

During much of the 20th century, there was a drive on from many quarters for the equivalent of the lingua franca of the 19th, a common worldwide language that all spoke and all understood, in addition to their native language, or as an alternative to it. Literally dozens of candidates were proposed. Among them was French from which the phrase lingua franca came, and English, due to their wide use in the colonial empires of France and England. There were also artificial ones such as Esperanto that derived their words and grammar from pre-existing languages to create an entirely new common language.

Something similar is occurring in the newly emerging world of net-centric computing, with Sun Microsystem's Java proposed as the lingua internetica for the Internet and the World Wide Web. But while Java may indeed become widely used, it will not become universal for many of the same reasons that artificial languages such as Esperanto or even natural ones such as English have not: individual and application- specific preferences and requirements. For the same reasons we do not all eat or like chocolate or vanilla ice cream, or use the same sized nuts and bolts in the machines we build, or exactly the same screw drivers and wrenches for all jobs, Java will coexist in the world of net-centric computing with a number of other programming languages and methods. The degree of its pervasiveness will depend on how well it satisfies a growing requirement for an object-oriented programming environment that reflects the way we deal with physical objects and distributed computing elements in a networked environment.

The Pros of Java

To understand the place of Java in the net-centric computing environment and when and where to use it, and not use it, it is first necessary to look at a brief history of this language and then analyze its pluses and minuses in various net-centric environments.

Originally, Java was developed as an object-oriented language to replace the complexities of such alternatives as C++, Pascal, and Smalltalk in embedded applications such as consumer electronics devices and consumer computing systems such as set-top boxes.

For many of the same reasons that telecommunications and data communications companies converged on C++ as the standard language for programming distributed applications, so, too, Sun developers discovered a much wider role for Java in the network-centric world of the Internet and World Wide Web.

Java, the language, must be discussed in the context of Java, the virtual machine; Java, the OS; and Java, the active content enabler. Much has been written about the "fact" that Java as a language is <u>platform-independent</u>. To some degree that is true, but on some architectures it runs better, and is more independent, than on others. This is because it is very processor specific. But the processor that it is specific to is a virtual processor maintained in software, a virtual machine. So the degree that a Java applet or application is processor independent is tied closely to how processor-independent the underlying virtual machine is.

Although now targeted at the desktop and network computing markets, Java has nevertheless generated considerable interest in other segments of the net-centric market for small footprint information appliances, Internet appliances, and Web-enabled set-top boxes. Why is Java attractive to all segments of the net-centric computing market?

First, it is a simple but powerful language. Some systems developers consider Java to be a better and easier to use version of C++. In fact, the developers of what was later to become Java at first chose to use the C++ language for their application, which was a classic networked device - a set-top box. Disappointed with C++, they then set out to design and develop their own language environment which subsequently morphed into today's Java. They essentially created a poor man's C++, with just enough of C++'s object orientation, but more approachable by being a much simpler language. For certain applications, where C++ may be too imposing on either the application or upon the application developers, a lighter weight C++ (i.e. Java) is very attractive.

Second, there is the feature that has received the most attention: dynamic loading and execution. Many systems must change their execution over time, adjusting to changes in the environment or perhaps updating certain parts of the application as the overall system continues to operate. Java's ability to download and execute applications on-the-fly can be very attractive for most net-centric applications, from network computers to set-top boxes and personal digital assistants (PDAs) linked to the Internet via wireless connections..

Very little serious use has been made of this capability as far as furthering and extending the concept of net-centric computing. In most NCs, Windows-based terminals, and NetPCs, most typical applications, such as word processing, spreadsheets, databases and even Web browsers, are written in monolithic form. They are either executed locally on a PC or NetPC with local storage for the program and the work generated, or on a server, where the input from a terminal or network computer is processed and stored.

The ultimate extension of the concept of dynamic loading and execution is for storage to remain on the server, but much of the computational chore shifted back down to the thin client in a sequential, modular sort of way. This concept takes advantage of the sequential nature of most PC operations. On a word processor, for example, you write and save, you format and save, you word and spell check and save. You seldom do all these operations at the same time. As a result, most of the time, the word-processing program is enormously underused.

Using Java's dynamic downloading and execution capability, this same function can be done in one of two ways. Using a monolithic approach, the entire program would be downloaded at the beginning of an operation, eating up bandwidth and local memory space, and turning an network computer from a "thin client" into a "fat client," as far as DRAM capacity is concerned. The other approach is to download modules, as needed to the local thin client. However, this approach, while attractive, will not be widely available for some time, since it will take time for major software companies to not only rewrite their applications in Java, but to do it in modular rather than monolithic form.

Third, there are advantages as far as safety and security are concerned. Memory leaks and bad pointers in an application are very difficult to detect and often take many hours or days of operation to be revealed. For many applications that involve net-centric computing, this can be very problematic for a number of reasons. One is that a memory leak can cause a system to fail, and for many applications such behavior is unacceptable. In addition, it may be very difficult to debug such a problem. For example, the client system may be located in a hand-held device linked to the Internet or intranet via some form of wireless connection. In this case, the primary indication of a problem is that the system simply crashed. It may be difficult if not impossible to have an engineer on site to watch and debug the failing system. Java's built-in memory manager with automatic garbage collection, which can inhibit memory leaks, and its restrictions on pointers which may eliminate bad memory references are therefore very desirable features.

Not to be ignored is Java's multithreading capability, with a strong model of how thread-critical code can be synchronized to avoid race and timing problems. Finally, there is the previously mentioned promise of the "write once, run everywhere" cross-platform capability of Java. While each underlying hardware and software platform has its own implementation of the Java Virtual Machine, there is only one virtual machine specification. Because of this, there is a standard, uniform programming interface to applets and applications on any hardware, making it ideal for net-centric computing on the Internet and World Wide Web, where the aim is to have one program capable of running on any computer in the world.

The JVM is the key to the independence of the underlying hardware and operating system — it is the platform that hides the underlying operating system from Java-based applets and applications. All that is necessary is to port the Virtual Machine to a variety of browsers and operating systems. The JVM defines a machine independent format for binary files called the class file format (.class). The format includes instructions for a virtual computer in the form of byte-code.

To create an applet in a Java execution environment (see Figure 5.1), the developer stores the byte-codes on an HTTP server and adds an <applet code= filename> tag to a Web page, which names the entry-point byte-code file.

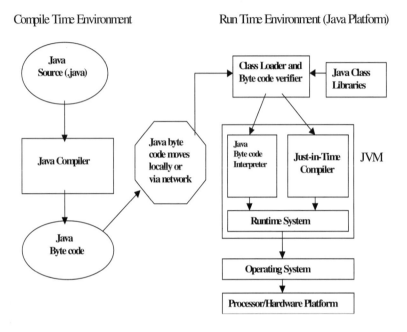

Figure 5.1 **Java Execution Environment.** In typical cross-platform application, Java applet created and compiled on a system can be sent across network where class-loader and byte-code verifier, and a Java interpreter or Just-In-Time compiler downloads and executes it.

When someone visits this page, the <applet> tag causes the byte-code files to be transported over the network and downloaded dynamically to the end user's browser on the Java-enabled platform. At this end, the byte codes are loaded into memory and then verified for security before they enter the virtual machine.

Once in the virtual machine, the byte-codes are interpreted, or turned into processor-specific machine code by a just-in-time code generator, also known as a JIT compiler. The interpreter and JIT compiler operate in the context of a runtime system. Any classes from the Java class libraries, or application programming interface (API) are dynamically loaded as needed by the applet.

The Java API framework (see Figure 5.2) is a set of standard interfaces to applets and applications, regardless of the underlying operating system, and include the Java base, or Java Applet API, and the Java Standard Extension APIs. The first defines the basic building blocks for creating fully functional Java-power applets and includes the following classes: java.lang, java.util, java.io, java.net, java.awt, and java.applet. The extension

APIs include published definitions for such things as 2D, 3D, audio, video, MIDI, telephone, server, animation, and security. The two newest, still in the approval stage, are the personalJava (pJava) and the embeddedJava (eJava) APIs. Once defined, they can be added to, but in order to maintain backward compatibility, cannot be changed in a way that calls to them would fail.

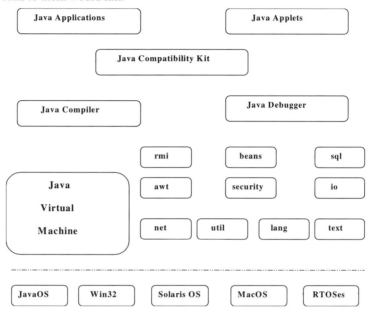

Figure 5.2 **Java Tools and APIs.** To speed the development of Java-enabled net-centric devices, Sun Microsystems has developed a complete tool kit of tools and interfaces to help programmers write applications.

Besides the obvious advantage of programs that can be moved around a network and run on any processor, this feature should be of enormous value to systems developers in the fast changing net-centric computing market.. As systems evolve, it is not uncommon to incorporate a higher-performance processor to increase system performance. If the new processor is of a different type from the previous one, the developers are faced with a porting task. This porting may involve not only the application, but also the underlying operating system. Presuming a Java implementation exists for the processor and that the old application was written in Java, the already compiled Java application may run as is, minimizing or even eliminating the porting effort.

Java solves a great number of portability problems. Under most conditions, it is no longer necessary to think about such things as the size of variables, the byte order in which data are stored by the processor, which functions are available or how they operate differently on different platforms. Nor is it necessary to worry about different ways to allocate memory or figuring out where to free that memory. Also of no concern are the different ways to process user events, to write things on the screen, to read from the keyboard, or to load additional libraries.

The Cons of Java

However, despite these advantages, Java has to be taken with a grain of salt, especially as the Internet and World Wide Web fragment into a number of different user populations. Although a significant step forward in platform independence, Java still has some problematic dependencies relating to hardware, feature sets, software interfaces and usability issues, none of which have anything to do with the language or the operating system.

The Java programmer will not be without resources in dealing with these dependencies, but they will be resources learned from programming in other languages. In many of today's "portable" programs written in the ubiquitous C language, platform dependencies are isolated into separate platform dependent and independent modules that can be reworked for any differences in the target machine. The remaining platform- independent code is usually portable enough that the code can be ported just by recompiling with the proper header files and options. Other approaches are to use entirely different platform-dependent libraries on each platform or use conditional compilation to handle the differences. Where the target machines, operating systems, and libraries are similar, these platform-dependent modules become a smaller proportion of the total code. And, of course, there is the old standby, hiding all platform dependencies in header files. That said, let's look at some of the dependencies that programmers will still have to concern themselves with when using Java.

Hardware Dependencies

Java programmers in the new net-centric computing environment will have to deal with differences in screen size, resolution, colors depth, if the device is always on the network, if it has power management, and the thickness of the pipe (network connection). The developer will also have to give some thought to the availability of sound output or input devices, printer, card reader, scanner, and other resources such as local storage.

 This is because Java, as originally written, assumed that the target would be a desktop or WebTop of a standard size with certain display and I/O characteristics. But in the rapidly proliferating world of net-centric computing such assumptions are no longer valid. With the new generation of Internet Appliances, you cannot take anything for granted. For example, the new HPC (handheld PC) devices that run Microsoft's Windows CE 2.0 have screens as small as 240x480 pixels with only four levels of grayscale, very limited memory, no disk, no cursor, and the equivalent of a single mouse button, are intermittently connected to the Internet by modem, and have sound output only.

Memory, Local File And Peripheral Issues

One difficult thing the net-centric system designer is going to have to face is that not all devices are equal as far as memory, local file, and peripheral support are concerned, and many do not match the assumptions made when Java was first proposed. Since many of these devices do not have a hard disk to provide virtual memory, the amount of memory with which the programmer has to work with is quite small. A wireless Internet telephone, for example, with 2 to 4 MB of RAM will have perhaps only 1 MB available for all Java programs. This means that the developer will have to write a program to work in a small amount of memory. This means not only that the programmer will have to write extremely efficient code but limit himself or herself to very small working sets of the data.

There is no platform independent way of dealing with this problem and stay within the rules Sun has set down for Java. If local files are used, the program would be non-portable and very un-Java-like. One alternative, practical only if there is a full-time network connection, is to store non-active working sets of data in a remote database on the server, swapping data in and out as appropriate. What this means, however, is that the program will run much slower, having to go out to the network more often, possibly across a slow network, which is certainly the case if the device is a wireless unit.

If the decision is made to write an applet or application that makes use of local files, it will not run on any of the devices that lack local storage, such as a diskless network computer. If the device has local storage, all of the problems relating to filenames emerge. A good case in point are the Java test cases shown in the Java security frequently asked questions (FAQ) file on the Sun Web site. They are not portable because the language and test cases make no attempt to handle anything but UNIX specific pathnames. Some of these test cases give a false "pass" reading even when security is circumvented because the applet attempts to read a file that doesn't exist on any platform but UNIX. It is necessary to know where the files can be kept on the target machine, and to do that requires the use of pathnames that can work on any platform.

Java, in its original implementation, has almost no support for peripherals, so there is, at the time this was written, no portable way to determine whether a device is available for sound output and input. Printer support was added in JDK 1.1, providing the ability to print text and graphics, and choose the printer and options. As of early 1998, the embedded Java API had not been released, so it is not clear yet how Sun will solve the problem of dealing with embedded hardware (analog and digital I/O, memory mapped devices, etc).

Java Applets And Feature Sets

Written originally when the assumption was that all net-centric computers would be PC-like, most Java applets are designed to take advantage of at least a VGA screen, lots of memory, high-quality sound and video capability, a full-sized keyboard, and a user operating the device with both hands while seated directly in front of it.

This is very much different than what the typical non-business net-centric computing device will have in the future, be it a hand-held device like a Personal Digital Assistant (PDA) or a set-top Internet access device. So the programmer is faced with a critical decision: write a Java program that will be the "lowest common denominator" operable on virtually any Internet device, or write one that meets the Java specs exactly, but can only run on a fraction of the available platforms. The venerable 80/20 rule applies here: it's likely that only 20% of the features designed into any given net-centric computer platform will be used by 80% of the total universe of users. It really is a trade-off between added and deleted features: adding capabilities in small devices that are not necessary in a PC, such as soft buttons to substitute for unsupported keyboard keys, but at the cost of some platform independence.

Real-Time Constraints

Another factor that will reduce Java's aim of "write once, run everywhere" ubiquity is the wide range of timing constraints that are imposed on the different types of net-centric devices on the World Wide Web.

If the net-centric hardware or software application must service real-time events, it may have timing constraints that cannot be met on all platforms running Java. A Java program that runs fast on your desktop might be unacceptably slow on a handheld PC companion or a Web-enabled set-top box. On a PC, just-in-time (JIT) compilers are capable of running code five to ten times faster than interpreted Java. Because of the scope of optimizations and the type of optimizations possible, JIT-compiled Java code will always be faster than a machine that is made to run Java as native code. One reason is that object-oriented code does a lot of copying to create instances. On a RISC machine like a JavaChip, each copy is a loop of instructions. On some CISC machines like the Pentium Processor or one of the earlier X-86 CPUs, this is a single instruction.

There are also a number of network issues that may affect Java portability. This is because, in order to operate efficiently, the Java applet needs to know about a number of network-related issues: (1) Is there a constant, or intermittent, connection to the net? (2) Is it a fast or slow connection? (3) Is the machine connected to the net through a proxy server? and (4) Are secure sockets (SSL) available, and how can these be used. At the time of this writing, there were no answers to these questions, at least none that are portable across all platforms.

personalJava (pJava)

Sun Microsystems, by late 1997, had begun to address some of these limitations first with pJava (personalJava). Then, in mid-1998, the eJava specification was released, not so much to provide a solution to the limitations of Java, as to provide a work-around alternative.. These are customized versions of the original Java API targeted at many consumer computing applications on Internet and information appliances, settop boxes, and Internet telephones and smart phones.

A brief review of the initial specification for pJava indicates that while the company has eliminated many of the packages available in the original developers tool kit for the full Java API, modified others and added several more specific to the requirements of consumer information appliances, very little has been done to the underlying virtual machine or operating system in terms of improving Java's interrupt response and its deterministic behavior.

Sun's argument is that the nature of the consumer market being what it is, many of the applications for which Java will be employed will not require either real-time nor deterministic behavior. In a typical implementation, the amount of program memory space available for the operating system kernel, the Java virtual machine, the Hot Java browser and the appropriate functions can fit into about 4 MBytes of ROM and 4 MB of DRAM. Not factored into this is the memory space required for downloaded HTML-based Web pages, which amounts to at least 1.5 to 2 MB, if you are ready to compromise on performance to some degree. At a minimum, without graphics code, no network support, no font support and no browser, the operating system kernel and the VM can fit into about 128 kbytes of RAM and 512 kB of read only memory. While that is more than adequate for many Internet appliances, such as Internet telephones, and portable Internet communicators, it is deaf, dumb and blind. Adding the additional features to make eJava useful could increase the memory requirements significantly.

With the pJava 1.0 API specification, first published in mid-summer 1997, a typical consumer implementation would require no more than 2 Mbytes of ROM and 1 to 2 MB of DRAM, but by the time the pJava specification is fully defined and approved it is expected that it will be possible to reduce this to less than 1 MB of ROM and no more than 512 kB of DRAM. To achieve this and make Java more appropriate for applications such as settop boxes, smart phones, and hand-held Internet appliances, a number of familiar packages from the original desktop SDK have been eliminated, several others modified, and some new ones added. Of the 21 components in the original SDK, only 12 have been retained.

The JDK 1.1 API packages included in pJava JDK 1.0 API are: java.applet; most of the Java abstract windows tool kit including java.awt.datatransfer; java.awt.event; java.awt.image; and java.awt.peer; java.beans; java.io; java.lang; java.lang.reflect; java.net; java.util; and java.util.zip. These packages provide the core capabilities of Java, as reflected in the many books on Java programming and in the capabilities of small devices and their users, insofar as it is possible to make generalizations about either.

Some of the functionality provided by the java.awt, java.io, and java.util packages was deemed by Javasoft programmers as not appropriate for inclusion in pJava. Examples of such inappropriate functionality are top-level windows and some features of internationalization. Deemed not appropriate for the kinds of applications for which pJava is targeted are a number of familiar packages, including java.math, java.rmi, java.rmi.dgc, java.rmi.registry, java.rmi.server, java.security.acl, java.sql, java.text, and java.text.resources.

A number of other features are included only as options. Among the most important of these are java.security and java.security.interfaces. A number of programmers and systems developers who have looked at the initial specification find it surprising that the security packages were not included, or at least an alternative, less memory intensive, albeit less effective method. Certainly the security features of Java are sophisticated and require considerable program memory space, so eliminating them was probably necessary. But users in this segment of the market are sensitive to anything that would render their Internet or information appliance inoperative. So some technique for providing at least minimal security should be included in the pJava API.

Finally, a number of features and APIs have been added. Among the most important are those relating to alternatives to the keyboard and mouse input common on the desktop. The pJava API includes four interfaces that allow developers to make it easier for their applets and applications to adapt to mouseless environments. Examples of such environments are keyboard-only systems and systems that are operated by remote control or that use a game control. In mouseless environments, users typically can navigate from one on-screen component to another by using keys or buttons on the system's input device. When the user navigates to a component, the visual representation of that component is modified in some way to indicate that it is the current component.

In the pJava API, the input device typically will provide a way for the user to select the current component, indicating that the user desires to interact with the component. For example, after navigating to an on-screen button component, the user might press the Go key on a remote control to indicate that the on-screen button is to be pressed. PersonalJava provides input preference interfaces to allow developers to specify the manner in which users navigate among components and the way that users interact with components.

Although the initial pJava spec does not address the issues of real-time embedded applications into which a portion of the net-centric computing market falls, it does address them indirectly. First, the timer utilities that have been added to the Java.util and Java.util.zip packages allow programmers to specify time periods with much finer granularity, from 100 or so milliseconds (msec) in the standard Java API to 10 msec. in the pJava specification.

Second, the amount of memory required for the application and to run the VM is smaller, which means that the impact of those features which limit Java in real-time deterministic environments - such as garbage collection and clear the memory - are reduced. Because the amount of DRAM required is reduced, the time required to do garbage collection is reduced. Not addressed is the lack of control the programmer has as to when garbage collection is done. In the development of the initial specification considerable effort was made by Javasoft developers to stay as true as possible to the "write once, run anywhere" dictum. This requires that the same language and dialect, the same APIs, and the same programming rules be used for all applications developed involving Java. With the pJava specification in its current form, all applications written using the pJava, with a few exceptions, will run on systems and environments built using the full Java API. These exceptions are the new packages, classes and APIs written specifically for pJava. However, consideration is being given to adding many of these new features to the full Java specification.

However, it looks as though downward compatibility has been compromised. Java applets are written with the full set of desktop Java APIs and it is a good bet that features such as mouse controls, CRT displays, etc., will not be readable or execute properly on an Internet device running in an environment based on pJava specifications. In this case, features and functions have been built into the pJava that would allow a device to go into a soft-fail or fail-over mode in which the device would not be able to run the applet, but would present to the user an innocuous message that the applet or function cannot be executed but does not indicate that the transfer has failed.

The pJava spec indicates the kinds of exception messages that would be delivered to the client system, but includes no details on how an application developer would make use of these in creating a soft-fail environment that does not threaten or panic the end user. For example, if a class is unsupported, calling its constructor or calling any of its static methods will result in the java.lang.message "NoPlatformMappingException." The same message will be sent for unsupported methods that are downloaded in a Java applet.

Interestingly, the most effective mechanism for allowing the information appliance to go into a graceful failure mode was not included in the current specification - letting downloaded applets declare their requirements, allowing browsers to avoid loading all or part of them if incapable of running them, is not included. This feature is still under investigation, and may be added as the specification matures and moves toward final ratification.

eJava Compromise?

The embeddedJava API specification was unveiled to the technical community in mid-1998 and represents an attempt at compromise with developers in the embedded and net-centric world who think it should be more real time.

Rather than modify standard Java in any way to make its interrupt response faster or to make it more deterministic, in eJava Sun chose to focus the API on features that would allow a programmer to squeeze the virtual machine and associated features down into a minimal memory requirement, so that it could ride atop a traditional RTOS, for those designers who need such capabilities.

To allow the eJava application environment to fit into the very tight memory footprints, Sun has developed a three-pronged strategy. First, it has reimplemented standard JDK classes to reduce their memory consumption. Second, in eJava, Sun has developed an efficient VM implementation optimized for the embedded environment, small enough to fit into a few hundred kilobytes of read-only-memory. And because the aim is also to reduce the DRAM footprint as well, it is intended to run without a JIT or virtual memory. Third, Sun is in the process of developing tools and and techniques with a variety of embedded tool vendors for configuring eJava in small memory applications. On its part is has developed special tools like . Java Filter and Java CodeCompact for shrinking the size of an executable image. Because it rides atop an RTOS, the memory requirement for the VM is even smaller, because it does not have to provide suppsort for a file system, a graphics/windowing system, or network support, all three of which Sun assumes will be provided by the underlying RTOS. For those applications where features are needed but cannot be fitted into the available memory, Sun has suggested in the eJava specification that the application developer store the additional necessary classes and applets somewhere on the network and download what is necessary for a particular service.

Do We Need Real-Time Java?

The answer to this question in the context of net-centric computing is "not yet, but soon", as bandwidths get wider and as Java programmers begin to take advantage of such language features as multithreading and active downloading of applications and applets. So the decision facing developers of net-centric applications, devices, and systems is whether to design for what exists now and modify it later as new capabilities emerge, or to plan ahead, building in capabilities that will take advantage of the likely new directions that the Internet and the Web will take?

First, we need to narrow down the definition of real time. The answers to this are as diverse as there are programmers and systems designers, but there seem to be two defining elements: interrupt time to external events and the degree of deterministic response.

What defines real time response varies widely, but broadly speaking response times can divided into hard and soft real time. Some engineers differentiate between the two extremes by the amount of response time necessary. Other differentiate by the amount of deterministic precision. According to the first measure, hard real time is typically is the low to medium microsecond response time, while soft real time is usually in the low millisecond to medium millisecond range. What they share in common is a need for reliable, deterministic responses to events; that is, just as important as responding fast to an interrupt is doing so in a reliable and predictable way so that the programmer can determine when, where and how the event occurs and incorporate the appropriate response into the system being designed. But there is considerable variation among engineers has to how precise the determinism should be: guaranteed to within a range of response times, "bounded" real time, or guaranteed to a specific response time. For a number of applications the response times can be as much as a second, 10 seconds or a minute but categorized a hard real time if the event required must occur at a specific point in the operation of the device and no other.

While most of today's net-centric devices (NCs, NetPCs, Internet appliances, etc.) fall into the non-real time or soft category, that is changing as the Internet and World Wide Web grows and changes. First, there is the bandwidth at which such devices are operating. Now, in the home or small office, data rates over the public telephone system range from 28 to 56 kbits/sec., and in the corporate networked environment they range from 10 to 100 million bits/sec. In the home and over the public switching system, data rates ranging from 128,000 to 3 or 4 million bits/sec. are in the offing. In corporate intranets and over the public switching system and Internet backbone, data rates will approach gigabits per second. A Java-enabled network computer or Internet appliance using Java as it is currently specified simply would not be able to respond adequately in systems operating at these higher data rates.

Second, as bandwidth and data rates increase, the Internet and World Wide Web will become much more multimedia-oriented, and the net-centric thin clients of various sorts will have to be able to handle such video, audio, and 3D transmissions. Current Internet-Web communications using the packet-based protocols on the public switching and private networks are more than able to support store-and-forward type communications, such as text and graphical email, file transfer, and most Web-based transactions between a browser and a Web page.

The problem is that video, especially when used in a two-way, interactive, and collaborative environment, requires that there be no delay or break in the two-way communications. Timing is also important so that video frames and audio arrive at the same time and are processed on the desktop or network computer in the right sequence and at the correct data rates.

Third, as bandwidth increases and programmers become more skilled, advantage will be taken of features of Java such as multithreading: downloading not one or two Java applets but dozens, and coordinating their interaction as the user of the net-centric devices accomplishes a task. This will require that the programmer be able to depend on a very precise, rather than a bounded deterministic response from the underlying software and the language used to write it, with nothing arbitrary, such as garbage collection, occurring to upset the devices' ability to handle multiple events and multiple threads of activity.

Three features of Java count against its ability to work in such a precisely deterministic and real-time, way: byte code, garbage collection, and numerous objects.

As discussed earlier, Java compiles to byte code, not machine code. This means that it can only run on a Java Virtual Machine (JVM) that interprets the byte codes. Since the interpreter will use several machine instructions to interpret each byte code, the interpreter is inherently about an order of magnitude slower than native code.

Java does not let the programmer allocate or free memory. This contributes greatly to Java's robustness and it makes it easier to program. But it means that the JVM must find and recover free memory. This is called garbage collection. The current version of Java will sometimes stop for a length of time that depends on the amount of RAM, but such interruptions can run to a second or more. From a real-time point of view this is the worst sort of performance problem. A program cannot easily know when Java is about to collect garbage code, and an unplanned delay of several milliseconds to a second is unacceptable. Also troubling to engineers and developers will be the fact that the garbage collection is under the control of the program: it occurs when and for how long the JVM or Java OS determines it is necessary. The developer has no control over when and how long the process will occur. If several of milliseconds to a second were required, a developer could live with that, provided he or she could specify when the garbage collection occurred.

Object invocation involves more overhead than the function calls with which most C, C++, or assembly-level programmers are familiar, and elegant object-oriented code tends to contain many objects. Java is elegant in this sense, and it encourages elegance on the part of Java programmers. And the more elegant a Java applet or application program is, the slower it is.

Working around to real-time Java

If the engineer or developer makes a decision to work with Java as it is, there are work-arounds that will help. First, a tuned interpreter or just-in-time compiler (JIT) can improve the basic performance of Java. The standard JVM's byte-code interpreter is written in C and forms a large, complex loop that is challenging for automatic optimization. Second, the byte code implements a stack machine. A human programmer will see that right off the bat and generate code with that in mind, but it is hard for a compiler to reach this conclusion. The result of applying this extra knowledge and a great deal of extra effort on the part of the programmer is between 30% and a factor of 2 performance improvement.

A conventional Java interpreter is not likely to get much better than a quarter the speed of native code. That is because of the way it allows code to operate: one instruction to actually execute the operation, one to pick up the byte code, one to select the operation based on the byte code, and one to branch back to the beginning of the loop. A JIT compiler can approach the performance of native code because the first time a JIT sees some byte code it converts it to native code. All subsequent times it executes that native code without looking at the byte code. This means the developer has to make a trade-off between converting the code fast, but generating poorly optimized code or converting the code slowly to well-optimized code. Each approach has a vulnerable point. All JITs will perform much worse than an interpreter if the code does not spend much of its time in loops. A fast, sloppy JIT will do better than a careful JIT in this case as well as when the time spent on the first pass through a function is an important measure. A slow, careful JIT will do well on benchmarks that measure the total time for many iterations of a loop.

A JIT would like to compile each method (function) once. This means that it needs to keep compiled code somewhere. The native code from a JIT can be expected to be between four and ten times the size of the corresponding byte code, so the RAM overhead of a JIT includes a code buffer several times the size of the byte code for the methods that the application uses.

Native methods and native processes are ways for an engineer to convert parts of a system into optimized native code long before run time. This fits the embedded real time model better than a JIT. The engineer knows what routines need the best performance and can focus optimization on these routines by converting them to C or assembly (and subsequently into native code). Java provides a smooth interface to methods in native code. Its interface to other processes is crude, but, in the worst case, a custom interprocess interface can be built with native methods.

Native methods can avoid garbage-collection delays entirely by not using garbage collected memory and not letting the thread scheduler execute. Native processes are only subject to garbage collection if they are written in a language that uses garbage collected memory. Java code is subject to garbage collection, but it can ask the JVM to collect garbage. This lets it schedule garbage collection at convenient intervals.

pERCing Up Java For Real-Time

What kind of functions would need to be added to the current standard Java to make it more deterministic and real time? To get a sense of what it would look like, one need look no further than the NewMonics Corp. pERC (see Figure 5. 3), a clean-room version of Java that incorporates real-time features. Because the extensions it includes are not part of Sun's standard Java specification, these capabilities can only be used by programmers and developers who do not need, or want, the "write-once, read everywhere" capabilities of the standard "pure Java" definition. But it is instructive to look at what Newmonics has done to get an idea of the direction in which mainstream Java may have to go if it is to remain the de facto "lingua almost-franca" of the Internet and World Wide Web.

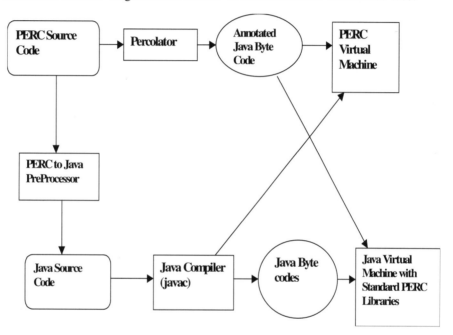

Figure 5.3 **PERCing up Java**. Newmonic's portable executive for reliable control (PERC) translates source code into Java byte with special nonstandard attributes that produce reliable real-time response.

pERC stands for Portable Executive for Reliable Control and is both a standard programming language notation and a virtual machine implementation. As a language, pERC is identical to Java, except that it supports two control structures that are not a part of the official Java definition, atomic and time control.

pERC can be translated directly to annotated Java byte codes using the NewMonics pERColator, or it can be converted to traditional Java by a pERC-to-Java preprocessor. The advantage of using pERColator is that it translates the pERC source into byte codes that are annotated using special attribute definitions to enable efficient and reliable compliance with real-time constraints. Traditional Java virtual machines do not understand these attributes. But without these attributes, compliance with programmer-specified real-time requirements can only be approximated.

The Atomic control structure is one of the two pERC syntactic extensions that allows programmers to specify whether certain code segments are to be executed in their entirety or not at all. The Time control structure allows programmers to specify a maximum amount of execution time for particular code segments. pERC provides a standard set of libraries to enable programmers to describe their real-time needs to the run-time environment. Two of the most important are those that relate to real time activities and real time tasks.

The RealTime package defines a RealTimeActivity class that pERC programmers can extend to create their own real-time activities. Each real-time activity consists of an arbitrary number of real-time tasks, a configure method, and a negotiate method. pERC supports a number of different kinds of real-time tasks. Most tasks are characterized in terms of their desired execution frequency and their periodic CPU-time requirements.

There are four types of pERC tasks: periodic, sporadic, spontaneous, and ongoing. Periodic tasks are automatically executed in each period. CPU time is reserved for execution of sporadic tasks, but the tasks are only invoked if certain conditions are satisfied at the time this task would normally be triggered. A spontaneous task can be triggered for executions if particular conditions are met and sufficient time and memory resources exist.

Periodic, sporadic, and spontaneous tasks are executed to completion each time that they are triggered. In pERC, these kinds of tasks are comprised of essential start-up and finalization code and an optional work component. In particular, the configure and negotiate methods are required for real-time execution. The pERC executive makes sure that each task has sufficient time to execute its essential components and provides a configurable amount of additional CPU time for execution of the task's optional components.

A fourth type of real-time task is identified by the ongoing label. Unlike the other three kinds of real-time tasks, ongoing tasks are suspended and resumed rather than being killed and restarted at the end of each execution period. Garbage collection is an example of an ongoing real-time task.

When a real-time activity is first introduced to the run-time system for execution, the pERC executive configures it. During initialization, the executive invokes the activity's configure method, which determines the total amount of memory that the activity requires and the rate at which memory must be garbage collected in order to reliably execute. The activity also determines the number of tasks that need to be executed and specifies an execution frequency, a minimum CPU-time allotment and a preferred CPU-time allotment for each task. The RealTime package provides services to assist in the analysis of the activity's memory and CPU-time needs. Once its resource needs have been determined, the configure method returns a representation of its needs to the run-time environment.

The Impact of CORBA

Whether it achieves adequate real-time performance and whether net-centric computing applications will need hard real-time performance, there is at least one other factor that may undermine Java's ubiquity as a net-centric computing enabler. As we shall learn in Chapter 8, that would be the rise of the Common Object Request Broker Architecture (CORBA). While Sun is active in its support for CORBA, the emergence of the so-called ObjectWeb based on CORBA could reduce Java to just another programming language, albeit an important one. Certainly, its ability to perform dynamic downloading of applets and its built-in reliability features could make it "first among equals" under the CORBA umbrella. But where Java allows the execution of applets remotely, as long as they are written in Java, the impact of CORBA on net-centric computing could be even more profound. This is because the CORBA specification allows heterogeneous computing; that is, programs written in virtually any language can be located on any system and invoked remotely as long as the program has the right object shell surrounding it, be it C, C++, FORTRAN, or for that matter, Java.

For all these reasons and because Java alone may not be enough to accomplish the goals a system designer may want to achieve in a net-centric device, it is probably not good policy to put all of one's programming eggs in one basket. As we shall see in the next chapter, there are alternatives to Java, some of which have been abandoned and some of which still survive, as well as complements to it that should also be considered.

Chapter 6

Alternatives & Companions to Java

There are alternatives to Java. Some come from the consumer electronics industry, where Sun is pushing hard to make its pJava a standard. Others are emerging out of the same telecom-datacom environment that led to the emergence of object- oriented languages such as C++. One alternative is embedded C++ (EC++), an effort by consumer electronics companies such as Hitachi, NEC and Toshiba to come up with a simpler, more compact and faster version of C++. Another is Limbo, that can best be described as a network-aware object-oriented version of C from AT&T. This is the same company and the same group of programmers who originally developed both C and C++. Still a third, is C+@, or CAT, that also got its start at AT&T, and was the victim of bad market timing. In many respects Java is its mirror image. But unlike Java, C+@ emerged at a time when it was perceived that there was no need for such a language, except among a few telecommunications equipment manufacturers. Beyond these that are a variety of portable and platform independent scripting and "small languages" that are as simple and easy to use as Java, if not more so.

This is not to say that any of these alternatives represent a challenge to Java as the linga almost-Internetica of net-centric computing, given the momentum and popularity that it has achieved. What I am saying is that Java is not the be-all and end-all of computer languages, and more often than not it will have to be used in combination with a number of other languages. In some instances, it may be appropriate to use other languages rather than Java.

Depending on where the industry goes with the Common Object Broker Architecture (CORBA), that allows programs written in a variety of languages to be accessed remotely, Java may be reduced to just another language. Certainly, it has some significant advantages, such as platform independence and dynamic downloadability, that will make it first among equals. But it must be balanced against the alternatives.

Alternative #1: Embedded C++

A new implementation of the venerable C++ object-based language, called Embedded C++ preserves many of its features, but without the code bloat. By omitting C++ features such as multiple inheritance, exceptions, and run-time type identification, EC++ hits the median between C and C++, preserving the object-oriented capability of C++ while achieving the memory and execution efficiency of C, an advantage that Java certainly does not share..

The primary thrust of EC++ will be embedded applications that utilize 32-bit microprocessors, but it could find significant use in net-centric computers where program size and performance are issues. Interestingly enough, the EC++ initiative had its beginnings in 1995 in Japan, a country not usually given much credit for its software prowess. Nonetheless, EC++ has already gained the support of most major Japanese CPU manufacturers, including NEC, Hitachi, Fujitsu, and Toshiba, all of whom are bent on dominating the newly emerging Internet appliance market. And U.S. manufacturers are beginning to take a serious look at the new specification.

The EC++ Technical Committee, as of late 1997, had already produced a draft specification for the new dialect. The committee was also working on a style guide that will help programmers maximize the efficiency, readability, reliability, and maintainability of their EC++ programs. Also, validation suites are available for a broad range of processors, including MIPS, PowerPC, and 68k families, Coldfire, Hitachi's SH family, and NECs' V800.

Despite the fact that the full-blown version of C++ is a powerful language whose object-oriented features enhance code reusability and simplify the partitioning and maintenance of complex code, it has numerous drawbacks. Unlike C, for example, where simple statements lead to simple code sequences, C++ uses constructors, destructors, and overloaded operators that often have the opposite effect. For example, moving an array declaration inside a loop can cause a dramatic increase in overhead in C++ programs, particularly if each element in the array must be constructed and destroyed during each pass through the loop.

Relative to C, C++ is also quite large and slow, but it is still faster than Java. Features such as multiple inheritance and exceptions, for example, result in considerable overhead in the run-time environment, even when not used by the target application. And library functions such as input/output often require much more code than is really needed for simple I/O operations.

EC++ looks very much like C++ did before a flood of "improvements" were added in 1990, though it incorporates all of the useful clarifications and minor enhancements that have been made subsequently as a result of refining and standardizing the language. For example, there are minor enhancements such as the addition of explicit statements that prevents single-argument constructors from being used for implicit conversions. Also, unused keywords such as mutable and template have been retained for compatibility. Small incompatibilities such as differences in floating point syntax (float_complex versus complex <float>) can be handled via conditional directives, macros and type definitions.

Among the C++ features that EC++ omits are multiple inheritance, virtual base classes, exceptions, run-time type identification, virtual function tables, and mutable specifiers. While each of these features is useful in itsr own right, none is compelling enough for a sufficiently broad range of applications. Eliminating them yields substantial reductions in the size of class libraries and improve run-time efficiency. These features also improve a programmer's chances of writing code that runs correctly the first time, one of the reasons many have turned to Java.

Multiple inheritance, exceptions, and run-time type identification are also the biggest contributors to C++ run-time overhead. Just as important, the overhead associated with these features is carried regardless of whether they are used in the target application. Features such as mutable specifiers, by contrast, add little overhead to the run-time environment, but needlessly complicate compiler design and systems programming.

Exception handling is the biggest of all C++'s overhead baggage. Though useful for dealing with errors, exception handling has two major drawbacks. First, exception handling makes it difficult to estimate the time that elapses between an exception and the transfer of control to the corresponding exception handler. As control passes from a throw point to a handler, destructors are invoked for all automatic objects that were constructed after the try block was entered. If an object has a destructor, then the object must be destroyed by calling its appropriate destructor, and it is difficult to determine the total time required to destroy all automatic objects.

The second problem with using exception handling is the difficulty associated with estimating the amount of memory consumed during the exception handling process. In general, the exception mechanism requires compiler-generated data structures and run-time support, both of that can add unexpectedly to the size of the program. Because of the difficulty associated with predicting the size and speed overhead associated with exception handling, it has been omitted from the EC++ specification.

Multiple inheritance can also add significant overhead to C++ programs. Designing a class hierarchy using multiple inheritance can make complex programs easy to partition, reuse, read and maintain. Unfortunately, it is difficult for even expert programmers to design such hierarchies using multiple inheritance. And when done improperly, the resulting programs tend to be less readable, less reusable, and more difficult to maintain than regular C programs. Given the overhead associated with multiple inheritance, and that the goal of the EC++ initiative is to simplify object-oriented programming, multiple inheritance has also been omitted from the specification. So too has virtual inheritance, that makes sense only when multiple inheritance is included.

Also eliminated was support for virtual functions, mainly because they do so at the cost of increasee run-time overhead in C++ programs The reason is that operations performed by such functions cannot be determined until the attributes of the object that they operate on are determined. And this is not done until runtime. This adds overhead to the target program, especially if the function is used in a loop that accounts for a large portion of the program's runtime.

Also eliminated are a number of features in C++ that add little or no overhead to the application, but needlessly complicate programming. For example, consider the specifier 'mutable', that determines whether members of a class can be modified. In many memory constrained environments in both net-centric computing and many embedded applications, objects that are specified 'const' are generally meant to be placed in ROM. However, class members that are declared 'mutable' can be modified, even if the object itself has been declared 'const'. As a result, objects that belong to a class that has a 'mutable' member cannot be placed in ROM. Moreover, 'mutable' is specified in the class definition for each class member, not in the object declaration.

The bottom line to all this is that it is impossible to determine whether an object is 'mutable' by looking at its declaration. To spare programmers this confusion, the "mutable" specifier has been omitted from the EC++ specification.

Some features that have been initially omitted from the EC++ specification may be restored upon further consideration. Name space support is one possibility. On the minus side, name conflicts avoided by name space support are more important in large programs, but are rarely encountered in the smaller programs typically used in many net-centric systems. Moreover, even if name conflicts do become a serious problem, they can be avoided can by using the static member of a class. However, on the plus side, name space support adds little or no run-time overhead. Moreover, namespace support is a plus for larger programs and very important when mixing libraries from multiple vendors.

New-style casts, that enhance flexibility by enabling programmers to direct the compiler to substitute one type for another (for example, to treat an integer as a pointer), have also been initially omitted from the EC++ specification. But new-style casts add little or no overhead to the run-time environment. Three of the four new-style casts (const_cast, static_cast, and reinterpret_cast) add almost no complexity or overhead to the compiler, and no overhead at all to the program when it executes. Only the fourth new-style cast (dynamic_cast) actually increases run-time overhead. So, why not add them back into the EC++ specification?

A direct consequence of simplifying the C++ language, of course, is a significant improvement in the size and efficiency of the associated run-time library. For example, eliminating exceptions from the EC++ language eliminates the need to provide exception handling functions or classes in the run-time library. And the lack of run-time type identification eliminates the need to provide a type-info class.

Even with these omissions, the EC++ run-time library supports a number of powerful C++ features that are not included with standard C, or for that matter, Java. For example, iostream operations for c-in and c-out are supported by providing istream, ostream, ios, and streambuf classes. String operations are supported for the class string; and math functions are overloaded for both double and float in both real and complex modes.

A number of additional EC++ run-time library features are also under consideration. These include the ability to provide input and output to strings; the use of allocators for string objects; and the addition of ifsteam, ofstream, and filebuf classes to support fopen and fclose. These additions add littles overhead and would broaden the appeal of EC++ beyond its initial target, embedded systems, to many equally constrained net-centric applications.

Alternative #2: Lucent's Limbo

Another alternative to Java, and specifically written for network applications, Limbo was developed at AT&T Bell Laboratories (now Lucent Technologies) as the programming companion to the Inferno operating system (see Figure 6.1). Whereas Java started as a programming language and spun off a VM and an OS, Inferno is a full network operating system including a kernel, communications protocols, libraries, security and authentication, naming protocols, and APIs, while Limbo is the programming language.

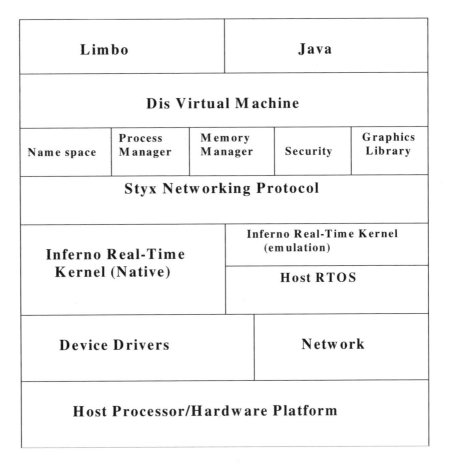

Figure 6.1 **The Limbo-Inferno Connection.** Derived from C with some object-oriented features, Limbo is a language designed specifically for telecommunications and networking applications.

When a Limbo program begins execution, it has a fixed interface to the OS services and is therefore inherently more portable than any language-only solution, even Java. One example is the network API, that is identical in all Inferno systems, independent of the host operating system or the network itself. Another example is security: features such as digesting and encryption are done in the lowest levels of the system, allowing an application to use them uniformly or even remain unaware of their presence, such as when a remote application wishes to establish a secure connection to a local one.

Moreover, Inferno was designed to manage the network, servers, and other network elements, not just client applications. Inferno provides the means to configure the execution of programs from pieces anywhere in the network, not just on the client.

Looking at just the language component, Java and Limbo have some significant similarities as well as crucial differences. Both use C and C++ like syntax, compile to a virtual machine that can be interpreted or compiled on-the-fly for portable execution, and use garbage collection to manage memory. On the other hand, if Java has any progenitors, it is C++ and it uses the object model to provide interfaces to system services. By comparison, Limbo is more C-like and avoids the complex object-oriented features of C++. Despite this it has more basic types (lists, strings, tuples, etc) and programming concepts (threads, communication channels) built into the language.

Limbo differs from Java in several other important ways. For Limbo, concurrency and communication are an inherent part of the language and the virtual machine and are used extensively in the programming model. Unlike Java, in which Sun developers have totally eliminated pointers, or at least hidden them from the casual programmer, Limbo has more general pointers, without sacrificing safety. Unlike Java, Limbo's garbage collector has constant overhead, so its operation does not conflict with real-time constraints, and, more importantly, to deterministic operation, is predictable . Also, the collector reclaims memory as soon as the last reference is released, minimizing the memory needed for execution. Limbo completely manages the lifetime of system resources by tying windows, network connections, and file descriptors to the garbage collector. Coupled with Limbo's "instantly free" property, this eliminates the need even to write special "free" routines, let alone call them.

Limbo programs are built as sets of modules that export interfaces whose implementation may be selected at run time. An application can configure dynamically, loading only those modules needed at any moment. It may load and even unload modules as its environment or execution dictates. The ease of exporting interfaces across the network allows modules to be run remotely; for example, a client can use the DNS (domain naming service) facilities of a server without the need to run DNS on the local machine.

Limbo has numerous libraries, packaged as modules. While the current collection covers some of the same ground as Java's libraries, there are significant differences. For one thing, Limbo's floating point package provides control of exceptions. For another, the graphics toolkit is based on the familiar model of the Tk tool kit, rather than built from scratch.

Limbo's virtual machine, Dis, is much more well suited to on-the-fly (just-in-time) compilation. The machine is a memory-to-memory architecture, not a stack machine, and for this reason translates much more easily than Java to native instruction sets. In fact, many Dis instructions translate to a single Pentium instruction. The on-the-fly compilers are small – for Intel 386 architectures it is only about 1, 300 lines of C code. This results in code that performs within about a factor of two of compiled C. For the same reason, the Dis interpreter is considerably smaller than the implementation of the Java virtual machine.

The combination of programs composed as modules and efficient compilation makes plug-ins unnecessary: codecs for playing things like AVI and QuickTime clips are written as Limbo modules. Although this means that existing C-language implementations must be rewritten, it also means that the modules are completely portable. Also, the multithreading in the language simplifies the implementation of scheduling in real-time codecs, a significant advantage in the future multimedia-based Internet.

In its details, Limbo has a number of features that should be of interest to programmers developing applications for net-centric computers. First, there is the chan (for channel) type. Unique to Limbo, a chan represents a synchronous bidirectional typed communications path between two threads. A communications operator is used to send and receive values along a channel. Also provided is an alt statement, similar in many respects to the well-known case statement. It allows a set of channels to be guaranteed a chance at performing, and completing a send-receive operation. What this does is prevent a single heavily used channel from hogging processor time, keeping less frequently used channels from communicating. Once a channel is created, any thread that has been referenced to it can use it to read and write. When writing or reading to a channel, a thread blocks contending activity until a corresponding read or write, respectively, takes place, making it much easier to synchronize threads.

Limbo also has a simple language element called a spawn statement that has been added to support multithreading. Similar to the select() and poll() functions in UNIX, it accepts a single parameter, which in turn provides a function for a new thread to execute. When used in combination with the alt statement, this allows an application thread to operate simultaneously on more than one channel. With the use of this statement, a single thread can block waiting on one of a number of channels to complete a read or a write. It can then perform an actin that depends on which channel is completed. This capability alone gives Limbo the ability to create uniquely robust and efficient concurrent applications.

Given all its advantages, will Limbo replace Java? Probably not, given the history of personal computing. Who would have thought that the original 8080 microprocessor, with its awkward instruction set and syntax, would give rise to offspring that have come to dominate the computer industry? One reason they have is that with each processor generation, Intel has responded to market and application requirements and moved the architecture away from the original implementation. Perhaps something of the sort will occur with Java. Maybe it will be possible to "Dis" the Java Virtual Machine, and move the Java language more in the direction of something like Limbo.

Alternative #3: C+@

The origin of C+@ was the same place that give birth to C, C++ and Limbo: AT&T. In may respects, it is identical to Java. Specifically, it was derived from software developed at AT&T Bell Laboratories by a group of telecommunications software engineers who wanted a true, dynamic, object-oriented programming environment similar to that in Smalltalk. It was licensed to a small company in Naperville, Illinois, called Unir Corp., that took on the job of commercializing the language.

Preserved in the resultant C+@ Programming Language was the syntax of C. It is designed as a companion language to C, a distinct difference from C++, which was designed as an extension to C, as indicated by the ++ in the name. The name C+@ reflects the syntax of C and the power of Smalltalk (the @ operator is used heavily in both C+@ and Smalltalk to create objects of class Point).

One of the most important characteristics of the C+@ programming language is that it is simple, even which compared to Java. Any C programmer can learn the syntax in less than a day because C+@ methods look just like ANSI C. C+@ syntax is easy to learn because it actually has fewer features than C. It solves the problem of bad pointers, the big reason many programmers have moved away from both C and +, but by taking the exact opposite approach to Sun, which eliminated pointers altogether from Java. Pointer mistakes are not likely to occur in C+@ because there are no explicit pointers in C+@. Rather, everyting is a pointer. Also, the infamous C preprocessor is not used in C+@. Thousands of lines of C+@ code have been written at AT&T Bell Labs and by other companies without using a single header file. This has dramatically improved the quality of the software.

But even though C+@ has the syntax of C, the semantics are very different. For one thing, in C+@ all operators and methods are handled as messages between objects, similar in some respects to Limbo. These objects contain state variables that are modified as a result of the methods being executed. When a method is executed, additional messages can be sent to other objects. Synchronous and asynchronous events are a natural part of the design process, because the C+@ programming language makes it possible to easily describe state variables and message handling. The entire system is modeled as objects and messages.

A major benefit of C+@ is how easy it makes it to reuse software, one of the most important aspects of object-oriented programming. In order to get widespread software reuse, classes have to be designed so that they perform isolated but well- documented functions, similar to the way ntegrated circuits are designed as a series of building blocks.

To make this as simple as possible, the C+@ programming language employs a simple plug-and-play syntax that allows designers to quickly develop classes, test the classes and deploy classes into working. A natural interface to C is provided so that designers can easily call compiled C functions, very similar to Java. These compiled routines can be used to access hardware registers, to perform system calls, or to handle time-consuming tasks. C+@ and C programs can pass data between each other as objects.

In order to effectively reuse software, a powerful development environment has been provided to allow designers to browse on existing classes and methods. Using it, C+@ programs can be developed just like C and C++ programs. Because of the emphasis on software reuse, the entire development environment is built around a powerful visual framework, called DoorStep, which itself was built using CAT. Similar to what Sun has done with JavaBeans and Microsoft with ActiveX, a DoorStep framework is a collection of classes that have been designed to work together and provide a cohesive set of functions. The DoorStep visual framework is used in all C+@ based systems that have a graphical user interface.

A primary advantage of C+@ over Smalltalk is that it promotes the concept of a ''unilanguage'' environment, similar to Java and Limbo. C+@ can be used for graphical user interface development, real-time telephone feature scripting, object-oriented database programming, and system administration programming. Programmers do not have to switch languages once they are working in a C+@ development environment.

Because the C+@ development environment is developed using this plug-and-play philosophy, it is possible to reuse classes from the development environment. These classes can be freely used in systems so that developers do not have to start from scratch each time they start a project. The attitude of reuse is pervasive throughout the entire development environment.

The C+@ Operating Environment

Similar to the relationship between Java and its Virtual Machine or Limbo and Dis, in order for C+@ to execute, an operating environment is required. This is both a plus and a minus. The plus is that once the operating environment is established, all projects can use it and this only needs to be done once for a particular processor architecture (68000, SPARC, X86, PowerPC, etc.). The minus is that this operating environment must exist to run C+@ programs and it takes space in the system memory. As the cost of memory decreases, this is less of a concern.

The C+@ operating environment is required to support the efficient creation (and automatic garbage collection) of objects (referred to as ''CAT Litter''). The operating environment also supports the message passing between the objects, as well as the interface to compiled C functions. The operating environment creates a virtual processor layer much similar to Java's virtual machine that C+@ programmers assume exists and use without being concerned about the internal operation.

Important in improving the reliability of net-centric applications, the operating environment can often be extended to include hooks for remote diagnostics, field software updates, and statistics collection. These capabilities can be used to help improve the quality of the software via better testing techniques and field diagnostics. With native compiled software systems (C and C++, for example), it is difficult to get between the program and the processor because the program runs directly on the processor. Similar to Java, with the C+@ operating environment, it is possible to peek at the operations from behind the scenes without affecting the running software and also without programmers changing their source code. This improves the quality of the testing and products.

Another advantage of C+@ is the concept of audits, used widely in telephone switching systems and other high reliability systems, and just now being used in net-centric computing. Audits typically execute behind the scenes and consist primarily of memory sanity checks that make sure that data structures that are supposed to be linked, are linked, and those that are not, are not. Range checks can also be done on the values in variables to make sure they are within a specified range. In native compiled systems based on C and C++, it is difficult to add these audits without involving the programmer. With C+@, these audits can be added behind the scenes because the operating environment has a view of the memory space of the applications. Unlike native compiled systems, in object-based programs using C++@, or Java for that matter, the objects can be informed that they have been audited and can take action. If they do not respond, the operating environment can monitor their behavior and escalate the auditing until the system repairs itself.

Similar to Java, C+@ programs are compiled and encoded in an architecture neutral format that allows the binary to be moved from one processor platform to another without being recompiled. The same C+@ binaries can be loaded into the different machines, and the C+@ operating environment tailors the binaries to the machine architecture as it is loaded. C+@ was developed on 68000 processors and evolved to the SPARC architecture without changing the binaries.

The binary format used by C+@ programs is very compact which makes it attractive in net-centric applications where Java has made the idea of platform independence so attractive. When these binaries are loaded they are expanded into DRAM as they are tailored to the machine architecture. Most of the execution occurs in the native instruction set of the machine. Therefore, there is very little performance penalty for C+@ programs. If performance becomes a major concern, then C can be used as a natural companion language.

Companions to Java: The Small Languages

No matter what the ultimate fate of Java, it will have a profound impact on programmers and programming as we enter the era of net-centric computing. If nothing else, it is opening up minds to other alternatives than the traditional C, C++, and other compiled languages. Already, programmers developing applications for net-centric computer systems are looking at all the alternatives that attention to Java has opened up: interpretive languages, object-oriented languages, and platform independent languages. Programmers, having gained some experience with the good and bad of Java are not only looking at EC++, Limbo and C+ @, but also a number of "companions to Java" as well.

While Java may now be the most well known interpretive language, it is not the only way (and maybe not even the easiest way) to add extensibility and portability to an application or to build simple graphical or textual applications for networked applications. Other, smaller interpretive scripting languages like Tcl (pronounced "tickle") offer similar functionalities, and may be even more appropriate to net-centric systems than Java. The well-known Tcl, with its roots in C, has unique strengths that make it a more credible choice for a for a scripting-extension language in many net-centric environments.

Examples of these alternatives abound on the Internet, due to the fact that it was originally an exclusively UNIX environment designed to link workstations, minicomputers and mainframes that used that OS exclusively. In addition to the OS, another legacy were the many "little languages" provided with it. Leveraging the flexible syntax of the UNIX environment, programs or "scripts" written in the little languages could be strung together to accomplish many useful user and system administration tasks. Some of these little languages, such as sed, awk, lex and yacc, are still in wide use in the engineering community. The yacc language, in particular, is still widely used as the parser generator for a host of UNIX tools and applications.

Scripting languages are particularly well suited to stringing together other self-contained programs to perform specialized tasks. These include such things as removing old, useless files from a directory, archiving email messages, or arranging for disks to be backed up on a certain schedule. In each of these cases, tools already exist to do the particular task (finding, removing, and backing up files). But what is lacking is the intelligence to do these tasks in a particular, customized way that suits the user. Such scripting languages come to the rescue by offering a simple environment for coordinating and sequencing the existing specific programs into a larger whole that accomplishes a goal.

In the desktop PC environment, under the original MSDOS, the only comparable scripting capability is the "batch" programming capability that allowed the relatively inexperienced user to string simple English-like commands together in ".BAT" files to perform a sequence of operations automatically. Traditional compiled languages are not very good for this kind of job because it is necessary to compile the code frequently during the debugging and testing phases of development. Since the compiled result will not be executable on machines different from the one where the tool was compiled, scripting languages, which were design to be portable, get around this problem.

Some of these scripting languages might even be useful to the systems designer using Java as his main programming vehicle, allowing him or her to side-step some of its limitations. For example, while Java has no facilities for local disk storage and other platform specific functions, a language such as Tcl could be used to complement it, providing the additional functions that Java does not, without compromising Java's "write-once, run-everywhere" ubiquity.

Today the Internet and World Wide Web is fecund with a variety of newer scripting languages, such as Perl, Python and REXX , that fill many of the roles previously handled by the aging sed and awk. They also run Web server CGI scripts, allowing the creation of customized Web pages in response to a search query, for example. Compared even to

Java, these scripting languages are much more precise and make possible much more rapid application-design cycles. Similar to Java, they also allow the developer to dispense entirely with native machine-code compilation, as C and C++ require, and all the attendant platform portability hassles that go with it.

The Tcl Alternative

One of the first of the newer scripting languages was Tcl. Coming from the world of UNIX and user interfaces such as the Bourne shell, Tcl corrected some of that shell's deficiencies, such as the difficulty in doing nesting quoted constructs, and the need to spawn an subprocess for almost every important operation.

Most importantly, Tcl provides a well-designed interface for joining a Tcl interpreter to an existing C program, giving what may have been a noninteractive program or one with a small, specific command language, access to Tcl's simple, general, extensible syntax. The idea its developers had was to create a universal extension language by providing a neutral, but powerful, langauge core that was meant to be extended with new commands representing an application's internal functions. In this way ,Tcl could serve as both a scripting and extension language at the same time.

Java's Internet connection almost eclipsed Tcl and the other little languages. Rather than being superseded by Java, Tcl and its partner, the graphical toolkit Tk, have experienced a renaissance of interest on the Web. The reason why is the ease with which existing C code, even legacy code, can be joined with a Tcl interpreter to provide scripting and extension capability simultaneously. And this same flexibility applies as well to Java. To bind a function in an existing application to a new verb in the Tcl language, all that is needed is to write a function that receives the arguments of the call from the Tcl program, invoke the existing application feature, and return the result. At initialization, a Tcl interpreter is notified as to the list of extensions provided by the application. This process can be made automatic by Tcl's dynamic loading features, allowing the writer of a Tcl script to load it in at run time.

Tcl also makes it possible for a new C or Java routine to receive its arguments as a vector of strings (in the familiar argc, argv format). However, it is up to the application to convert these strings to a more efficient internal representation and , after the work is finished, report the result in the form of another string.

In the abstract, it would seem that this approach could create a serious performance bottleneck. Indeed, some small, simple, mathematically oriented programs run quite slowly under Tcl when compared to other interpreted languages, to say nothing of compiled or semicompiled languages. But if used to provide a set of services for a distributed net-centric application, the performance penalty associated with the change in format to and from strings is quite low, if any at all. In most cases, in such applications the time spent by the call depends mostly on the efficiency of the communication path and the time required by the server to accomplish the request, not on the amount of time spent by the scripting language in gathering arguments and results.

Although it may sound contradictory, while Tcl is a simple language, Tcl has the flexiblity of more powerful ones. A case in point is its exception handling facility. Like everything else in Tcl, an exception (or error) is simply a descriptive string. But it is easy to create handlers for specific error conditions that can retry failed operations or take a more user-friendly action than just reporting low-level errors to the user. Differentiating Tcl from other modern scripting languages is Tk, the Tcl-based graphical user interface system. After binding an API to Tcl, it is possible to exploit Tk's powerful interface-building capability with the old application. Tcl and Tk work across a wide variety of environments, including Windows 95 and Windows NT.

If the two great strengths of Tcl/Tk are the ease with which large bodies of application code can be exploited in a new, scriptable environment and the simplicity of the Tk graphics extension, Tcl's greatest flaw is its lack of software support. That may be changing for — paradoxically — Sun has thrown its support behind Tcl and maintains a Web page on its site as extensive as that devoted to Java. There are also other flaws. For example, Tcl has a global name space; if two extensions define procedures with the same name, that collision has to be resolved; ditto for global variables. Unlike C or Java, there are no such things as static or private procedures. Unlike Java, Tcl has no internal support for object-oriented design, although Tk is somewhat more object oriented. This is because of the language's syntactic simplicity. Standards for the creation of packages (self contained bodies of Tcl code) or new Tcl application bindings meant to be dynamically loaded and exploited as a unit, have only recently been introduced.

Tcl and the Web

Say what one will about its lack of "heavy" language features such as native arithmetic, pointer-based data structures, and global modular structure, Tcl is being used successfully in a large number of varied environments on the Internet. The astonishing speed with which Tcl has found usefulness in the creation of Internet applets and as a CGI language extension is a testament to the simplicity of its design and the quality of its implementation.

Compared to a similar Java implementation, a Tcl/Tk applet (Ticklet) is extremely compact. A ticklet can be produced in a few broad strokes, whereas Java insists that the applet designer spend considerably more time on the details of the interface and implementation. A ticklet can be generally implemented with only about 150 lines of code, taking advantage of Tk's simplicity of expression, allowing the programmer to completely ignore the deeper software engineering that would be needed to construct a similar Java applet. A Tcl-Tk plug-in is now available for the Netscape and Internet Explorer Web browsers, that in itself is a demonstration of Tcl's unique flexibility.

In short, Tcl-Tk is a handy partner for engineers who want to create simple graphical or textual applications, or create a programming-prototyping environment for a new or old application. With just a small amount of planning, a new system can be given a full-featured language binding via Tcl that inherits Tk's graphical presentation capabilities automatically, and Tcl can give an old legacy application a new look and new life with only a modest amount of interface code.

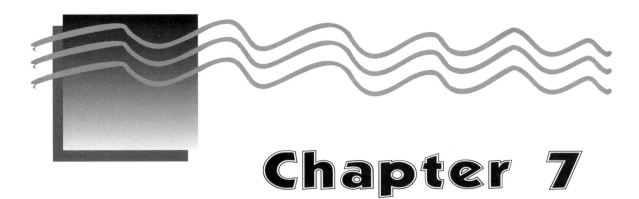

Chapter 7

Selecting a Net-Centric Operating System

In the 20 or so years since the introduction of the first microprocessors, systems designers have gotten used to the fact that if they were building a desktop computer they could use any operating system they wanted, as long as it was from Microsoft or Apple. If they were building a workstation, they had a little more flexibility, with a half dozen or so choices, all of which were just dialects of UNIX. Admittedly, there were a number of contenders for center stage in the marketplace. Some, such as CP/M from Digital Research, fell by the wayside. Others, such as the IBM's OS/2 became established in their own niches, but did not make it into the mainstream.

By comparison, in the world of net-centric computing, designers of connected desktop, set-top or hand-held Internet devices will have a wide range of options from which to choose. Not only will they have the use of scaled-down versions of the Microsoft Windows operating system, but several real-time versions as well. Additionally, there are a number of operating systems developed for use in the embedded systems market, real-time OSes, that are candidates as well. Eventually, by the turn of the century, some standardization will emerge, most likely around Sun Microsystem.'s Java OS. This trend will be somewhat dependent on how successful Sun is in moving the Java paradigm to silicon in the form of pico-, micro- and ultraJava 32-bit microcontrollers. But because of the diverse forms the Internet appliance will take, and the degree to which these devices will operate independently of the network, there will likely be diverse dialects, as software vendors develop different combinations of the Java OS and their own RTOSes.

Because the world of net-centric computing is not likely to settle down soon to one or a few paradigms, the engineer and systems designer will be faced with some difficult decisions about the most appropriate OS to use. The designer of an Internet appliance such as a Web-enabled screen phone or a wireless email phone, for example, might need the small kernel size of a real time OS. But because of the limited number of features and limited functionality might not need such sophisticated capabilities as multithreading or the ability to handle multiple processes. And depending on the network requirements, such an application might not need the deterministic, microsecond response time of an RTOS, but would need its ability to operate with a minimum of features and to operate from application code stored in a few hundred kilobytes of RAM or ROM. A more sophisticated system such as a network computer or NetPC for use in a corporate environment, might need real- time operation but not the sparse memory space. This is because it may be necessary to respond to multiple network inquiries quickly on the one hand and update screen information quickly on the other.

What And Why Of OSes

Everyone is familiar with conventional operating systems such as DOS and Windows 95 for PCs and UNIX and Windows NT for workstations and servers. The responsibility of such OSes is to manage the resources of the system within which they operate: disk drives, dynamic memory (RAM), printers, modems, keyboard, mouse, and the graphical user interface. The familiar desktop OS is typically a monolithic program, ranging from about 10 to several hundred megabytes in size. It handles all these functions using carefully constructed logic to sequence and service the interrupts that drive some of the input-output operations, such as data in and out of the communications ports.

From a real-time perspective, the limiting factor in both Windows 95 and Windows NT is not overall throughput, but worst-case performance and predictability. In some cases, these shortcomings can be resolved using facilities already available in the operating system. For developers of net-centric computers coming from the desktop world, the familiarity of the Windows development environment and the many low cost tools must be balanced against the less than optimal performance of this desktop OS vis-à-vis real-time deterministic operation.

For example, both Windows 95 and Windows NT use paging to shuffle processes between the hard disk and main memory. While this technique improves overall performance by ensuring that only the most active processes are stored in system memory, it also introduces unpredictable latencies (page faults) that complicate real-time programming. However, they both have page-locking mechanisms that a programmer can use to override this mechanism for critical code and data segments. Another feature of both OSes that can also complicate real-time programming is how they handle dynamic priority assignment. In both operating systems, thread priorities are dynamically raised and lowered in order to resolve deadlocks (like priority inversion) and prevent high-priority threads from commandeering the CPU 100% of the time. But in doing this, it is difficult for the OS to ensure timely response for critical threads. To circumvent this mechanism, both operating systems allocate their 32 priority levels equally among non-real-time threads, whose priority can be altered, and real-time threads, which have fixed priority. While features like paging and dynamic priority assignment can be a real pain to real-time programmers, they can be circumvented if the programmer has a good understanding of the OSes and the facilities for modification they have available.

More of a problem, however, are the interrupt handling mechanisms of both operating systems, which are not conducive to deterministic scheduling and are difficult to work around without the help of third-party tools or programs. In Windows NT, interrupts trigger service routines (ISRs), which run for a short period of time at the kernel level. In turn, ISRs trigger Deferred Procedure Calls (DPCs), which implement the bulk of the time-critical driver code, also at the kernel level. Finally, DPCs are used to launch application threads (scheduled with other Windows NT processes) that perform less critical driver functions. With respect to real-time applications, the drawback to this is that all DPCs run at the kernel level, where they have the same priority and run until completion on a first in, first out basis. So, while DPCs take precedence over application threads, they cannot pre-empt each other. This leaves the programmer with no way to establish priority-based hierarchy of DPCs. As a result, application and driver latency can vary widely depending on the number and type of DPCs that are scheduled at any give moment, from a millisecond when no DPCs are scheduled, to tens or even hundreds of milliseconds when DPCs for slow devices have been scheduled. Windows 95 handles things in a similar way except that events provide the functionality associated with DPCs, and both interrupts and events are encapsulated as VxDs, which run at the kernel level (Ring 0). Here again, the problem is that all events run at the same priority. Programmers can control the scheduling order somewhat. But once events are scheduled, forget about changing the order of execution. That is set in concrete.

At the other end of the spectrum are the so-called real time operating systems which are written modularly allowing them to operate in environments of no more than 50 kbytes all the way up to 10 Mbytes or more. This is due to the fact that they often do not need to manage such resources as disk drives and keyboards, much less GUIs. Their basic responsibility is to manage the response of a microprocessor's resources to requests from the outside, such as a communications device or a sensor requesting information, and then make a decision on the use or validity of the data collected. Unlike large monolithic OSes where response time can typically range from hundreds of milliseconds to a few seconds, an RTOS must respond in as little as 10 microseconds or less to no more than 100 microseconds. And where a desktop OS must handle perhaps half a dozen different operations (threads or processes in the parlance) an RTOS must handle many, many more. And whereas a desktop OS makes decisions on which request to honor on the basis of a priority list without regard to the exact time such a response will occur, an RTOS must be much more deterministic; that is, to handle the many tasks clamoring for management, it imposes restraints on when such requests occur. If they do not occur during a specific period that the RTOS and the programmer have determined is most appropriate, they do not get serviced.

The real time OSes that have emerged to aid embedded controllers in such operations exhibit some unique features that differentiate them from more monolithic desktop OSes. One is that they normally require very little memory space for storage of the kernel, usually ranging from a few tens of kilobytes to at most one or two megabytes. They are also able to execute programs in very small amounts of dynamic memory space, no more than a few hundred kilobytes to, again, one or two megabytes. Finally, where the monolithic OSes switch between different operations in tenths of seconds to several hundred milliseconds, RTOSes, because they must manage and execute what amount to several programs simultaneously, need to be able to respond in 5 to 10 microseconds.

To get an idea of the capabilities that an RTOS would offer the designer of a net-centric computer, it is instructive to look at some of the alternatives available. Virtually every major and minor RTOS vendor is targeting the net-centric environment, so it is impossible to review and compare each. But some offerings are interesting to look at because they are either typical of the choices available, or they have managed to make some headway in one or more net-centric computing segments, based either on unique RTOS capabilities or the marketing efforts of the companies that developed them.

The QNX/Neutrino Solution

If you like UNIX, you will like the QNX/Neutrino RTOS from QNX Software Systems Ltd. Neutrino (see Figure 7.1) is the successor to the most recent version of the company's QNX 4 real-time operating system. But where the original RTOS was dedicated to the X86 architecture and all of the many variants that have come to market, Neutrino has been ported to not only the X86, but the PowerPC, MIPS, SH3, and ARM architectures as well.

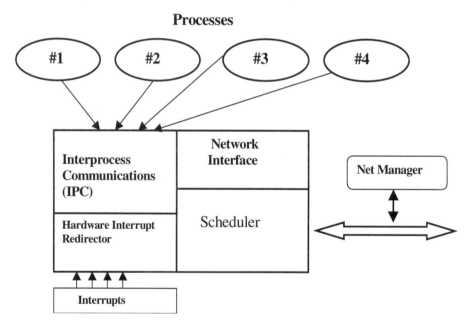

Figure 7.1 **Memory Protected.** QNX/Neutrino is a real-time operating system with configurable memory protection that manages two fundamental operations message passing and scheduling.

Programmers who got their introduction to programming on UNIX will be very comfortable with Neutrino because, like its predecessor, it has incorporated virtually every feature of the industry-standard POSIX specification, an effort to come up with a version of UNIX that is portable across virtually every processor platform. A preemptive scheduling, multitasking OS with one of the industry's most complete set of memory protection features, the Neutrino microkernel occupies only 32 Kbytes of memory. The kernel is written from the ground up around POSIX real time standards and extensions.

Based on a true microkernel architecture, the Neutrino kernel implements only a minimal set of core services: interprocess communications, interrupt handling, and thread scheduling) and can optionally manage a team of cooperating, add-on processes that are also available or which can be written by the developer. The add-on processes use the core set of services to implement other, higher-level OS services (such as filesystems, networking, and so on). Each of these processes can run in its own separate, memory management unit-protected address space, allowing a net-centric system developer to build in only the amount of protection from crashes that he or she needs. This protection ranges from no protection (for systems without MMU hardware) to full user-to-user mode, where user processes (including drivers) are protected from each other.

The kernel's interprocess communications and optional network services handle message passing, while a interrupt redirector and scheduler oversee thread-level scheduling, interrupt handling and timing. However, the kernel is itself never scheduled for execution; it is only entered through kernel calls invoked by a process or from a hardware interrupt. The kernel's main job is to manage the operation of two types of cooperating processes: system processes and ordinary processes. The main difference between the two is that the first manage services that are traditionally associated with the OS, while the second type handle everything else. The system processes include the Process Manager which coordinates process creation, the File System Manager which handles file system support, the Network Manager and the Device Manager which handles everything associated with I/O. The Process Manager is the only process that actually operates inside the microkernel's address space, adding only 32 Kbytes of additional memory to the footprint. But in this small amount of space features such as memory allocation, process contexts, resource manager namespaces and other extensions have been added to the functionality of the kernel.

Because it supports execute-in-place (XIP) code, the OS and its applications can execute out of a minimal amount of ROM or flash memory. Because of its message-passing, process-oriented design, it is possible to build precisely the OS that the developer requires. This can be done either through the many modules provided by QNX or by writing customized service-providing modules that extend the OS for a variety of specialized applications and devices, features that would be very useful in the diverse net-centric computing environment that is emerging.

Among the specialized applications and services provided by QNX for support of network-based computing is an implementation of the Microsoft Common Internet File System standard that allows a client workstation to perform transparent file access over a network to a Windows NT/95 server. Another module that can be added is the Tiny TCP/IP Manager (Ttcpip), which provides a small, modular implementation of TCP/IP that supports PPP and 802.3 networking. Although it requires under 40 kilobytes of code, Ttcpip supports most of the common Internet protocols, including the UDP and TCP protocols. A small set of useful TCP/IP utilities and daemons (such as ftp and ping) are included.

QNX was one of the first RTOS vendors to develop its own graphical user interface: the optional Photon microGUI Manager. It implements a scalable graphical architecture for a full-featured, embeddable GUI that can run in less than 500 Kbytes of memory.

Microware's OS-9, OS-9000 and DAVID

OS-9 is a real-time, multiuser, multitasking operating system developed by Microware Systems Corp. It provides synchronization and mutual exclusion primitives in the form of events, which are similar to semaphores. It also allows communication between processes in the form of named and unnamed pipes, as well as shared memory in the form of data modules.

OS-9 is modular, allowing new devices to be added to the system simply by writing new device drivers or, if a similar device already exists, by simply creating a new device descriptor. All I/O devices can be treated as files, which unifies the I/O system. In addition, the kernel and all user programs are ROMable. Thus, OS-9 can run on any 680x0-based hardware platform from simple diskless embedded control systems to large multiuser minicomputers. Originally developed for the 6809 microprocessor, OS-9 was a joint effort between Microware and Motorola. The original version of OS-9 (OS-9 Level I) was capable of addressing 64 kilobytes of memory. OS-9 Level II took advantage of dynamic address translation hardware, and allowed a mapped address space of one megabyte on most systems.

In the 1980s, Microware ported OS-9 to the 68000 family of processors, creating OS-9/ 68000, which is used in a variety of industrial and commercial arenas, including the Philips CD-I, set-top boxes for interactive television, and a variety of net-centric computing devices. More recently, Microware has developed OS-9000, a portable version of OS-9, written primarily in C, more oriented to small-footprint applications common in net-centric computing devices. It is available for the Intel (386 and higher) and PowerPC processors. Code is portable across OS-9000 platforms and between OS-9 at the source code level. Theoretically , OS-9000 can be ported to any processor architecture. DAVID is a configuration of OS-9/OS-9000 targeted toward the interactive TV set-top box (STB) market. DAVID stands for Digital Audio Video Interactive Decoder. The unique characteristics of DAVID are that it will always include the following I/O subsystems: (1) SPF (Serial Protocol File Manager) which manages high speed, packet based, streaming networks, using protocol modules to add support for X.25, UDP/IP, Q.2931, etc.; (2) MPFM (Motion Picture File Manager); and (3) MAUI, the Multimedia Application User Interface. DAVID is currently at v2.0 and being shipped for 68xxx, Power PC and 80x86 processor families.

To provide complete Internet access support for all of its OS configurations, Microware also provides an ISP, or Internet Support Package, which is a complete TCP/IP package that includes telnet and ftp client and server applications. ISP also provides a C lannguage BSD 4.2 compatible socket library. ISP supports both Ethernet and SLIP.

ISI's pSOS+

Another good operating system for developers who need the small memory size, but not necessarily the real time, but want to keep their options open as far as the latter is concerned is Integrated System Inc.'s pSOS+ (see Figure 7.2). It is a real-time, multitasking operating system kernel that acts as a supervisor, performing services on demand and schedules, manages and allocates resources, as well as generally coordinating multiple, asynchronous activities. The pSOS+ kernel maintains a simplified view of application software, where applications consist of three classes of program elements: Tasks, I/O Device Drivers, and Interrupt Servic Routines (ISRs). pSOS+ is small (17 to - 40kbytes)and fast, requiring less than 10 microseconds for context switching, fully preemptible, reentrant, and deterministic.

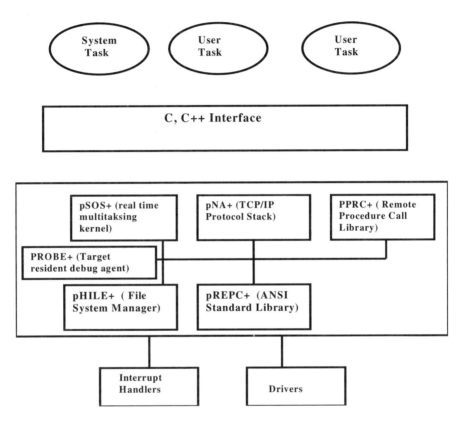

Figure 7.2 **Device Independent.** PSOS+ is an RTOS built around a small 17- to 40- kbyte kernel that handles only three classes of program elements: tasks, I/O device drivers and interrupt service routines.

pSOS+ employs a priority-based task scheduler that supports time-based and preemptive scheduling, specified on a per-task basis. The scheduling policy is dynamic and can be changed at runtime. pSOS+ services include task management, semaphores, events, timers, fixed- and variable-length queues, and asynchronous signals. pSOS+ is dynamic in its operation with all operating system objects being created, modified and/or deleted at runtime. Tasks can be dynamically loaded into a running target system, and added or removed from the schedule. Important in the still rapidly changing world of network computing, pSOS+ is completely device independent. All hardware-specific operating system elements, such as processor initialization and device drivers, exist outside the pSOS+ kernel, decoupling the kernel from the hardware operation. This results in interrupt and exception handling that is fast, highly deterministic, and under developer control. It also means that the programmer is presented with unchanging kernel image that vastly improves application reliability. At the same time, it allows the RTOS to be customized for a wide range of environments.

Wind River's VxWorks

Available on a wide number of microprocessors, including the PowerPC, 68K, CPU32, Sparc, MIPS, i960, x86, and 29k Wind River's VxWorks embedded real-time OS has been adapted and complemented for use in a wide range of net-centric computing applications.

Supporting a wide range of industry standards, including POSIX 1003.1b Real-Time Extensions, ANSI C (including floating-point support) and TCP/IP networking, VxWork's wind microkernel incorporates a full range of real-time features, including fast multitasking, interrupt support, and both preemptive and round-robin scheduling. The microkernel design allows VxWorks to minimize system overhead and respond quickly to external events. Kernel operations are fast and deterministic; for example, context switching requires only 3.8 microseconds on a 68K processor and interrupt latency is less than three microseconds. VxWorks also provides efficient intertask communication mechanisms, permitting independent tasks to coordinate their actions within a real-time system. The developer may design applications using shared memory (for simple sharing of data), message queues and pipes (for intertask messaging within a CPU), sockets and remote procedure calls (for network transparent communication), and signals (for exception handling).

Important to developers of net-centric computing devices is VxWorks' scalability, which allows developers to allocate scarce memory resources to their application, rather than to the operating system. OS configurations range from a few kilobytes of memory to several megabytes. Highly modular, the developer may choose from 80 different options to create hundreds of configurations. Furthermore, these individual modules are themselves scalable, allowing the developer to configure VxWorks run-time software optimally for the widest range of applications. For example, individual functions may be removed from the ANSI C run-time library, or specific kernel synchronization objects may be omitted if they are not required by the application.

VxWorks integrates not only the TCP/IP networking protocols, but a wide range of other networking facilities: sockets, NFS client and server, Zbuf (a.k.a. No-Copy TCP), CSLIP, RPC, FTP, rlogin, telnet, BOOTP, and SNMP. Through its third-party partnerships, a designer can implement virtually any networking solution: STREAMS, ATM, ISDN, SS7, Frame Relay, X.25, and OSI communications protocols; CMIP/GDMO and SMARTS for distributed network management; and CASE and CORBA for distributed computing environments.

Microtec's VRTX

A powerful entry in the net-centric OS arena is the latest implementation of Microtec's VRTX, Version 5.0, which was updated with telecommunications, networking, and network computing in mind. More than most of its RTOS competitors and certainly more than desktop alternatives such as Windows, this current version of VRTX represents a convergence of the desktop and embedded technology in net-centric computing devices.

Important to not only embedded systems designers, but those working on a variety of net-centric devices is VRTX's componentized architecture which includes (1)VRTXsa, a high-performance, multitasking kernel; (2) VRTXmc, a multitasking kernel optimized for minimal RAM & ROM usage; and (3) a range of networking modules, including a TCP/IP protocol stack, SNMP, PPP, SLIP, NFS, and RPC. For devices with stringent memory and resource requirements, such as Internet appliances, display telephones, cellular phones and pagers, the VRTX product family also includes the VRTXmc kernel. VRTXmc, designed specifically for such applications, is optimized to minimize RAM and ROM usage.

Based on a proven microkernel architecture, all the VRTXsa system calls are deterministic and preemptible. This allows context switches to occur even during a system call. And, because VRTXsa supports priority inheritance, it eliminates the problem of priority inversion. In many real-time operating systems, if an interrupt occurs in the middle of a system call, it is necessary to complete execution of the system call before scheduling a new task. This results in a delayed response to critical events. Since VRTX has preemptible system calls, it does not suffer from this problem. As a result, VRTX can schedule a new task as soon as it is ready to execute, ensuring an immediate response.

PowerTV: An Application Specific OS

In many respects, Web-enabled set-top boxes represent the future of network computing as a mass consumer phenomenon. And many of the requirements for real time and determinism that many hardware and software suppliers do not think are necessary in many other net-centric computing applications are an absolute requirement in this particular segment.

So, when evaluating RTOSes for this environment, the designer should keep in mind the weaknesses of traditional RTOSes as they relates to set-top boxes. First, a complete software solution is required for the set-top, not simply the kernel, hardware abstraction layer, device drivers, and TCP/IP stack provided by most RTOS vendors. The operating system must also support additional protocols for connecting to the network, such as DAVIC (Digital Audio-Visual Council) and DSM-CC (Digital Storage Media, Command and Control). These specify how a set-top signs on to the network establishes sessions, and receives data over mechanisms such as data carousels. A comprehensive 2D imaging model, including support for bit-blit operations (rectangular pixel operations), is required as is support for high-quality antialiased text and graphics rendering.

Many of the OS vendors seeking to provide such capabilities do so by dynamically linking applications at run time and patch the operating system (when it resides in ROM) using RAM or FLASH-based tables. However, this functionality needs to be extended to include the ability to decrypt the applications and patches sent over the network and to digitally authenticate them upon receipt by the set-top. Also, most general-purpose RTOS solutions do not provide a multiple-application framework and resource-sharing environment, which is critical for set-top solutions running in a constrained memory environment. And with respect to providing an optimized low-latency real-time system, many existing RTOS systems are adequate, but typically incorporate predefined kernel objects that encompass a high level of functionality and perhaps do not give the application programmer or system software developer the ability to optimize their implementation for performance.

The introduction of the Web and Internet and the use of multiple IP interfaces on the set-top requires modifications to standard TCP/IP stack implementations to support path associations with DAVIC signaling connections. In addition, the TCP/IP stack must be optimized to operate effectively in a constrained memory environment. In providing a solution for the set-top market, PowerTV has developed its own in-house compact operating system that incorporates all the necessary functionality to support set-top operation and enable compelling applications in a small footprint. Why a specialized kernel, rather than adapting one of the existing OSes? First, from PowerTV's point of view, it was important to be able to adapt the OS to the platforms as they were presented by its customers. This is a difficult job, because more than in many other segments of net-centric computing, most set-top platforms use exotic processors with a lot of silicon integration, not typically supported by existing kernel providers. Second, there was a need to incorporate directly into the heart of the OS the interfaces that enable system and application software developers to access the lowest granularity objects. This is necessary to build a system that operates efficiently in the given memory constraints.

At the heart of the PowerOS kernel (see Figure 7.3) is a unified events system that enables low-latency delivery of data to either single or multiple sources. Applications and drivers merely register interest in the type of events that they are interested in and the generated events are then directed to the registered receivers. This model is enhanced by enabling applications to register interest with a mask-compare system, allowing applications to listen for all key events; for example, whether they come from an IR remote, keyboard, front panel, or external device. By providing application developers with access to the lowest level kernel object, they can build their own optimized kernel objects on top. The PowerTV operating system incorporates a range of additional modules not normally associated with an RTOS, such as Application Manager, Resource Manager, Audio Player, Crypto Manager, PowerDraw, TV Manager, and Purchase Manager. They provide the additional functionality required for the set-top environment, but would mean additional development using an existing RTOS.

PowerTV, Inc., is also involved in both system design and ASIC development, two key competencies that have enabled it to develop an integrated solution with the right balance of hardware and software to address issues such as the throughput of Internet-based IP data. For example, the set-top environment supports true broadcasting and multicasting of IP packets, making Webcasting a viable service. However, the overhead in processing all these packets in order to retrieve the 1% of packets of interest takes a lot of processor cycles. To eliminate this overhead, PowerTV has extended its system software offering to include key Internet and Web technologies such as HTML, HTTP, and

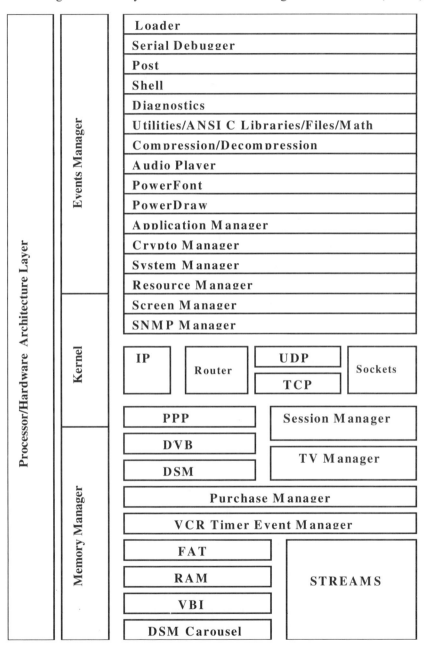

Figure 7.3 **All-in-One RTOS.** Unlike general purpose OSes, the PowerTV OS is specific to the real-time video, audio, and imaging requirements of the Web-enabled digital set-top box environment.

SSL to enable a wider range of content to be supported in broadband networks. To provide an optimized implementation, such extensions are architected in conjunction with end-to-end network designs as opposed to a client-centric focus. Applications such as email and Web browsers are provided, and PowerTV was one of the first companies to demonstrate an RF broadband Web Browser environment, as well as the operation of a Java Virtual Machine and Applet API running on the PowerTV operating system.

The many RTOS alternatives, however, face significant competition, not so much on the basis of more advanced and better features, but because of the sheer market presence from two alternatives: Java OS from Sun Microsystems, Inc., and Windows CE 2.0 from Microsoft Corp. Both offer small footprint OSes that many net-centric computing applications will require and are learning fast about OS design and the requirements of these applications. Both offerings are credible alternatives, particularly for those designers who prefer the familiarity of the Java and Windows programming environments and whose applications that are not particularly memory constrained and do not require real-time or deterministic response.

If there is any area in which the traditional embedded RTOSes have a weakness it has to do with the graphical user interface. Because of the nature of their traditional embedded market there was little need to interface with a human. In net-centric devices, things are much different. However, many of these companies have made up for lost time. QNX, ISI, and Wind River have all developed the capability to offer GUIs with their OSes. And unlike Java OS and Windows CE, these GUIs come with tools that allow the designer of a net-centric system to build an interface that precisely meets the requirements of the end user. And as we shall see in the following survey, while neither JavaOS nor Windows CE can compete with the RTOSes on performance, the one area that is attractive to developers is the ability to build one quickly, as in the case of JavaOS, or which comes included, as in the case of Windows CE.

JavaOS

Sun Microsystem Inc.'s Java, both as a language and an operating system, changes things altogether for developers of Internet appliances and network computers. It was designed originally to provide a platform-independent means by which applications developers can write programs independent of the underlying hardware. Of more interest and concern is Java's distributed object orientation and dynamic loading mechanisms ,which make it possible to distribute applications and programs throughout the network on both the client and the server. In the net-centric computing model initially conceived by Java's developers, an application may be stored on one system, but downloaded and run on an entirely different one. Internet devices can download pieces of programs from different locations to run on one processor, and different devices and processors can run a single application, passing data amongst each other across the network. Depending on how it is implemented on both the server and the client, Java could make a "thin client" such as an Internet phone considerably less thin. A concern for Internet appliance developers who need rea- time, deterministic operation is the still evolving nature of Java and its less than adequate capabilities in this area.

Sun's first effort at resolving these concerns is the JavaOS (see Figure 7.4), developed and introduced to the market in early 1997. It consists of two parts, the Java run-time code or Virtual Machine, which is processor independent, and the Java kernel, which can be optimized for specific architectures. To achieve the goal of platform independence even in the operating system, JavaOS was created with as many of the platform dependencies removed as possible, migrating them to platform-independent code. This minimalist kernel and the virtual machine are then used to implement all the other services such as GUI support, networking, I/O drivers, and the file system. Designed specifically to sup-

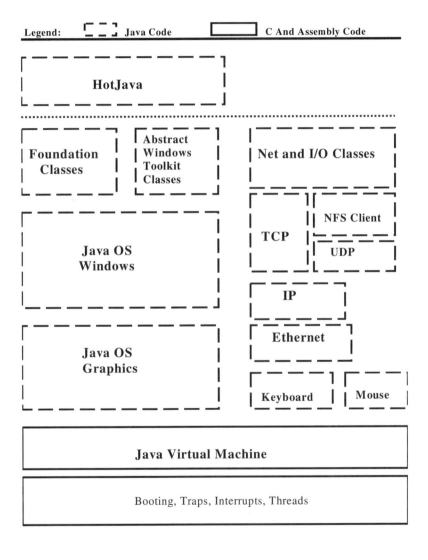

Figure 7.4 **JavaOS.** Built around the Java virtual machine that executes machine- independent byte code, this OS can be adapted to virtually any net-centric environment.

port the JavaOS, the kernel provides three basic types of services: booting, traps and interrupts, and thread support. interrupts, and thread support.

The bootstrap code for the kernel allocates memory for the Java heap, DMA registers and I/O device registers as well as detects hardware devices, mapping them to the appropriate drivers. The trap-interrupt code in this kernel provides all the traditional I/O services, interacting with hardware devices and passing necessary information to and from the device drivers. The kernel also supports context switching for the virtual machine, allowing it to handle as many as a dozen or so threads at any one time. While it is processor independent, the JavaOS does require that the host system support several mechanisms: context switching, GUI tool kit, networking and I/O. Because it supports only those features that can be implemented independent of the underlying platform, it requires a very small amount of memory. However, adding processor specific features, either directly or via an underlying RTOS, contributes considerably to the memory size. Because memory protection is provided by the Java programming language, all applications can run in a single address space, minimizing considerably the context-switching overhead. Because of this, the system can run almost entirely in the supervisor mode, which again minimizes system overhead and speeds up execution. Java OS also does not require a memory management unit; but if is one available, JavaOS uses it to make disjointed address spaces appear contiguous. Without optimizing or porting to specific architectures to improve performance, the JavaOS, at the time of this writing, performs more than adequately for most network computing and Internet appliance applications. Its TCP/IP throughput is about 500 kbits/sec., more than adequate for most Web browsing.

It is the memory requirements that should give Internet appliance designers some pause. At a minimum, without graphics code, no network support, no font support and now browser, the JavaOS can fit into about 128 kbytes of RAM and 512 kbytes of read only memory. While this is more than adequate for many Internet appliances such as Internet telephones and portable Internet communicators, in this minimalist form there is no way for it to communicate with the human operator. Adding those features to make it workable could increase the memory requirements significantly. Also, the designer must factor in the additional RAM needed for applications and run time requirements. At the high end, a complete JavaOS environment, consisting of the OS and the HotJava browser, can fit into about 4 Mbytes of ROM and 4 Mbytes of DRAM. Not factored into this is the memory space required for downloaded HTML-based Web pages, which amounts to at least 1.5 Mbytes, if you are ready to compromise on performance to some degree.

Integrating Java And Real Time

For designers looking to the future, it is not only the small memory footprint of the OS that must be considered, but its support for real time deterministic response as well. While many of the first and second generation of net-centric devices may not need this capability, future designs, such as for all digital set-top boxes with cable modem connectivity, certainly will. To combine the "write once, run everywhere" capabilities of Java with real time, vendors of many of the leading RTOSes have licensed the Java Virtual Machine and integrated it into their operating systems. The most direct way to do this is incorporate the VM in toto and run it under their own RTOSes as just another task to be managed.

Alternatively, companies such as Wind River Systems Inc. have taken a much more aggressive approach. Wind River's Java support is a highly scalable port of the JAE (Java Application Environment) to VxWorks (see Figure 7.5). JAE includes the Java virtual machine and all the Java classes. Any Java program following the Java specifications defined by Sun can be executed on a device running VxWorks. The port of the Java virtual machine to VxWorks maps Java's multithreading architecture to VxWorks' multitasking structure. Multiple Java threads operate as separate VxWorks tasks, allowing full concurrency and control. Like any VxWorks task, Java threads can send and receive messages, operate on semaphores, and control tasks via native methods. Java threads can even communicate and synchronize with other VxWorks tasks. Java's flexible memory model provides real advantages but its garbage collection mechanism, during which all Java threads are blocked, prohibits it from being fully deterministic. In real-time devices this is not acceptable. Wind River's Java port deals with the problem by reserving a separate memory partition for Java dynamic memory requirements and by mapping Java threads to VxWorks tasks whose priority can be set appropriately. Native, high priority VxWorks tasks outside Java are not influenced by Java execution, including the garbage collection thread.

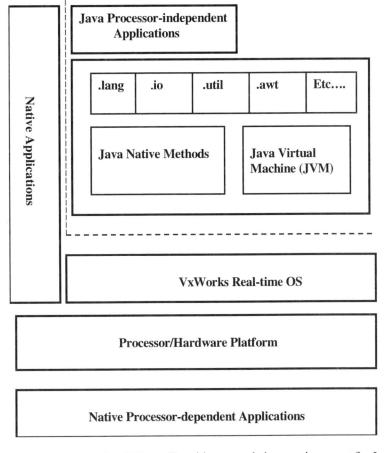

Figure 7.5 **Integrating Real-Time.** To achieve a real-time environment for Java, Wind River maps its multithreading architecture into VxWorks multitasking structure in each which thread operates as a separate VxWorks task.

Java applets may be loaded from the network, a local disk or ROM. To make memory usage more efficient, a unique tool has been added to the company's Tornado tool suite that allows execution of Java classes directly out of ROM, without first loading them into RAM. This arrangement allows the potential reduction of a net-centric device's RAM requirements. As discussed earlier, VxWorks is a true real-time system, and remains so even when Java is running. The mapping of Java tasks to VxWorks threads allows Java to be used in real-time applications. Nontime-critical parts of an application such as configuration, graphics, and user interface, as well as nonreal-time applications, can use the advantages of Java, while VxWorks supports time-critical code, usually written in C or C++.

Windows CE 2.0

Windows CE 2.0 is Microsoft's second attempt at developing an operating system based on its server-based Windows NT that would be useful in a wide variety of memory-efficient consumer computing applications. This category includes hand-held PCs and mobile communication devices as well as many wireless net-centric computing devices. It comes after long years of effort by the company to develop an operating system appropriate for applications other than the desktop.

Despite the fact that Microsoft is playing catch up with small-footprint OSes for embedded applications in consumer electronics, the marketing muscle it has put behind the effort should ensure that it will play a significant role in the net-centric computing market. This is despite fundamental limitations when compared to the many small-footprint embedded and real-time operating systems that are its competitors.

One of the first things Micrsoft learned about RTOSes soon after the introduction of Version 1.0 of WinCE was the value of modularity. In addition to requiring little memory space, many of the more popular RTOSes are designed modularly, so that systems designers can customize the OS to specific system requirements. Designed as a 32 bit preemptive multitasking operating system, WinCE 2.0's main features (see Figure 7.6) are provided as a set of modular components: the OS kernel; the OEM abstraction layer (OAL), which adapts the OS kernel to a specific hardware platform; a graphics, windowing and event subsystem (GWES); the object store, containing the registry, file system, and data bases; the device manager; the communications stack; and the shell.

Except for the OAL, all of these components, while customizable, are identical regardless of the hardware platform on which the operating system is running. Also called the hardware abstraction layer, the OAL is a thin layer of code that provides an interface between the kernel and the device hardware. The system developer receives a version of the kernel from Microsoft tailored for a specific processor and then finishes the job of tailoring by implementing hardware-specific power management, serial and parallel port, interval timing, real time clocking, and interrupt handling functions appropriate to the application.

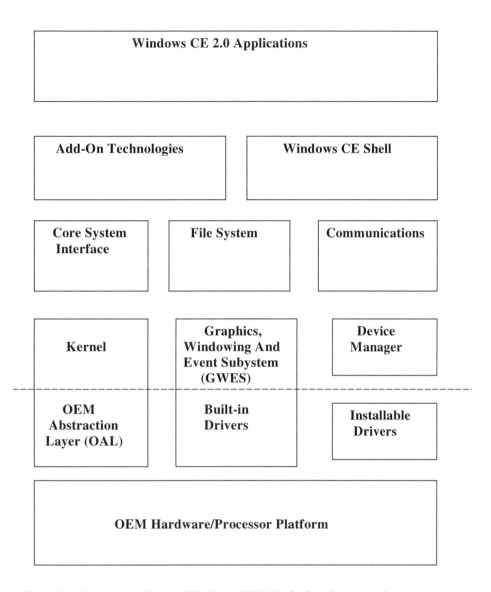

Figure 7.6 **Modularity Rules.** Windows CE 2.0 is designed as a set of modular components from which the developer can select the features appropriate to his or her application.

Although not platform independent as Java and JavaOS are, the above features of WinCE 2.0 make it easily portable to a variety of processors. As of the mid-1998, WinCE 2.0 ports had been made to the AMD X86-derived ElanSC400; the Hitachi SH3; the 486DX from Texas Instruments, National Semiconductor, Intel, and ST Microelectronics; Motorola's PowerPC821; NEC's MIPs-derived VR4101, VR4102, and VR4300; Philips MIPS-based PR31500; and Toshiba's TX3912. It also supports a number of ARM variations, including the DEC SA1100 StrongARM, the IBMPPC 403GCX, and the Motorola 860.   W in C E 2.0 also supports dynamic applet and application downloading in two ways. First, it supports the ActiveX controls, DCOM and OLE features of its desktop parents. Second, Microsoft offers the WinCE Toolkit for Visual J++, as well as for C++ and Visual Basic. To support Internet and Web access, some of the communications options available for WinCE 2.0 are the WinINET API for HTTP and FTP support, and SLIP as well as PPP for serial networking.

Microsoft's aim with this OS is to support not only deeply embedded applications in process control, telecommunications, instrumentation and high-speed data acquisition, but also the real-time needs of future net-centric computing devices as the Internet band-width increases from its present 9600 to 56,000 bit per second requirements to several million or so bits per second. But unlike the hard real-time features of most traditional RTOSes, WinCE 2.0 still falls into the category of "soft" real time because of its depen-dence on bounded interrupts, and worst-case latency times; that is, the total time to the start of the interrupt service routine (ISR) and to the start of the interrupt service thread (IST). Windows CE 2.0 splits interrupt processing into two steps: first an ISR and then an IST. Each hardware interrupt request line is associated with one ISR and, when interrupt occurs, the kernel calls a registered ISR for that interrupt line. Most of the interrupt handling is done within the IST, where the two highest priority levels can be assigned to real time ISTs. This ensures that these threads run as quickly as possible. At the highest priority level, the ISTs are not preemptible by any other threads. These threads continue execution until they either yield or are blocked. WinCE 2.0 supports multitasking in much the same way as RTOSes such as VRTX. VxWorks and pSOS+ do. But where these RTOses support 256 levels of task priority, CE 2.0 supports only seven. But even so, the task prioritizing scheme it uses gives programmers much greater flexibility in applying those priorities to application threads and device drivers. This is because the bulk of the device driver functionality is implemented with interrupt service threads, which are sched-uled along with user processes. This enables programmers to prioritize the functions performed by device drivers in a way that enables either higher-priority drivers or applica-tion threads to preempt lower-priority drivers

A credible effort that will fit the requirements of many, but not all, of the net-centric computing systems now emerging, Windows CE 2.0 still has far to go as far as both footprint and deterministic real-time response is concerned. In its minimalist configura-tion, running only one thread and without a GUI, WinCE 2.0 fits into about 128k bytes of ROM and 32 kbytes of RAM. A full-blown implementation typical of an Internet appli-ance or network computer in which many of the components are used is much larger. For a hand held PC, for example, Win CE 2.0 requires about 2 Mbytes of ROM and 1 Mbyte of RAM. With applications typical for a net-centric application (TCP/IP stack, browser, alphanumeric display, and address book), a WinCE 2.0 implementation would fit in about 512k bytes of RAM and ROM, respectively. A full blown GUI would double the memory space required.

What about similar configurations using one of the above mentioned RTOSes? While the applications and user interface requirements are about the same, with an RTOS the de-signer can start with a much smaller footprint. The pSOS+ and VRTX kernels start at about 2 kbytes of ROM and 800 bytes of RAM. VxWorks can range from no more than 15 kbytes of ROM for a deterministic stand-alone microkernel to about 1 Mbyte with Java, graphics, browser and TCP/IP networking support. As far as performance is concerned, as of this writing, the range of ISR latency times for Win CE 2.0 ranged from 1.3 to 7.2 microseconds on an 60 MHz Hitachi SH3 reference platform. More importantly, the range of IST latency times was in the 80 to 190 microsecond range. While it is difficult to make comparisons unless the measurements on done on the same platform, the estimated IST latency time on a roughly comparable platform is typically in the 5 to 10 microsecond range for both VxWorks and pSOS+.

Is Real-time Windows Real?

For some of today's net-centric computing applications, such as within many Internet/ set-top box combinations, especially those using a fully digital approach, the soft real-time features of WinCE 2.0 may not be enough. But the wealth of development tools, and programmers, available within the Windows environment is seductive. For those who want the familiarity of the Windows programming environment, but feel their net-centric computing application needs, now or in the future, will have to be more real time, there are a number of options available.

Though not optimized for real-time applications out of the box, Windows 95 and Windows NT can be readily adapted through the use of third-party products. Anasoft Systems for example, offers kits that enable designers to greatly reduce the Windows footprint by eliminating facilities such as help files, multimedia extensions, and fonts that are not required in the target system. The disk space required to run Windows NT, for example, can be reduced from well over 100 Mbytes to as little as 10 Mbytes, with RAM requirements reduced from 32 Mbytes to just 8 Mbytes. The footprint for Windows 95 can also be cut in half, with a minimal configuration requiring just 12 Mbytes of storage (RAM plus hard drive).

Meanwhile, enabling software from companies such as Spectron and VentureCom makes it possible to significantly improve Windows real-time performance. IA-SPOX from Spectron Microsystems Inc., for example, enhances the real-time response and predictability of Windows 95 by relieving the Windows 95 kernel of scheduling responsibility for critical real-time interrupts and drivers. VentureCom's Real-Time Extensions (RTX) does the same for Windows NT. Both Windows 95 and Windows NT were designed to provide good overall performance for a broad range of applications, and not to ensure fast, predictable CPU access for any particular thread. In order to adapt Windows 95 and Windows NT for real-time applications, then, a number of issues must be resolved.

For one thing, though it provides a rich set of services, the Win32 API does not specifically address real-time applications. In particular, it does not have services for interacting efficiently with hardware devices (such as performing efficient port I/O and configuring interrupt handling). So such services have got to be provided through the use of a separate API that either supplements or replaces the Win32 API. While both Windows 95 and Windows NT provide priority-based pre-emptive multitasking, a must for real-time applications, it is no match for an RTOS. Typical thread-to-thread context switching is 2 to 5 milliseconds and interrupt latency is 10 to 20 microseconds. While not up to the performance of most RTOSes, it may be more than enough for many net-centric computing applications. From a real-time perspective, the limiting factor in both Windows 95 and Windows NT is not overall throughput, but worst-case performance and predictability. In some cases, the developer can work around these shortcomings by using features built in to the operating system. In others, it will require real-time add-ons such as those supplied by Spectron and VentureCom.

Though hard real-time response is not possible with either Windows 95 or Windows NT, substantial improvements are possible. One technique, typical of the way early real-time DOS and Windows 3.1 implementations were adapted, is to run Windows as a low-priority task under a real-time operating system. In this way, high-priority real-time tasks get first call on CPU cycles, with the remaining time allocated to Windows and its nonreal-time tasks. This is similar to the way some RTOS vendors are integrating Java into their environment: running the JVM as a low-priority task under the RTOS. An alternative strategy is to integrate the RTOS as a device driver under Windows. Such a configuration allows the systems developer to intercept critical real-time interrupts and direct them to a separate scheduler that provides faster response and greater determinism.

For designers who require advanced features such as client-server support and built-in remote access that Windows NT provides, but need real time also, there are a number of options. For example, there is Real-Time Extensions (RTX) from VentureCom Inc. that leverages a modified Windows NT HAL (Hardware Adaptation Layer). This modified HAL improves real-time response and determinism by creating a firewall between non-critical interrupts (which are handled by Windows NT), and real-time interrupts, which are handled by a separate real-time scheduler (RTSS). The modified HAL also prevents Windows NT from masking RTSS interrupts and intercepts blue screen events (Windows NT crashes), allowing real-time processes to continue executing until the Windows NT application can be recovered. Also, it provides fast timer services (100 microseconds versus 1 millisecond) that can be used to reduce thread switching latencies. RTSS can handle Win32 threads as well as those developed using VentureCom's own RT API. RTSS also provides an interface between RTX and Windows NT processes. To improve scheduling flexibility and determinism, RTSS incorporates a pre-emptive thread scheduler with 128 fixed, non-degrading priority levels and support for priority inversion management, a substantial improvement over the original OS. RTSS processes are automatically locked into memory to avoid page faults. RTSS user processes can perform I/O, enable/disable interrupts, and access physical memory, eliminating the need for device drivers.

What about making Windows 95 real time? A real-time multitasking environment known as IA-SPOX from Spectron Microsystems runs as a VxD at Windows Ring 0. Like RTX, IA-SPOX boosts real-time response and determinism by intercepting time-critical interrupts and routing them to a separate real-time scheduler. The IA-SPOX kernel provides preemptive multitasking with 16 priority levels, deterministic interrupt handling, and high-speed data streaming. It also provides memory management facilities that allow IA-SPOX tasks to access the Pentium's floating-point and MMX instructions from Ring 0, which could be important in applications where real-time response to networked multimedia is required. To invoke such services, Windows applications simply make calls to the WinSPOX API, which, in response, dynamically downloads the appropriate real-time tasks to Ring 0 for execution under the IA-SPOX real-time kernel.

A key advantage of using SPOX in a Pentium-based real- time system is that IA-SPOX also runs native on digital signal processors. This enables programmers who require hard real-time response and higher performance levels to utilize DSPs as front-end multimedia coprocessors, offloading the host of compute-intensive real-time tasks, a real possibility in future World Wide Web applications that make use of new networked multimedia standards such as MPEG 4. Another benefit of this approach is that using an RTOS that runs native on the host and the front-end processor makes it much easier to partition the application into hard and soft real-time components and migrate code between the two environments. This is much the same as the scheme that many RTOS vendors are proposing with the Java VM.

QNX's Windows Solution

For developers who are reluctant to commit everything to one of the previously described Windows 95-NT real-time fixes, QNX also offers an alternative without giving up any of the features associated with traditional RTOSes and UNIX.. As its starting point the QNX Windows solution makes use of the European Computer Manufacturers Association's Application Programming Interface for Windows (APIW). The intent of APIW is to define a standard Win32 API for developing open systems, cross-platform applications that don't necessarily run on Microsoft- or Intel-only operating systems.

In a general sense, the approach that QNX took to integrating the Win32 API is similar to the way that Wind River Systems integrated the Java Virtual Machine into its RTOS, and offers a number of advantages over the earlier described approaches. First, the RTOS execution model is at least as robust as Windows NT, since every process (including those that make up the OS itself) runs in a separate, MMU-protected address space. Thus, the environment seen by Win32 applications under the RTOS is much "tougher" than running real-time processes as HAL-resident tasks without memory protection. Second, source code portability is maintained because real-time applications can be written to the full Win32 API, including having access to the graphical user interface. This is quite unlike being limited to the API within the HAL, or only that of a supervisory OS. Developers can continue to use the Win32 API, for both real-time and nonreal-time applications. Existing Win32 source code can then be immediately compiled and run in a real-time Win32 environment. Third, because the QNX RTOS is a purely event-driven and doesn't need to poll for real-time events, hardware interrupts are responded to only as they occur, executing interrupt handlers and scheduling processes with the low latency expected of a real-time OS. Latencies and overheads incurred from polling for real-time events are also avoided.

Fourth, implementing the Win32 API on an RTOS such as QNX produces a runtime system smaller than Windows 95, and significantly smaller than Windows NT. Despite the small memory requirements, the environment still provides a fully memory protected environment, real-time determinism, the Win32 API, and a fully-compliant POSIX API.

In the case of the QNX Windows offering a further benefit also exists: because of its POSIX and UNIX API compatibility, a programmer can readily compile and run source code from most UNIX OSes This allows developers of real-time systems to readily adopt source code from both the Win32 world and the UNIX world, promoting code reuse and increasing code reliability.

Toughening Up Windows CE

There are still some real concerns about Windows CE's hard real-time capabilities. While not necessary in the kind of net-centric applications developers are looking at right now, as bandwidths increase this will be a real requirement. If a developer wants to be sure of his or her ground as a net-centric project develops it is important to be aware of some of Windows CE's limitations in this regard and plan accordingly.

Developers who have looked at Windows CE 2.0 in detail have some real concerns about its real-time capabilities. One reservation about WinCE 2.0 is with reference to threads and thread priority and the way in which the OS manages them. For one thing, while CE is preemptive and multitasking, it is limited to supporting only 32 processes simultaneously. While enough for small systems but this limitation can be a problem for applications in larger systems like those found in the telecom/datacom or office automation markets. With the current implementation, while Microsoft says there are eight priority levels available, only seven of them are really useful. The lower levels are idle and barely usable by applications and threads since they can only execute after all other priorities are complete. This makes it only usable for soft real-time applications, and then only if an engineer is very careful in the way he or she writes the code or implements the hardware. By way of contrast, most RTOSes not only have as many as 256 levels, but all can be used in any hard or soft real-time situation. In addition, having only eight priority levels requires that multiple threads share a single priority level. For those threads sharing a priority level, Win CE 2.0 uses a round-robin scheduler. The problem is that this scheduling technique is hard on system performance and makes the system determinism extremely difficult to evaluate. Also, according to Microsoft specs, levels 2-4 are dedicated to kernel threads and normal applications. But what the documentation does not say is that applications cannot be prioritized against each other or against kernel threads. Since there are only eight priority levels, the number of threads per level will usually be bigger than the number of priorities.

This situation will make the time-slice scheduler predominant over the priority-based scheduler. Consequently the overall system will act more like a time-slice base scheduler rather than a priority-based system, limiting its effectiveness even in soft real-time applications. By comparison, with an exclusively priority-based scheduler the only requirement is the need to establish priority between threads. Once established the scheduler will use these priorities to run the threads. Since thread completion (by blocking or dying) is requested before switching to another one, no additional context switches are needed. Engineers targeting net-centric applications in the near future where real-time operation will be necessary should also be concerned not only about the way Win CE 2.0 handles interrupts, but in its lack of nested interrupts, almost de rigour in embedded systems.

What this lack means is that an interrupt can not be serviced while a previous interrupt is being processed. So, when an interrupt is raised in WinCE 2.0 while the kernel is within an ISR, execution continues to the end of that ISR before starting the ISR for a new IRQ. This can introduce latencies, or delays between the time of the hardware interrupt and the start of the ISR.

The bottom line is that the WinCE 2.0's ISR/IST model, no matter what improvements Microsoft makes to the underlying kernel, is generally not an efficient way to handle interrupts and not well-suited for real time. First, interrupt prioritization is not respected. Interrupts are taken and processed by the kernel in the time order of their arrival. Once an interrupt is accepted by the kernel, this interrupt cannot be preempted by another one arriving later. The only case where an interrupt can preempt another one is when the interrupt IST has a higher priority than the other interrupt IST. But WinCE provides only two levels for IST priority. This really limits IST prioritization. Programmers thus only have the choice between high and low for their ISTs priority. For system with more than two devices generating interrupts, ISTs will have to share priority levels and in that situation priority between devices cannot be guaranteed. A second problem with WinCE interrupt model is in the number of context switches requested to process an interrupt. Once an interrupt is generated, at least three context switches are performed by the kernel. A first context switch happens when the interrupt is received by the processor. A second context switch occurs when the IST is started and finally a third one happens when the thread processing the data is activated. For system with low interrupt rate this kernel will work perfectly but once the rate increases the system will completely fall down.

Microsoft has taken some of the criticisms of Windows CE 2.0 to heart and has incorporated some improvements in its next-generation Windows CE 3.0, including faster thread switching, nested interrupts, more priority levels and semaphores.

The new 3.0 kernel will require no more than 50 microseconds when switching between high-priority threads. It will also include the ability to nest interrupts, making it possible to respond to higher-priority interrupts while servicing lower-priority interrupts. The new kernel will also support more than WinCE 2.0's eight priority levels. The new kernel will also support semaphores, a Win32 thread synchronization construct that controls access to a fixed-size pool of objects. If a thread requests an object from the pool when none is available, a thread is placed in a queue by the semaphore where it waits for an object to be released. Via this mechanism, the semaphore allows applications to limit the total usage of critical resources such as network connections.

Inferno: The Best Net-Centric OS Ever

My candidate for the best net-centric OS ever is Inferno from Lucent Technologies Inc., the spinoff from Bell Labs, the research institution that gave the computing industry C, C++ and C@+ and had a major role in the development of UNIX and POSIX. Primarily because of poor timing, Inferno will most likely never be a major player in network computers, Web-enabled set-top boxes or Internet appliances, although it is ideally configured for all of those applications. But it will be behind the scenes, used in the bridges, routers and switches, where reliability counts for more than popularity or ubiquity.

Emerging from the laboratories about the same time as Sun's Java, but without the public relations fanfare, Inferno and its sidekick language Limbo (see Chapter 6) are intended for use in a variety of network environments. Whereas for most efficient operation Java may require Java-specific silicon, Inferno is designed to be portable across a wide range of processors and currently runs on Intel, Sparc, MIPS, ARM, HP-PA, and AMD 29K architectures and is readily portable to others. The purpose of most Inferno applications is to present information or media to the user; thus applications must locate the information sources in the network and construct a local representation of them. The information flow is not one-way: the user's terminal (whether a network computer, TV set-top, PC, or videophone) is also an information source and its devices represent resources to applications.

The model upon which Inferno was built has three basic principles. First, all resources are named and accessed like files in a forest of hierarchical file systems. Second, the disjointed resource hierarchies provided by different services are joined together into a single private hierarchical name space. Third, a communication protocol, called Styx, is applied uniformly to access these resources, whether local or remote. In practice, most applications see a fixed set of files organized as a directory tree. Some of the files contain ordinary data, but others represent more active resources. Devices are represented as files, and device drivers (such as a modem, an MPEG decoder, a network interface, or the TV screen) attached to a particular hardware box present themselves as small directories. These directories typically containing two files, data and ctl, which respectively perform actual device I/O and control and status operations. System services also live behind file names. For example, an Internet domain name server might be attached to an agreed-upon name (say /net/dns); after writing to this file a string representing a symbolic Internet domain name, a subsequent read from the file would return the corresponding numeric Internet address.

The glue that connects the separate parts of the resource name space together is the Styx protocol. Within an instance of Inferno, all the device drivers and other internal resources respond to the procedural version of Styx. The Inferno kernel implements a mount driver that transforms file system operations into remote procedure calls for transport over a network. On the other side of the connection, a server unwraps the Styx messages and implements them using resources local to it. Thus, it is possible to import parts of the name space (and thus resources) from other machines. To extend the example above, it is unlikely that a set-top box would store the code needed for an Internet domain nameserver within itself. Instead, an Internet browser would import the /net/dns resource into its own name space from a server machine across a network. The Styx protocol lies above and is independent of the communications transport layer; it is readily carried over TCP/IP, PPP, ATM or various modem transport protocols.

Inferno creates a standard environment for applications. Identical application programs can run under any instance of this environment, even in distributed fashion, and see the same resources. Depending on the environment in which Inferno itself is implemented, there are several versions of the Inferno kernel, Dis/Limbo interpreter, and device driver set.

When running as the native operating system, the kernel includes all the low-level glue (interrupt handlers, graphics and other device drivers) needed to implement the abstractions presented to applications. For a hosted system, for example under Unix, Windows NT or Windows 95, Inferno runs as a set of ordinary processes. Instead of mapping its device-control functionality to real hardware, it adapts to the resources provided by the operating system under which it runs. For example, under Unix, the graphics library might be implemented using the X-window system and the networking using the socket interface; under Windows, it uses the native Windows graphics and Winsock calls. Inferno is, to the extent possible, written in standard C and most of its components are independent of the many operating systems that can host it.

Despite the elegance of design, efficiency of operation and ease of implementation, Inferno, other things being equal, would end up being the dominant net-centric OS of the future. But things are not equal. RTOSes, because they were in the market first and have the right mix of small size and real-time operation will likely find a place in the net-centric future. But both the RTOSes and Inferno will have to compete for the attention of developers with JavaOS, which will be a player because of the market momentum and enthusiasm for Java and with WinCE 2.0 from Microsoft, simply because of that company's sheer size and market presence.

Chapter 8

Distributed Objects & Net-Centric Computing

The ultimate goal of net-centric computing is to turn the network into the computer, to shift the burden of computing from the client to the servers. At the very least, the aim is to balance the compute burden through out the network among many clients and servers, with a minimum of resources expended on each specific client.

Right now, the closest one can come to this ultimate dream is in the traditional "dumb terminal" or its more recent manifestations, the X-windows terminal under UNIX or the Microsoft Windows-based smart terminal. Ultimately, as more functionality, such as remote procedure calls are added to Java, and the common object broker architecture (CORBA) becomes more ubiquitous, such remote computing capabilities will be employed much more widely. But because of the particular requirements of the clients, the servers, and the connections between them, it will still be far from the ultimate dream of many systems designers of a platform-independent means of accessing programs and applications remotely on the network. A significant gating factor will be the availability of appropriately designed modular application software that can use the dynamic downloading features of Java more efficiently.

The issues facing the systems designer in the future are diverse: Does the application justify or even require such remote computing capabilities? If the engineer is designing in a relatively closed environment, such as a private intranet, a virtual private network or an extranet, are the capabilities of Java or CORBA needed? Indeed, might there not be other alternatives that are less costly and more suitable in terms of real-time deterministic response and memory requirements? Answering such questions will be important for the designer of the net-centric computing device, as well as for the one who is responsible for implementing the network infrastructure within which it will operate.

To get a clear idea of the choices available and the decisions that will have to be made, it is necessary to look at how "remote computing" has been done in the past, what Java brings to the game, and how CORBA will change the way such capabilities are added to a net-centric computing environment. Not only will the decisions made affect the ultimate cost and configuration of the network appliance and the system that supports it, but the alternatives present other difficult issues relating to safe, reliable, and secure computing.

Remote Procedures and Dumb Terminals

In the days of minicomputers and mainframes, the desktop system available to the user was, by necessity, no more than a monitor with some sort of communications interface to either a local area network or a modem. The intelligence that was incorporated in such systems was originally nothing more than some basic functions, just enough to (1) handle the interface to and from the keyboard; (2) generate simple text-oriented displays; and (3) respond to commands from the central mainframe or minicomputer, asking for additional keyboard input and to generate responses from the computer on the remote terminal screen. All program activity - word processing, database compilation and searching, and a wide range of basic applications - was, and still is, performed remotely on the central computer or server, with only the results delivered to the terminal and displayed.

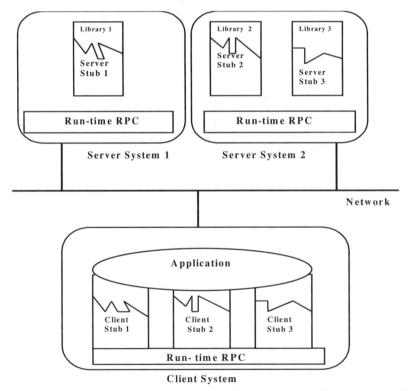

Figure 8.1 **Remote Procedures**. Until recently most net-centric mechanisms have depended on the use of the remote procedure call to initiate functions on a remote computer, usually a server.

Based as they were mainly on various flavors of the UNIX operating system, the basic mechanism for such operations was the remote procedure call (RPC), an extension of the common UNIX procedure call.

Locally, within a UNIX-based computer, procedure calls are the interprocess communication protocol used in the UNIX environment to allow program-to-program communication. RPCs are simply an extension of the procedural call to allow similar communications between programs on two different computers or to allow a computer to communicate and interact with a user on a remote terminal.

In the client, in this case a network computer or Internet appliance, a process calls for a function on a remote server, telling it that it wishes to execute a particular program or subroutine on the server. While the process is issuing the call to the server, the RPC run-time software on the client collects the parameters, forms a message with the information to be processed and sends it to the remote server. It then goes into a state of suspension until it receives the results. The server receives the request, unpacks the parameters, calls the procedure requested, and then sends the reply back to the client.

In the past, the use of RPCs was limited to systems using the same version of UNIX. But with consolidation in the UNIX world into just two or three basic variants, remote computing is achievable with the RPCs implemented as a compiler feature and supported by platform-specific RPC run-time libraries. These allow a program located in ROM on a remote terminal to initiate remote procedure calls in the same way as a local procedural call would be made to a run-time library. Similar to the way it interacts with the local procedural call, the run time library on the mainframe, minicomputer, and now the microprocessor-based server, is charged with finding the procedure or program, establishing the connection, and handling all the communications, transparent to the application.

Essentially a command-line and text-oriented technique, the most wide usage of the RPC technique outside the closed environment of a local area network and UNIX has been via Telnet on the pre-WWW Internet . It was the means by which early users accessed a variety of public and private information services such as the Well and CompuServe as well as the thousands of computer bulletin board services that have proliferated. It is also the means by which users still access library databases, as well as numerous program and news group repositories.

Until recently, most servers in wide use on the Internet have been UNIX-based. To allow access to a variety of information resources at university and other locations via the Telnet scheme using modems and remotely connected PCs, advantage was taken the emergence of a set of standard terminal interfaces. In addition to their native means of displaying data, most PCs via the communications programs available for early modems, were and still are able to emulate a number of terminal interfaces: Digital Equipment's VT 100 series, Hewlett-Packard's 700/94, IBM's 3101, and the Hazeltine 1500, among others. To a limited degree, such Telnet and terminal emulations allowed a user to do some limited remote computing. Via simple command line sequences it was possible to run a number of simple, single-function applications to help search a remote database. This provided, to some degree, a means by which it was possible to use the network as a computer regardless of the platform from which the requests came.

GUI-Based Network Computing

More recently, the RPC mechanism has been given a GUI look, by integrating this mechanism with graphical user interfaces such as Microsoft's Windows NT and X-windows/OpenLook, and the opportunities for remote computing via the network have expanded.

Microsoft's implementation of RPC in WindowsNT is based on the Open Software Foundation's (OSF) Distributed Computing Environment (DCE) standard, which allows it to work with any system that adheres to the DCE's Cell Directory Service as well as Microsoft's Name Service Provider. Under WindowsNT, all applications on the system are local DLLs (dynamic link libraries) that contain stub procedures, one for each remote procedure. These stub procedures take parameters passed to them and organize them for transmission across the network: organizing and packaging according to the requirements of the particular communications link, resolving references and making copies of any data structures to which a pointer refers. The stub then calls RPC run-time procedures that locate the computer where the remote procedure resides, determines the transport mechanism to be used and sends a request to it. When the remote Windows NT server receives the RPC request, it unpackages the parameters, reconstructing the original procedure call and then calls the procedures. When the server completes the requested function, it follows the same sequence of operations in reverse, returning the results to the client.

The RPC run-time mechanism makes use of a generic RPC transport provider interface to talk to various communication protocols, which act as a thin layer between the RPC functions and the transport. It maps the RPC operations onto the functions provided by the communications transport mechanism. Under Windows NT, the RPC supports transport provider DLLs for named pipes, NetBIOs, TCP/IP, and DECNet. And writing new provider DLLs can support additional transport links.

When one Windows NT is talking to another NT remote execution of applications using RPC occurs automatically, except for determining who in a particular connection will be the server and who will be the client. However, what about the situation where the client is a UNIX-based workstation or X-windows terminal or is a Windows 3.1- or Windows CE-based terminal or Internet appliance? Fortunately, WinNT's RPC has two features that resolve this dilemma.

With regard to a situation in which the client is an X-windows system, WinNT's RPC conforms to the RPC standard defined by the OSF in its DCE, to which virtually all UNIX systems vendors conform as well. What this means is that not only can applications written using Microsoft's RPC facility call remote procedures on other systems using the DCE standard, but the reverse is true as well: any DCE-conforming RPC on a remote system or terminal can access and run programs on the Windows NT.

In the case of a Windows 3.1/95- or Windows CE-based client, which in their original implementations do not have RPC capabilities, advantage can be taken of the fact that as part of its WinNT RPC package Microsoft includes a compiler, called the Microsoft Interface Definition Language compiler (MIDL). This compiler considerably simplifies the writing and creation of an RPC application. All the programmer need do is write a series of function prototypes in C or C++ describing the remote routines. These routines are then placed in a file to which is added information such as a network-unique identifier for the package of routines, a version number as well as a set of attributes that describe whether the parameters are input, output or both. The programmer then compiles this IDL with the MIDL compiler, which creates both client-side and server-side stub routines as well as a set of header files that are included on the server-based application.

Extending Windows' RPC

Where the equation breaks down is in linking Windows NT to remote NetPCs or Windows terminals running Windows 3.1, Windows 95 or Windows CE, since none of these consumer-oriented GUI-based operating systems support RPC. Using the interface creation capabilities of WinNT, a number of companies have created a considerable market since the mid-1980s with proprietary run-time programs resident on the terminals. Their function is to emulate the RPC capabilities of Windows NT, allowing them to access programs running on servers.

The Intelligent Console Architecture, originated by Citrix Corp. is a general-purpose presentation services protocol (see Figure 8.2) for Microsoft Windows that is very similar to the UNIX X-windows protocol. In essence, it allows an application to be executed on a WinFrame multi-user WindowsNT application server with only the client system's user interface. Only the keystrokes and mouse movement are transferred between the server and the client device over any network or communications protocol.

The ICA protocol presents only the user interface from an executing machine

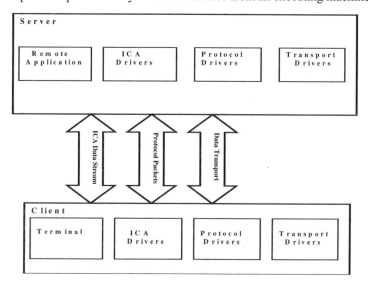

Figure 8.2 **Extending Windows.** Using remote procedure call methods as a starting point, the Citrix Intelligent Console Architecture to link remote terminal to applications that that are executed on a WindowsNT-based network server.

on the display of the client, be it a Windows-based terminal, a NetPC, a desktop PC, or a network computer. This distributed Windows architecture allows Win 16 and Win 32 API-based client-server applications to perform at very high speed over low-bandwidth connections. It also allows 16- and 32-bit applications to run on legacy PCs, as well as on new-generation, lightweight client devices.

ICA has been licensed to a wide range of vendors as an embeddable object in new hardware and software products. Tektronix, Insignia Solutions and NCD have developed turnkey software solutions for running Windows applications on UNIX workstations and X-terminals. Hardware vendors such as Wyse Technology and TransPhone offer ICA-based Windows terminals that connect to WinFrame servers over dial-up or network connections. There are also companies such as TransPhone that makes an integrated telephone, fax machine and Internet terminal device that use the remote computing capabilities of the ICA protocol.

Perhaps the biggest vote of approval is from Microsoft Corp. itself, which in 1997 entered into an agreement with Citrix to integrate its ICA protocol into the follow-on to Windows 95 as well as the PDA-oriented Windows CE.

How ICA Works

ICA is the physical line protocol used for communication between the client PC (or any other ICA-enabled client device) and the Citrix's NT-based WinFrame application server. Thinwire is the name of the data protocol that exports an application's graphical screen image (see Figure 8.3). It is not a physical protocol, but rather a logical data stream that flows encapsulated in an ICA packet. The physical protocol, ICA, must guarantee the delivery of the data stream with no errors and no missing or out-of-order data.

From the perspective of the application server, the major portion of the thinwire component is part of the GUI and video driver subsystem. In cooperation with key elements in the WinFrame Win32 subsystem, a series of highly optimized drawing primitives is generated, called the thinwire protocol. The output of the thinwire protocol driver is a logical data stream that is sent back up a virtual channel API, which takes the data stream and encapsulates it into an ICA packet. Once a packet is formed, it can be passed through a series of protocol drivers to add functionality like encryption, compression, and framing. It is then put on the transport layer and sent to the client. Once at the ICA client, the data packet passes through the same layers in opposite order, resulting in the graphical display of the remote application user interface on the client.

Each ICA packet consists of a required 1-byte command, followed by optional data. This packet can be prefixed by optional preambles, negotiated at connection time, to manage the transmission of the packet. The nature of the transmission medium (LAN versus Async) and user-defined options (e.g.,compression) influence the total packet definition.

ICA Protocol Stack

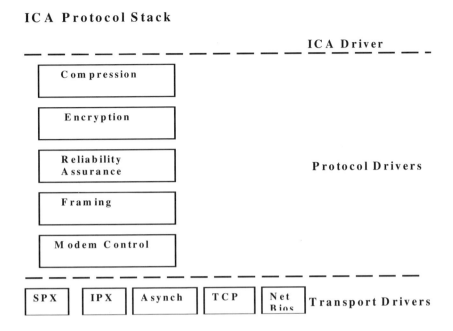

Figure 8.3 **Thinwire.** Key to the Intelligent Console Architecture is Citrix's proprietary Thinwire protocol, which exports an application's graphical screen image from the server containing the program to the thin client-network computer on which data are entered.

Beneath the ICA data packets, there are several optional protocol driver layers. Their existence and use is negotiated during the ICA handshaking that occurs at the start of a session. These layers sit below ICA and can be removed or replaced. The ICA protocol stack is dynamically configured to meet the needs of each transport protocol. For example, IPX is not reliable, so a reliable protocol driver is added above the IPX transport driver. However, since IPX is a frame-based protocol, a frame driver is not included in this particular ICA protocol stack. On the other hand, TCP is a stream protocol, so a frame driver is incorporated. And since TCP is reasonably reliable, this driver is not added to the stack.

Distributing Objects, Not Applications

The ICA approach is a short-term solution to the problem of distributing computing throughout the network and will continue to be viable until the alternative, based on distributed objects, becomes more widespread and until some sort of standard emerges. Even then, alternatives such as ICA will remain in use because of the overhead in memory and processing power imposed by such object-oriented paradigms.

The closest thing to a standard, right now, is the Common Object Request Broker Architecture (CORBA), the creation of the Object Management Group, an organization of over 700 companies that developed the original specification. The two most popular alternatives to CORBA are the object mechanism inherent in the Java language and the distributed common object model (DCOM) from Microsoft Corp., a distributed extension of the common object model that underlies the Windows operating environment.

Regardless of the approach, objects are nothing more than software components or building blocks that provide a specific set of services. With objects, the designer need only know the interface to the object to access its services. The code and data that implement the services within an object and allow them to be executed over the network are invisible outside an object; that is, they are encapsulated within the object. By encapsulating various software functions within an object, complex applications and software systems can be decomposed into a set of simpler components, each of which is implemented as an object and interacts through a common set of object interfaces. Object software is organized into hierarchical classes and subclasses. Starting with a library of object classes, an engineer or system developer can create new classes or, more importantly, extend existing ones by inheriting properties from previously defined classes, whenever new services are needed.

Because the code and data that implement a service or set of services are encapsulated with the object, the code that actually executes an application can change over time and can be altered or upgraded without any impact on the use of that object by other objects, as long as the object interface has not changed. As a result, software upgrades can be done on particular parts of a system without affecting others. Important to developers of applications and the main reason that there is gaining momentum toward such an alternative is that legacy code (old code written in older languages) can be given a new "paint job" by wrapping it with a software layer that provides the object interface. Many object-oriented languages and development environments have been designed to support such an object model, of which Java is the most recent: C++, C+@, Eiffel, Ada, and Objective-C among others. But object-oriented shells can be built around virtually any software language.

The Java Alternative

Java as a language and an applications programming interface grew out of applications such as set-top boxes long before it found its place in the networking and Internet environment as a platform-independent way of running and distributing programs.

Since then, it has moved in directions that in some respects are diametrically opposite to actions taken by Wintel (Microsoft and Intel) with the Windows-based terminal approach. The Windows terminal approach is server centric in that the majority of applications are executed on the server with only the results sent back to the thin client. By comparison, the ultimate aim of Java was to download applets and applications to a client only as needed, with little actual computing performed on the remote server. However, the Java model is, ironically, becoming almost as server centric as the Windows terminal approach.

The basic idea behind the Java language is "write once, run everywhere," no matter what the operating system and no matter what the underlying processor. The ultimate effect of this is that the network computer is no longer a "thin" client. Such peripheral functions as hard drives, floppy drivers, printers, and other devices are eliminated, but the network computer is "fat" as far as the burden placed on it by downloaded Java applets and applications and in terms of the additional dynamic DRAM that must be used.

The client-centric model originally adopted by Sun Microsystems and Javasoft for Java is gradually giving way, at least temporarily, to a more server-centric orientation with the incorporation into the Java API in late 1997 of remote procedural calls (RPC) and remote method invocation (RMI).

RMI enables the programmer to create distributed Java-to-Java applications, which allows remote Java objects to be invoked from other Java virtual machines, possibly on different hosts (see Figure 8.4). A Java program can make a call on a remote object once it obtains a reference to the remote object either by looking up the remote object in the bootstrap-naming service provided by RMI or by receiving the reference as an argument or a return value. A client can call a remote object in a server, and that server can also be a client of other remote objects. RMI uses object serialization to marshal and unmarshal (organize and disperse) parameters .

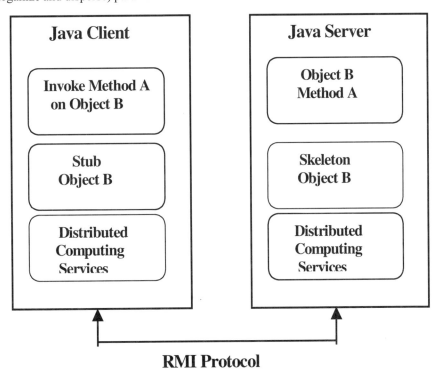

Figure 8.4 **Remote Invocation.** Adapting remote procedure calls to the Java environment, Sun has added remote method invocation to its repertoire of techniques to initiate functions on a remote computer on a network.

RMI, however, is strictly for Java-to-Java communication, complementing CORBA (Common Object Request Broker Architecture), which can communicate among multiple languages. RMI takes its place firmly on top of the Java VM, leveraging the strengths of code portability and security as principal reasons for an all-Java environment. A good way to look at RMI is to view it as a collection of layers just underneath a developer's code and above the Java VM. Incoming invocations (requests) from other Java objects pass through the VM to a particular object. Similarly, return values pass from this object through the VM to other objects. Because RMI is dynamically extensible, it is ideal for command objects, strategy objects, and agents. Objects can be of a core class, a JavaBean, or a custom object and can communicate in client-server, peer-to-peer, or behavior-based fashion (as with agents). Client objects can call methods in remote server objects, which in turn can be the clients of other remote objects.

The underlying problem with Java is that it is a programming technology that, by definition, only works within the boundaries of the language itself. That is, if an engineer or systems designer were thinking about using it as the means by which to do distributed computing it could only be done between programs written in Java. If it is necessary for a Java applet or client to communicate with another language, it must be done by using a Java intermediary with an interface that is identical to its mirror image on another system written in another language, but also acting as an intermediary.

What is needed in the emerging net-centric environment for computing is an interaction technology, something designed specifically to be the glue that links different programming methodologies together. In other words, what is needed is something other than just another programming language, something that occupies the spaces between languages and between the systems on the network that connect their hosts together. This is where the common object request broker architecture (CORBA), its interface definition language and the Internet Inter-ORB Protocol (IIOP) comes in.

The CORBA Alternative

CORBA dovetails neatly with the Java paradigm in many respects. Both are object-oriented, but where Java is created to allow object-oriented distributed computing in a homogeneous environment — that is between programs that have been written in Java — CORBA is designed to provide distributed computing in a heterogeneous environment. In other words, it is an interface by which programs written in any language for which an object request broker (ORB) has been defined can interoperate. Currently, ORBs have been developed for a wide range of languages, including Java, C, C++, Smalltalk, Ada, and COBOL among others.

Within the CORBA standard (see Figure 8.5), a common interface definition language has been defined. Its job is to express interfaces to services provided by an object. This IDL is independent of the specific programming language used to write the code of the client and server objects. Right now I can almost hear the engineers and programmers reading

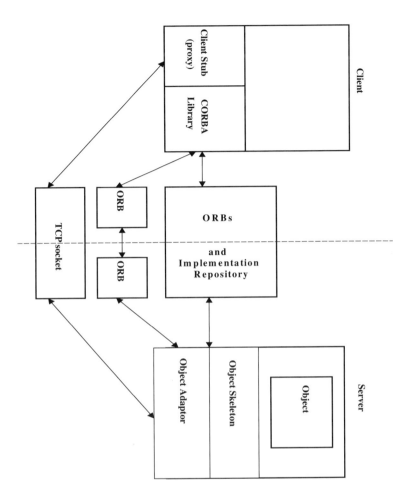

Figure 8.5 **Distributing Objects**. Key to the ability of the Common Object request broker architecture is encapsulation of software functions within objects that can be organized and acted upon.

this saying to themselves "Not another language!" So it is important to note that the IDL is not a programming language in the strictest sense of the term. Rather, it is more of a description of the interfaces between objects that is independent of any particular programming language. But the aspects of it that are similar to a language are very simple indeed. Because its only job is to describe interfaces, virtually all the issues that complicate programmer's life, such as control flow, functional composition, and memory management, are absent.

To create the appropriate interface does not mean that the programmer or system developer has to abandon the language with which he or she has the most experience. This is because the IDL is so constructed that IDL-specific objects methods can be written in and invoked from any language that provides bindings, or mechanisms, to link CORBA to that specific language. As of this writing, many of the languages in wide use have such bindings available, allowing IDLs to be generated from the language's native interface description: ADA, C, C++, COBOL, Smalltalk, and, of course, Java.

If there is any piece missing that will permit CORBA to be used as the ubiquitous environment within which to do net-centric computing it is the absence, until recently ,of a common distributed -bject communications protocol. IIOP is slowly emerging as the de facto interobject communications protocol. Already it is required on all CORBA 2.0 compliant systems. Netscape and Sun have placed implementations and the code for IIOP in the public domain, and a number of CORBA vendors, such as Iona and Visigenic, have been shipping free versions as well.

With all the pieces in place, how is a net-centric computation performed over the network? When a network-connected computer boots up at run time, access to object services located at other sites is supported by an ORB, which allows objects to interact with each other, even though they may not have detailed information about each other. The ORB's job is to translate service invocation requests from the client object into a format that is understandable by the server object. These requests and the replies are then transported back and forth via the network and IIOP. By means of the IDL-based interface, a client running on an ORB can make requests of other objects and other ORBs in two ways.

The most common way, because it is similar to the way things are done in the current UNIX-Windows environment, is to use a compiler generated stub. Using this stub, a request can talk in the syntax of a native programming language object invocation. The alternative to this requires that the system developer commit himself to the CORBA environment and become familiar with concepts such as object references, object adapters, servants, skeletons and bindings. An object reference provides the information needed to uniquely specify and object within a distributed ORB environment, a unique name or identifier. The client uses an object reference together with what is called a dynamic invocation interface (DII) to make invocations without compile time knowledge of the object type. Using the DII, an ORB directs a request to the server identified by a specific object reference.

When the server receives a request, it identifies what is called the object adapter, which is responsible for locating, and even creating, servants. Servants are anything executable - a script, a subroutine or more typically, a programming language object. The object adapter must locate the skeleton, or framework, within which a servant can be invoked. Using a mechanism called binding, the application code is executed on the server, and the server registers the CORBA objects in an object adapter. Finally, the server invokes the servant through the skeleton, completing the transaction.

The integration of CORBA with the Java-based object methodology provides most of what is needed for ubiquitous distributed computing. What CORBA provides is the ability to access legacy applications on their native platforms, while Java provides the mechanism to develop an object once and install it on any platform. Also, Java's platform binary independence allows an object to relocate at run time from one platform to another.

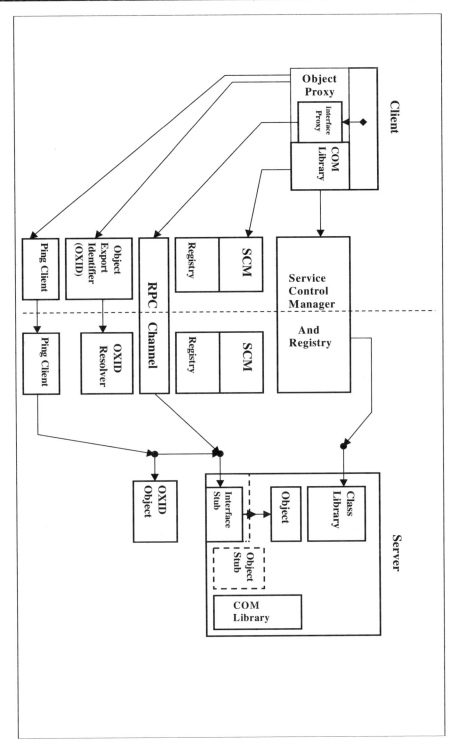

Figure 8.6 **Microsoft's DCOM**. The alternative to CORBA, DCOM is built on the Window's OLE and tools such as ActiveX to distribute applications over a network.

Microsoft's Way

The one main obstacle to the emergence of the CORBA-Web is Microsoft and its vested interest in an existing object-based mechanism, the common object model (COM) and its distributed extension, DCOM, upon which Window's OLE and tools such as ActiveX are based (see Figure 8.6) Like CORBA, DCOM is a complete object system and in many ways is a mirror image of the industry-wide standard, at least in terms of functionality. Rather than adopt the CORBA standard protocol, DCOM uses the remote procedure call methodology that underlies Windows as the main mechanism for linking objects and locations on a network. DCOM also has its own interface definition language called MIDL (guess what the M stands for?) and its associated compiler.

Similar to CORBA, DCOM defines a clear separation between the object interface and its implementation and likewise requires that all interfaces be declared using MIDL. Where CORBA is based on a standard object model, DCOM is not, nor does it support multiple inheritance, at least directly. By using the ActiveX component builder and encapsulating the interfaces of the inner components and representing them to a client, DCOM can support multiple interfaces and reuse, the main uses for inheritance in a traditional object environment.

Unlike CORBA, DCOM objects are not objects in the traditional sense. Among other things, DCOM interfaces do not have state, which means that an object cannot have a unique object reference. Without a specific object ID, DCOM clients must obtain a pointer to an interface, rather than a pointer to an object with a state. Because this pointer is not related to the state information, a DCOM object cannot maintain state between connections and cannot reconnect to exactly the same object with the same state at a later time. It can only reconnect to an interface pointer in the same class. Because DCOM objects are not able to maintain state between connections, it is operating at a disadvantage in environments — such as the Internet — where faulty connections are endemic.

JavaBeans Versus ActiveX

As part of its strategy to blunt the move toward Java as a development language and environment and as a means to distribute applets and applications throughout a network of servers and clients, Microsoft has developed ActiveX. It is a fully Web enabled component architecture, creating an alternative network-computing model. In response, Sun has introduced its own component architecture, called Java Beans.

In the component approach, programmers can work at a higher level of abstraction, one step removed from the details of a specific language to create, assemble and use dynamic applets or applications. A software component is a reusable software module supplied in binary form. It is somewhat like an integrated circuit in that it can be plugged into an application and connected to other components.

ActiveX is both a component-based approach to applications building as well as a way to create an environment within which applets and applications can be dynamically downloadable. It is an implementation of the Distributed COM model that serves as the basis for the Window's distributed-object scheme. ActiveX defines software components that can be written in any of the languages supported by Microsoft.

In the Microsoft model, the ActiveX controls are the main mechanism for distributing applications, not any particular language. Unlike the Java paradigm, where all applets must be written in Java, under ActiveX a number of languages can be used: C, C++, Visual Basic, as well as Java, or at least Microsoft's particular Windows-optimized and non-standard extension. With ActiveX, any part of an ActiveX component written in any language with the correct links to the controls can be downloaded to a client machine as often as is necessary in compiled form to known processors. If the developer does not want to download the application to the client, the component can be written to run entirely from the server.

Whereas ActiveX was a response to Java's dynamic downloadability, Sun's Java Beans was a response to the componentized approach to programming that ActiveX introduced. The Java architecture has no mechanism by which programmers can build components that can be assembled to create applications or make applications work. So, taking a page from Microsoft's game book, Sun has come up with the JavaBean API. A Java Bean is a reusable software component that can be manipulated visually in a builder tool. Similar to ActiveX, but specific to Java, Beans can be used to create a variety of functions ranging from special-purpose controls, such as buttons and boxes to components with no GUI, such as communications interfaces and data-type converters.

The strength of both component-building approaches is that they considerably simplify the building of user interfaces, which is likely to constitute a considerable portion of a programmer's time during the development of a net-centric product. Using traditional coding techniques, the creation of a user interface is much more complex. Each time a component needs to be added it is necessary to write code setting properties and invoke the right event handling methods. The problem is that it is not easy to see immediately how a particular addition or modification affects the user interface. This is because it is necessary to go through a laborious trial and error process, first coding, then compiling, and then running it, to see if it achieves the desirable look. Also, each time a component is created a lot of repetitious code must be written.. It is also uninteresting and in combination with the boredom of the formrer, leads to errors in programming, further drawing out the development process.

What these component builders do for the system developer is eliminate much of the repetitiveness and shorten, and in some cases eliminate, the lengthy code-compile-run process. In a visual building environment, programmers can pick precoded components from a tool box, move them to a container and lay them out with other components. The builder automatically generates code and provides mechanisms for inserting code for event handling.

There are tradeoffs that the system designer will have to make. In exchange for rapid prototype development, important in the rapidly evolving net-centric computing arena, the designer will have settle with somewhat more bloated code. Because prewritten code is more generic, the code sparseness that is possible with hand coding is not available.

For the programmers who have gotten used to the class libraries available in Java and in ActiveX, there will also be somewhat less flexibility. Whereas class libraries are built for the purpose of solving broad categories of problems and give the programmer a lot of flexibility, components, on the other hand, are typically self-contained and focused on the performance of a particular task. While a component is usually bundled with the necessary class files and resources so that its appearance and behavior can be changed, the structure is fixed.

Which is the best way to go? The easiest answer is both, even it you are not working in a traditional desktop environment. With the porting of Windows CE and Java to a widening variety of hardware platforms, the choice is no longer dependent on the underlying processor.

If the engineer wants flexibility and a choice of languages, probably the ActiveX route is the way to go. Java Beans is limited, of course, to Java, while ActiveX can be used with C. C++, Java, and Visual Basic to name a few. But in exchange for this flexibility, the developer will have to accept the complexity of ActiveX. Because it is built on the use of object linking and embedding (OLE) and dynamic link libraries (DLL) through the Visual Basic- Java - C- C++ custom controls, it carries a lot of baggage. Because both DLL and OLE evolved over time, they are not well designed and are needlessly complicated. And compared to Java, both C++ and Visual Basic are inherently complex and hard to maintain, even if the program is well designed. In the net-centric computing environment, where software reliability will be a make or break issue for the system provider, these are considerations to take seriously.

If the developer wants to retain the flexibility of building applications that run in any operating system environment, Java and Java Beans is the way to go. Although Windows is now available on a wider number of hardware platforms, ActiveX and DCOM does require that the programmer limit himself to a single operating system. The one big advantage of ActiveX - the availability of programming languages other than Java - may not be the deciding factor in the future, especially for developers who are looking in the direction of CORBA.

At one point the members of the CORBA initiative had agreed to its own component building architecture, called OpenDoc. Now, many of its major proponents (IBM, Netscape and Oracle, among them) are moving to a component architecture based on Java Beans. Because of CORBA's independence from the underlying language implementation, this derivative of JavaBeans (or CorbaBeans?) may be the way to go if the system developer plans to be a long-term player in net-centric computing. It makes it possible for a system designer to develop applications not only in Java, but in a wider array of programming languages than even Microsoft offers with ActiveX.

Making Choices

In the net-centric world of Internet appliances and network computers, no one solution to achieving the ultimate ideal of "the network as computer" is likely to dominate. This is because the variety of devices that is likely to emerge will have different requirements as to memory space, interrupt response time, degree of determinism and mix of local versus remote execution of applications.

In an information appliance such as an Internet telephone with limited memory space - no more than one/two megabytes or so for application programs, operating system and graphical user interface - there are important questions the designer must answer: How many remote computing capabilities does he or she want or really need? And are there alternative ways to achieve them.?

Using HTML Commands.

For example, in a smart telephone with Internet access, it is not necessary to move beyond the traditional HTML and HTTP protocols in order to give the user the illusion of looking up, say, a telephone number, or accessing some sort of simple database, such as a recipe list, a TV program listing, or a yellow pages listing of categories.

At a very primitive level, some degree of interactivity could be added by using HTML 2.0 and 3.0's FORM capability in which an area in an HTML page is made available for user input, combining this with the GET and POST commands. Both commands construct a query URL, which is sent by the browser to an executable script on a server identified in the URL, directing it to take an action specified in the ACTION attribute. With both methods, anything that the script writes to a standard output is sent back to the reader's browser as a new HTML page, which can be the old HTML page with new information entered or inserted in response to the input. The interactions between the browser and the server are mediated by a collection of interactions called the Common Gateway Interface (CGI). Each server has its own set of scripts and programs needed to process information from Web pages on the server. These scripts have in the past been written in the Perl or the C-like TCL scripting languages on UNIX machines, and something like ActiveX onWindows machines and they produce responses that are relatively static in nature.

Right now, with the FORMS/GET/POST/ACTION combo, it is possible to generate a response from a server that illustrates a point in the text of a Web page. While interactive in the general sense of the term, the response generated would be relatively static and limited to a specific set of inputs with predicable responses. The reader would not be interacting with the actual processor, program, or tool, but with a TCL or Perl script that would return information based on the reader input. In most cases, since it is not the actual program that is generating a response, but a TCL script, the reader input is usually confined to a multiple-choice set of options that nearest approximate the reader's requirements. The actions taken on the basis of these inputs are then returned to an updated HTML page with the new information - for example, a graphic simulating the response from a program or instrument or an an actual image.

Server-based Computing.

If the engineer has much more sophisticated remote computing applications in mind, much of the small memory footprint of an Internet telephone can be preserved by using the Windows terminal solution. Depending on the sophistication of the programs to be executed on the server and the degree of control that the user of the Internet appliance will have over its operation, a run-time remote procedure call program containing the client side stub routines can be generatde which ranges from 60 to 100 kBytes, over and above the program memory space required for the operating system kernel and the graphical user interface.

The majority of the system overhead is borne by the server. In a Windows-based terminal configuration, the load can be considerable. Assuming a normal user's desktop workload in a business environment, as much as 16 Megabytes of DRAM and 100 MB of disk storage may be required to do, say, a standard word-processing, database or spreadsheet operation. The way this load is apportioned in an Intranet is much the same as is done in environments where dumb or smart terminals are used now: limit the number of users to match the capabilities of the server or servers; carefully control what applications each user can execute; or allocate each user a precise budget on the server of memory, disk and CPU time adequate to the operations each needs to perform.

In the broader Internet environment, such alternatives are difficult to implement. In an Internet application, where thousands or hundreds of thousands of users may be accessing the network, the problem is one of controlling the number of users accessing the system. There are number of ways to mollify the situation. The most obvious way is to limit the functionality of the device and the infrastructure that supports it to the present Web-Net paradigm: using the Internet only as a mechanism for sending and receiving messages and the Web as a way of accessing information.

However, if the idea is to create a personal Internet device in which most of the computing is performed remotely, other techniques must be employed. One way is by subscription, creating a members-only subset of the Internet, as with America OnLine, or by means of a virtual private network (VPN), a closed set of users and servers within the general Internet, where access is limited to those with the appropriate passwords. A second alternative is to use special servers as translators. In its interaction with the client appliances, the server is limited to the capabilities of the client. But when interacting with the broader Internet "outside" the VPN, the server uses the full range of capabilities. A third alternative is to limit the number and sophistication of the functions that the Internet appliance user can perform on the servers.

Of course, one could use the traditional Java paradigm of "write once, run everywhere" and remove most of the load from the server. But this makes the appliance much fatter, and much more expensive in terms of local memory, display and storage. However, the modifications that Sun has made to Java in the form of RPC and RMI capabilities will allow developers to more carefully balance the compute load dynamically between the servers and the clients: using a server-centric remote computing configuration when usage load is high and then shifting to a pure Java-based RMI environment where compute chores are distributed amongst not only servers and clients, but between clients as well, when usage is low.

Long term, an engineer would be better served by adhering more closely to the CORBA specification if his or her system has the memory and performance to spare. Because CORBA imposes an additional level of complexity and coding on top the application, only connected desktops that are likely to have the performance and memory to spare. In other network-connected systems, such as Internet appliances and Web-enabled set-top boxes, a more conservative approach is in order until processor performance and DRAM capacity increase.

If a designer wants to design today in a format that is adaptable to the CORBA-based ObjectWeb of the future, the best route to take is to stick to a purely Java-based methodology, since Sun has undertaken a number of initiatives to ensure Java's compatibility with CORBA.

The only barrier to the eventual emergence of CORBA as the predominant mechanism for distributed computing on the network is Microsoft Corp. and its DCOM specification. Ironically, in its opposition to CORBA in favor of its own DCOM Microsoft may be undermining its efforts to reduce Java from the status of lingua Internetica to that of just another programming language. If the company were serious about eliminating Sun as a potential competitor and as the primary source for languages and programs for the Internet, its best bet would be to back the CORBA effort with all its might. However, this would also present the company with another dilemma, for in the CORBA-based ObjectWeb environment, it would only be one of hundreds of vendors supplying software.

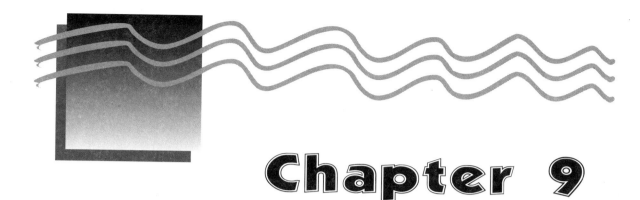

Chapter 9

Choosing the Right Processor

When choosing a processor for a net-centric computing device, the system designer must use a methodology and make decisions that differ significantly from those employed to build a desktop PC.

As far as the mainstream of the desktop computer market is concerned there have been only two choices: the Intel approach which is built around the X86 architecture and the Apple approach which depended on the Motorola 68X00 and PowerPC architecture. In workstations and servers, the choices are a little better; in addition to the x86 and PowerPC, the choices are among five or six high-end 64-bit CPUs: the Alpha, the MIPS architecture and a variety of specialized designs. Life is much different in the world of net-centric devices. Not only do you have the above choices but there are literally a dozen alternatives, and even more customized variations.

Because of the size and cost limits placed on net-centric devices, as well as performance and memory constraints, the processors that are to be used are not the familiar desktop alternatives, but new architectures that have come into common use in the embedded world.

Many vendors of reduced instruction set computer architectures designed their processors for the workstation market. Because of the number of competitors and the smaller than anticipated size of the market due to the inroads of the desktop PC, many of these 32 bit RISC CPUs have been adapted to the needs of the embedded market. About half a dozen architectures dominate this segment, all of which the designer of an Internet-connected system has available.

The MIPS family, for example, was initially used by Silicon Graphics Inc. to build its high end multimedia workstations. It has since them been adopted by a wide range of companies to build devices targeted at both embedded and net-centric computing applications.

The company that designed the architecture, MIPS Computer Systems Inc., has licensed the architecture to a variety of CPU vendors.. Among the vendors who have licensed the architecture are Philips Electronics, Integrated Device Technology, LSI Logic, Inc., Quantum Effect Devices, Inc., and NEC.

Another implementation that has found wide use is the ARM architecture from Advanced RISC Machines Ltd., which has licensed it to a wide variety of manufacturers, including Alcatel, Mietec, Asahi Kasei Microsystems, Atmel/ES2, Cirrus Logic, Digital Semiconductor, GEC Plessey Semiconducctors, LG Semicon, NEC, Oki, Samsung, Sharp, Symbios Logic, Texas Instruments, VLSI Technology. and Yamaha.

The X86, while an older complex instruction set computer architecture, has also managed to gain a stake in the embedded market, not because of its size, cost, or performance, but because of the wide availability of processors. Advanced Micro Devices, Inc., and National Semiconductor Corp. have taken the lead in adapting older 386 and 486 designs to the requirements of net-centric computing. National in particular is looking to use later designs in the X86 family, including the Pentium, to gain a performance boost in a variety of net-centric applications. To thisend, it has acquired Cyrix Corp., which has built Pentium and Pentium Pro equivalents for the desktop market. National's plan is to continue with that desktop effort, but also focus on the needs of net-centric applications where low cost, small form factor and performance are required.

Because of the small size of the die on which the CPUs are fabricated, RISC CPUs have managed to squeeze out every bit of cost to bring down component prices to the $5 to $50 range compared the $500 to $1,000 per chip cost of state of the art 100 MHz-plus x86 designs. such as the Pentium, Pentium II and Pentium Pro and their competitors from Advanced Micro Devices Inc., Cyrix Computer Systems, ST Microelectronics, Inc., and IBM..

The problem with the embedded alternatives to the X86, and why, at first glance, they may not be suitable for building Internet appliances and network computers has to do with tool availability and cost. In the desktop environment, tools such as compilers, debuggers, emulators, and linkers are within the price range of even the most shoestring of operations- from a few hundred dollars to about $1,000. By comparison, tools for embedded processors are usually not only more sophisticated but more expensive, from $5,000 to $10.000 for a minimum configuration. And, unlike the desktop, where the program or system being built has a potential market of tens of millions of units, most embedded processor tools are used to develop programs that have a very specific use in one environment with just one or two specific functions to perform and in volumes that seldom match those of the desktop world.

Java Opens Up the Network

So, on the face of it, the advantage would seem to lie with the desktop CPU vendors in the world of net-centric computing. But the entry of Java and the concept of platform-independent programs into the World Wide Web changes the equation. It means that virtually any processor can be used as the main CPU in a net-centric computing platform. The only question now is choosing among the alternatives which are best at executing Java-based programs.

The value of a stack-oriented virtual machine that is independent of the underlying hardware architecture is that the system programmer can write applications for the VM and not the underlying hardware. The Java language itself introduces another level of abstraction since it is a high level interpretive rather than a compiled language, such as C or C++. While this allows a system designer to develop code for any application regardless of the underlying hardware environment, this approach does not guarantee equivalent performance. While the virtual machine and its associated program code will run on any processor, they will not run equally fast.

The least efficient environments for running Java programs and the Java VM are the majority of RISC processors, which moved away from the stack orientation of earlier CISC processors and toward more efficient register-based architectures. Even the newer generation desktop PC processors such as the Pentium and Pentium Pro have moved away from their CISC origins. Internally they are register-intensive and very RISC like. To maintain compatibility with earlier CPUs, these newer architectures convert the original X86 machine code into more efficient RISC code before execution. In workstations and desktop computers, it is possible to hide the limitation relating to the inefficient execution of Java code by using brute force: depending on the 200-500 MHz plus clock rates of these advanced machines to overcome and hide the performance hit.

However, this is not a luxury available in most cost and memory constrained net-centric computing environments. There, the aim is to keep costs down to an attractive sub-$500 range. This requires making sacrifices in terms of processor performance and memory availability, both of which rule out the desktop solution.

As we have learned from earlier chapters, there are software workarounds to this problem. One is to compile the byte code into architecture specific C-code or machine code. Another is to use one of the JVM/RTOS combinations to provide a degree of machine specificity. The third is the use of just-in-time compilation that recompiles to the native code of the host processor on the fly. The first sacrifices Java's machine independence, while the other two require significantly more memory.

To avoid such sacrifices, the obvious solution is to look for a hardware platform that maps the Java virtual machine onto the underlying architecture more efficiently and which can perform stack-oriented operations faster than general purpose register-based RISC architectures. Fortunately, there are numerous alternatives emerging.

Sun's picoJava

To overcome the performance limitations inherent in executing Java code on a non-optimized hardware architecture, Sun Microsystems's solution is direct and to the point: implement the Java Virtual Machine directly in silicon. Direct execution using the Java VM in silicon overcomes much of the performance hit of the other alternatives. The first implementation of this approach is the picoJava I core, a stack-based RISC architecture featuring a pipelined implementation of the Java Virtual Machine byte code instruction set (see Figure 19.1). Sun is not building actual silicon with the core, but licensing the core to the many consumer electronics companies in Japan, Taiwan and Europe, who use it as the basis of customized designs for their own uses.

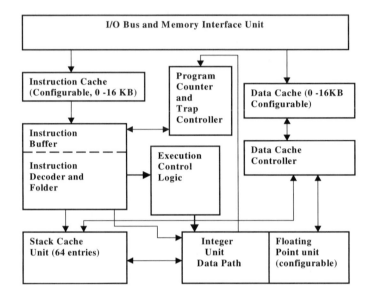

Figure 9.1 **picoJava.** Combining both stack and RISC-based structures, the picoJava I is essentially the Java Virtual Machine implemented directly in silicon.

Streamlined for maximum run-time performance, the picoJava I architecture is simple, offering designers variable size caches, a removable floating point unit, and a basic integer unit.. The instruction cache used for on-chip storage of Java byte codes can be sized anywhere from 0 (i.e., entirely eliminated) up to 16 Kbytes. This cache is directly mapped with a line size of 8 bytes (about the same as a 16 to 24 byte line in a RISC architecture). A 12-byte instruction buffer decouples the instruction cache from the main execution unit. A maximum of 4 bytes can be written into the buffer at one time, while 5 bytes can be read out. Since the instructions average less than 2 bytes in length, this means that the processor usually reads more than one instruction in a single cycle. All instructions consist of an 8-bit opcode and zero or more operand bytes, as defined in the Java Virtual Machine platform specification. Up to 5 bytes at the head of the instruction buffer can be decoded and sent to the next stage of the pipeline for execution (in one or more cycles). An on-chip stack cache is used by the execution unit to access operands needed by the functional units, which is the equivalent of a register file used in most RISC designs. The stack contains both integer and floating-point data, and there is no overhead for passing operands to and from the stack for either unit.

The simple RISC-style pipeline of the picoJava I core architecture breaks instruction execution into its four fundamental stages: fetch, decode, execute, and write back. The data cache is accessed during the execute phase, if necessary. Since, in a stack-based architecture, the instruction following a load almost always depends on the data returned by the load, access to the data cache is not implemented as a separate (fifth) pipeline stage. In standard RISC fashion, all compute instructions take their operands off the stack cache and return their results to the stack cache. This facilitates pipelining of compute and memory instructions. The picoJava I core pipeline is enhanced to accelerate the throughput of object-oriented programs. It includes support for method invocations, along with support for hiding loads from local variables. Moreover, to improve run-time, the pipeline has support for thread synchronization and write barriers that allow the use of a variety of high-performance garbage collection algorithms.

The data cache can range in size from zero to 16 Kbytes. To improve the hit rate it is designed as a two-way, set-associative memory. The data path between the picoJava I core data cache and the pipeline is 32 bits. An I/O bus-memory interface unit serves as the link between the picoJava I core and any other interface logic that could reside on the same die. The picoJava I core technology implements a hardware stack directly supporting the Java Virtual Machine stack-based architecture. The top 64 entries on the stack are contained in the on-chip stack cache, organized in a circular buffer. All operations on data in the core are performed through this stack cache. Data from the constant pool and from local variables are first pushed onto the stack; all instructions then access their operands from the stack.

The picoJava I core stack is used as a repository of information for method calls (see Figure 9.2). The Java Virtual Machine platform creates a method frame for each method called at execution time. Each method frame is composed of three types of entries: parameters and local variables, frame state, and operand stack. Parameters, local variables, and operands are just what their names suggest. The frame state documents pertinent information needed when returning after the method is completed. This information includes return program counter, return variables, return frame, return constant pool, current method vector, and the current method monitor address. The stack cache relies on a background dribbling mechanism (see Figure 9.3) that tracks when dirty entries should be written out to the data cache and when valid entries should be read from the data cache into the stack cache. This is controlled by programmable high- and low-water-mark values. When the

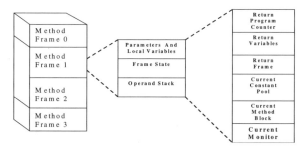

Figure 9.2 **Stack-based.** Used as a repository for method calls, the Java core stack creates a method frame containing parameters and local entries, frame state, and operand stack data.

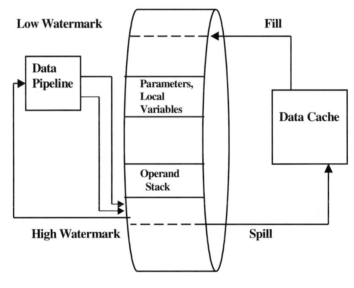

picoJava Dribbling Mechanism

Figure 9.3 **Dribbling Mechanism.** The picoJava stack relies on a dribbling mechanism that tracks when dirty entries should be written out to the data cache.

number of valid dirty entries in the stack cache exceeds the set high-water-mark value, the dribbling mechanism begins scrubbing out the dirty entries to the data cache. When the number of valid dirty entries falls below the set low-water-mark value, it begins reading additional valid entries back into the stack cache.

Stack operations are also accelerated in the core with a random, single-cycle access folding operation. Frequently, an instruction that copies data from a local variable to the top of the stack is followed immediately by an instruction that consumes these data. The instruction decoder detects this situation and folds these two instructions together, effectively performing both in a single cycle. This compound instruction performs the operation as if the local variable were already located at the top of the stack. In the unlikely event that the local variable is not contained on the stack cache, this folding cannot occur

Simulations show that the folding operation eliminates up to 60% of the local loads required by a stack architecture, or about 15% of all the instructions executed in a typical program. The result is an instruction execution profile for picoJava I core technology that more closely resembles that of a RISC machine. Without folding, many more stack operations are performed - to manipulate data, duplicate values, set up constants, and move local variables to and from the stack - than compute operations. After folding, the instruction profile is dominated by the actual compute operations themselves.

The picoJava I core architecture has been designed to be simple and flexible, providing developers with a number of trade-offs for optimal system performance. For instance, while picoJava I core technology is capable of executing basic "simple" instructions in one cycle, including arithmetic, shift and logical instructions, as well as constants, conversions, and stack functions, medium complexity functions rely on microcode-state machines for quick invocation and return. Some infrequently used complex instructions are trapped and emulated in software. The idea is to allow developers to utilize the core in the most effective way for their target market. As mentioned above, the caches can be scaled up or down in size, depending on the need to increase performance or decrease cost. The floating-point unit can be eliminated, if floating-point code is not used. The I/O bus can be freely specified according to the requirements of the target market. For example, I/O might be provided via a local bus, PCI, or some other standard, running at whatever frequency is appropriate. In the same way, the memory solution is customizable, be it RAM, ROM, DRAM, EPROM, or Flash PROM, or some combination of these.

Another dimension of design freedom involves the amount of peripheral logic integrated on chip with the CPU core. Thus, the picoJava I core can be implemented as a bare-bones Java CPU, with a simple (or high-speed) local interconnect to a second chip containing all the application-specific peripheral logic. This approach provides the best time-to-market and the most flexibility for the CPU. Alternatively, the core logic can be implemented on one die with all the necessary peripheral logic as a highly integrated, single-chip computer.

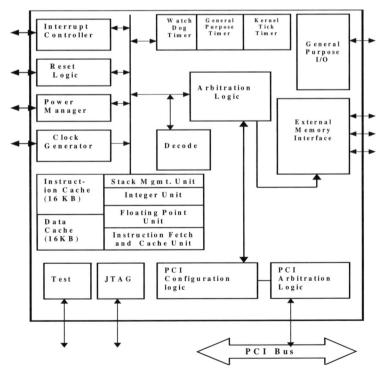

Figure 9.4 **microJava.** Containing the picoJava core, this follow-on chip from Sun incorporates peripheral core logic designed to execute both Java and non-Java code faster.

The microJava Compromise

A pure Java hardware implementation would be the ideal solution in an environment where one expects to run only Java code. However, most network computing applications will for a long time have to run legacy code; that is, existing applications written in languages such as C and C++, among others. To address this issue, Sun has introduced a second chip, the microJava 701 (see Figure 9.4). It is designed to run non-Java code about as efficiently as comparable RISC CPU architectures. While it offers no tangible price-performance advantage when executing non-Java code, it executes Java byte code faster than the alternative: using a general purpose chip optimized for other high level languages.

The 701 is designed to be a CPU able to facilitate the creation of inexpensive "thin client" machines of all types and do so in minimal time for minimal cost. . At the same time, the 701 attempts to minimize system-level cost by building the core logic chipset into the CPU itself. It integrates in only those basic functions that are needed in any type of system. And even there, it attempts to preserve choice wherever possible. In the 701, the basic picoJava core technology has been configured with a floating point unit that supports the Java Virtual Machine floating-point specification, and 32 KBytes of cache memory, evenly divided between instruction and data storage.

Other core features worth mentioning are a power down idle mode that cuts chip power consumption by stopping the core clock; a performance counter that can be used to accumulate any performance statistic of interest, such as pipeline stalls, cache misses, page misses, and so forth; and a hardware breakpoint, for halting program execution when a specified code or data address is fetched. In addition, for memory and I/O, the 701 includes a 16-bit general purpose I/O (GPIO) port, a general-purpose timer, a kernel tick timer, a watchdog timer, and an interrupt controller that handles 11 external and 6 internal sources. Also on chip is a built-in self-test module to ensure the integrity of the on-chip cache RAMs and an IEEE-standard JTAG interface that supports full boundary scan. The memory interface on the microJava 701 is quite versatile. A wide 64-bit data interface is provided, but the option exists to use only 32 bits for smaller memory configurations. The bus cycle timing is programmable, and supports three different timing configurations: for EDO DRAMs as fast as 50 nsec., for SDRAM up to 100 MHz, and for Flash/ROM/SRAM. The maximum block copy data rate is 128 MB/sec. in and 128 MB/sec. out.

The memory controller also has a local bus that can be configured to attach 8-, 16-, or 32-bit slave I/O devices. The local bus has 8 dedicated data pins, intended for attaching an inexpensive boot PROM. Wider devices are attached by borrowing additional pins from the memory data bus. In a 32-bit wide memory configuration, ample extra memory data pins are freely available; in a 64-bit wide memory configuration, any extra pins needed must be multiplexed. The PCI interface is compliant with the 2.1 standard. It is 32-bits wide, and can operate either at 33 MHz or 66 MHz, using 3.3-V signaling, The PCI clock runs asynchronously with the CPU clock. The CPU can be configured to arbitrate the bus as a PCI Host, or to work with a bus arbiter as a PCI Agent, and can either actively initiate accesses as a Master or passively receive them as a Target.

Stack-Oriented Alternatives

In addition to Sun, at least two other companies are offering stack-based CPUs which offer the promise of executing Java code and the Java virtual machine efficiently: Patriot Scientific and Rockwell Collins.

Patriot's PSC1000

Retaining a high degree of general-purpose functionality is the PSC1000 from Patriot Scientific Corp. in San Diego, California (see Figure 9.5).. On the PSC1000, the Java VM maps so closely onto the architecture that 38% of the Java VM byte codes translate directly into the same or fewer bytes of opcodes on the PSC1000. Since, like the Java VM, it is also a stack architecture, the byte code to machine code translation is direct; no register mappings or complex optimizations are required. Byte code expansion is only 20% compared to 300% to 500% on X86 chips.

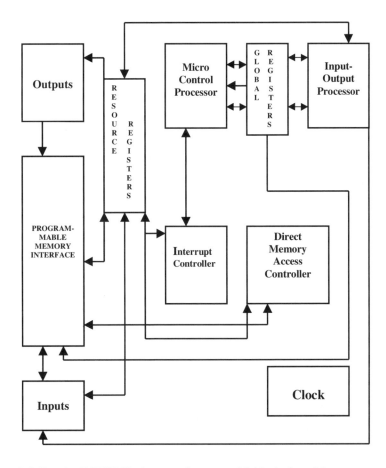

Figure 9.5 **Patriot PSC1000.** A general purpose 32-bit design, this processor contains both a stack-based microcontrol unit as well as a companion input-output processor.

The PSC1000 CPU was designed from the ground up to execute languages like C, C++, Forth, and Postscript efficiently. While picoJava and the PSC1000 have comparable performance, picoJava is much more complex. Compared to the picoJava, which a 4-KB instructions cache and an 8-KB data cache, the PSC1000 gets away with using much less. Other than separate caches for a 16-deep local-register stack and 18-deep operand stack (both implemented mostly as RAM) and a 32-bit instruction-prefetch register, the PSC1000 has no caches. And where the picoJava also uses instruction pipelining and register forwarding, the PSC1000 uses neither. Both factors greatly reduces chip area, complexity, and core power dissipation compared to picoJava.

The PSC1000, though using different machine instructions, is amazingly close to a Java processor. Like the Java VM, the PSC1000 is a 32-bit stack architecture, uses 8-bit instructions, and has local variables (registers) and floating-point support. Interestingly, the underlying hardware design is significantly different from existing RISC-based microcontroller architectures. Instead of a single processor core the PSC1000 includes two processors within one package: a microprocessor unit (MPU) for performing conventional processing tasks and an input/output processor (IOP) for performing time-synchronous input-output operations. The IOP executes deterministically and can, with appropriate software, replace many dedicated peripheral functions.

Aside from this dual-processor architecture, what makes the design unique is the use of a merged register-dual stack architecture. Also known as a zero-operand architecture, operand sources and destinations are assumed to be on the top of the operand stack, which is also the accumulator. An operation such as ADD uses both source operands from the top of the operand stack, sums them, and returns the result to the top of the operand stack, thus causing a net reduction of one in the operand stack depth. Both stacks contain an index register in the top element to minimize the data movement typical of register-based architectures, and also minimize memory accesses during the procedure. Not only does this mean that it can process the interpreted Java code efficiently, but it is also good as a general purpose processor because most of not all popular programming languages are designed to use stack data-passing and expression-evaluation mechanisms within the compiled code. Thus, the stack architecture is very efficient for processing languages such as C and C++, and even more so for stack-based languages such as Forth, Postscript, and Java.

Unlike most RISC designs, PSC1000 CPU instruction set is hardwired, allowing most instructions to execute in a single cycle without the use of pipelines or superscalar architecture. A flow-through design allows the next instruction to start before the prior instruction completes, thus increasing performance. The stack architecture eliminates the need to specify source and destination operands in every instruction.

Rockwell's JEM1

The other Java-specific alternative to the pico and microJava chips from Sun is a new family of stack-based CPUs based on the Advanced Architecture Micro Processor (AAMP) from Rockwell Collins, which allows a variety of derivatives to be constructed from an off-the-shelf library of standard cells. Virtually identical architecturally to the Java devices, the first Java-specific 32-bit processor to be generated is the JEM1, a 50-MHz design that combines the ALU, registers, and other CPU functions with an interrupt controller, two programmable timers, logic for external data bus support, and power management, as well as a JTAG test interface.

Like the Java chips, the AAMP devices do not rely on the use of big register sets, but instead perform all internal operations on top of an accumulator stack. To get around the performance issue relating to the use of interpreted Java on more general purpose architectures, Rockwell has in the JEM1 opted for the same strategy that Sun has in its Java chips: a hardware-based virtual machine to directly execute Java byte code. Like Sun's picoJava, Rockwell's Java implementation of the AAMP has incorporated a number of extended byte-code instructions in hardware that are not part of the standard Java instruction set. These extended byte codes correspond directly with such low-level chip control operations such as register accesses, cache control, and load and store operations. While the aim of such extended byte-code instructions is to make it easier to control such things as modems and network interface devices, the additions could make code developed for use on such devices nonstandard and not portable to other architectures. In order to take advantage of the extended byte codes and the resources that they control, it is necessary to use the Java class libraries to translate high level programming commands directly into extended byte codes.

Looking For General Purpose Alternatives

Unless Sun and its partners in implementing the picoJava and microJava in silicon are successful in coming down the fabrication and manufacturing learning curve rather quickly, it will be several years before such Java hardware implementations are able to match the low cost of many other, more mature CPU alternatives. A better strategy for the system designer who is building a net-centric system now is to determine which general purpose architectures, RISC or CISC, are most compatible with the stack-based, object-oriented Java VM. A number of the general purpose processors now on the market fill the bill to varying degrees.

X86

For designers familiar with earlier generations of the X86 architecture, especially its more recent 32-bit 386 and 486 implementations, it should come as no surprise that this processor family is one candidate. Firs, aside from the Sparc, the X86 was the first architecture to which native Java compilers were developed. Indeed, in its effort to preempt Sun's efforts to make Java platform-independent, Microsoft has introduced tools and a version of the Java virtual machine that is optimized for the X86 architecture. This means that developers can write compiled X86-specific code for this architecture that will run significantly faster than the interpreted platform-independent alternative. Beyond this, the 386/486 architecture, as a stack-based CISC architecture, can execute Java byte code much more efficiently that some of the register-based RISC alternatives.

SPARC

Nor should come as a surprise that one of the RISC processors most suitable for executing Java byte code is Sun's Sparc architecture. Similar to other RISC CPUs, Sparc supports function calls and parameter passing with fixed-sized register windows, which are usually partitioned into three segments of eight registers each. Fortunately for the Java programmer, Sparc maintains a conventional stack in memory as a backup store for window overflows and parameters or local variables exceeding the respective window partition sizes. Although not designed for this specific purpose, this feature can be used to advantage in the execution of Java code.

ARM

Another possible alternative for running Java code much more efficiently is the ARM architecture from Advanced RISC Machines Ltd. in Cambridge, England. The ARM is a good candidate because of its origins as an architecture originally designed to support not only C-like structured languages, but interpreted languages like BASIC, Lisp, Forth and Smalltalk as well. Based on this lineage, the ARM architecture has instructions and memory management support that work together to make a very efficient Java environment. If attention is paid to coding and to mapping of Java's stacks to the underlying registers of the ARM, it is possible to execute Java code efficiently and quickly. In the matter of stack handling, the ARM architecture is able to perform stack post- and pre-increment and decrement during a stack access operation, saving a great deal of execution time. Also useful are the ARM's shift and add functions, which are an advantage when moving from byte codes to the native 32-bit instructions of ARM processors. The ARM method of mapping pages and handling complex protection checks for operating systems is also beneficial in running Java code, requiring only modest optimizations for added performance.

If this is the route the designer of a net-centric system is going to take, one of the first things that must be done is to construct an instruction set simulator using the Java byte-code interpreter and to decide how the primary data storage of the simulated machine will be mapped into the data storage of the host machine. This must be done with some care because, whereas, the Java virtual machine provides an operand stack, a stack pointer into that stack, a program counter and main memory, the ARM processor does things slightly differently. It provides 15 general-purpose registers, a program counter and a condition code flags register.

The best strategy is to use a single ARM register to hold the Java Program Counter and another to hold the Stack Pointer. The ARM assembler allows the programmer to assign symbolic names to registers, so one thing that could be done is to give each register a unique name: one called "jpc" to hold the Java Program Counter and another called "jsp" to hold the Java Stack Pointer. Holding the Java stack in ARM registers is a bit more of a problem, but easily solvable. Without doing detailed analysis of the Java program, efficiently mapping the Java stack onto ARM registers will be too costly for a simple emulator , and this register allocation is in fact one of the major tasks for a compiler. Instead, the best approach is let the Java stack reside in memory and access it in memory. As the ARM processor is a RISC machine, it does not support data-processing operations (like Add or Compare) directly on data in memory. So data must first be loaded into ARM registers, the processing done on these registers, and the results stored back to memory.

One of the things that makes ARM a particularly efficient Java engine is its ability to access a stack location and update the stack pointer in a single instruction. Most RISC machines need a separate instruction to update the stack pointer after each stack access instruction, making the program bigger and slower. Also useful is its ability to efficiently load Java byte codes and interpret them with ARM code sequences. A good way to access the correct sequence of ARM code to interpret a Java byte code is to use the byte-code itself as the index into a C-like switch statement of ARM code.

One of the things that can be done in the ARM architecture is to load a byte code and increment the Java Virtual Machine Program Counter with a single instruction. This is done by making use of the ARM's ability to perform a memory access and update the address register value in a single instruction. This sequence automatically fetches the byte code and increments the Program Counter. A switch statement then uses the byte-code as an index into a jump table to find the address of the ARM code sequence corresponding to that byte code, and then jump to that ARM code. When done, the ARM code jumps back to load the next byte-code.

Two optimizations can be made to this approach. Both are code size versus interpreter performance trade-offs. The first optimization resevers a fixed number of ARM instructions for each byte-code interpretation sequence. This allows the programmer to just use the byte code as an index directly into the interpreter code, rather than through a jump table. With two instructions it is possible to load a byte code, increment the Java program counter, and jump to the correct ARM code to interpret the byte code. In more traditional RISC machines as many as five instructions would be needed to perform this complex operation. A second optimization folds the byte-code interpretation and next byte-code fetch into one ARM code sequence. Instead of having every ARM code sequence interpret a byte code to the end with a branch back to the byte-code fetch, this optimization involves putting the bytecode fetch instructions into every sequence of ARM code to interpret a byte code. It costs one more ARM instruction per code sequence, but saves one (branch) instruction in execution time, a reasonable trade-off. The five instruction sequence that other microprocessors need to fetch and dispatch a byte code make this optimization much more costly in code size.

ARM processors incorporate special MMU hardware for speeding up garbage collection, negating one of the biggest criticisms that programmers have about Java.. First, every virtual page (there are three page sizes, 1 Mbyte, 4 Kbyte and 64 Kbyte) can be individually marked as cachable and write buffered, allowing the operating system to specify exactly how pages are used. The 64-K and 4-K pages also have individual access permissions on 16-Kbyte and 1-Kbyte subpages (respectively), allowing very fine control over access to data within that page. But with the MMU there is one more possibility:

every page belongs to one of 16 domains, and access to an entire domain can be controlled with a single MMU register write. This is specifically designed to speed up garbage collection by allowing a whole group of pages (a domain) to be instantaneously made "no access" while they are garbage collected.

StrongARM

As efficient as the original ARM architecture is in executing Java code, the one variation that has gathered a lot of interest from network computer builders is the Digital Equipment Corp.'s StrongARM architecture, which was acquired by Intel Corp. after the dissolution of the Massachusetts-based computer pioneer Published benchmark (CaffeineMark) ratings clock the 233 MHz StrongARM running Java interpreted code at nearly double that other architectures at similar speeds, even in the case where StrongARM is running the virtual machine, rather than code compiled into native instructions on the fly.

A number of characteristics make the StrongARM architecture so much better at executing Java than most other processors, including other ARM architectures. For one thing, while it uses the same instruction set as other ARM family CPUs, the StrongARM is unique in its use of Harvard architecture caches instead of the von Neumann architectures. In the Harvard architecture cache separates the instruction and data streams, whereas the von Neumann maintains a single stream for both. One of StrongARM's biggest advantages when running Java is its ability to contain the Java Virtual Machine entirely within its 16-KB instruction cache. In addition, a knowledgeable code designer can place a subset of the execution code in the independent on-chip 16-KB data cache. By using cache this way, most Java components execute at the high clock rates of the on-chip memory. When running code inside the StrongARM at 200 or 250 MHz, it is being executed at about three to four times the clock rate of the external environment, which in most present designs is in the 33-MHz to 66-MHz range. With that kind of performance differential, it really matters little whether the code is interpreted or compiled.

In conjunction with separate instruction and data streams, the StrongARM microprocessor incorporates separate memory management units dedicated to each stream. To maintain high performance while providing operating systems such as Java with robust memory management features, a few small tweaks were made to the on-chip translation look-aside buffers (TLBs). These changes ensure that the complex protection checks and translation lookups that simplify operating systems occur in a single cycle. Another architectural advantage for StrongARM is its instruction pipeline. In one of the first implementations of the architecture in the StrongARM SA-110, use is made of a RISC-like five-stage pipeline rather than a three-stage pipeline of the traditional ARM CPU. The number of stages in a pipeline is not critical for Java, but the extra two stages allow on-chip clock rates to exceed 100 MHz without significant code stalls.

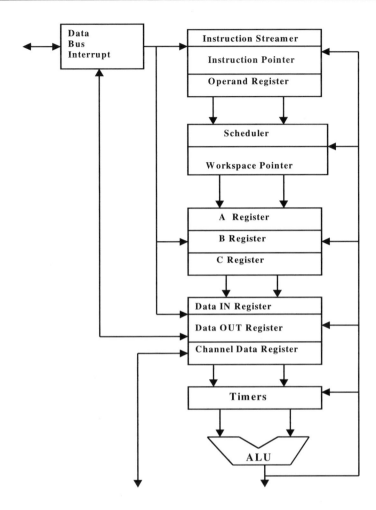

Figure 9.6 **Java-friendly core.** Based on the original Inmos Transputer architecture, the ST-20 core used in ST Microelectronics' family of consumer-oriented microprocessor designs contains a stack-structure that processes Java code efficiently.

The SGS Transputer

An architecture from the past that engineers have found is a good stack-oriented architecture for executing Java instructions directly is the 32 bit Transputer architecture, introduced in the early 1980s as a multiprocessor architecture by Inmos Ltd. for supercomputer applications. The architecture is now owned by ST Microelectronics, Inc. and has been retargeted as a general-purpose controller family, the ST20 series, for use in many net-centric applications. The first of these, the ST20450, is in production as a standard part. Additionally, the architecture is available as a core and is being used by a number of companies in application-specific designs.

Making the Transputer architecture (see Figure 9.6) particularly attractive for stack-based applications running Java is its very compact object code: the average instruction is only 10 bits long, due to the stack-based approach. With no directly addressable registers, instructions need only specify the operation, not the operands, considerably speeding up the execution of Java code. Unlike traditional RISC architectures, the Transputer architecture sports a small number of registers. There are only six: a workspace pointer for storing information about where variables are kept; an instruction pointer that points to the next instruction to be executed; an operant register; and most importantly for executing high-level languages such as Java, three registers (A, B and C) which form an evaluation stack.

The A, B and C stack registers serve as the sources and destinations for most arithmetic and logic operations. Loading a value into the stack pushes B to C, and A to B before loading A. Storing a value from A pops B into A and C into B. Expressions are analyzed on the evaluation stack, and instructions refer to the stack implicitly. For example, the ADD instruction adds the top two values in the stack and places the result on the top of the stack. The use of the stack removes the need for instructions to specify the location of their operands. An interesting feature of this design is the floating-point execution unit, which uses a microcoded computing engine with a three deep floating point evaluation stack for manipulation of the floating point numbers used in many 3D operations. Because of its structure, the Transputer executes a number of high level languages directly without recompilation: C, C++, Pascal and, of course, Java. This should not be surprising because the original Transputer CPUs were designed to execute not machine instructions, but high level commands written in a company-proprietary language, called Occam.

AMD 29000

Advanced Micro Device's 29000 family of RISC processors is another general-purpose architecture that could execute Java byte-code efficiently. Although it never established a foothold in the workstation market, it became widely used as an embedded processor, especially in high-end printers, which used Postscript, another stack-oriented, interpretive language. Specifically, the 29k has 128 local registers that have been organized in stack-like fashion. The 29k family, unfortunately, is a victim of bad timing. While it supports current members of the family, AMD is no longer funding efforts into expanding the family. It is instead devoting itself to the X86/Pentium/Pentium Pro market, a decision made about a year before the Web-Internet entered the mainstream consciousness and Java was released for use by outside developers.

Move To Customizable Solutions

To meet the sometimes conflicting requirements for not only better performance, but low cost and better reliability, the emergence of net-centric computing may well be accompanied by a totally new way of selecting processors and implementing them in a design.

Traditionally, in the desktop world, systems vendors usually started with a reference design, an evaluation board developed by the CPU vendor with recommended peripherals, memory devices, and even board and package designs. Often it is the reference design that the systems designer went to market with, with a few modifications to further reduce the cost of manufacturing. Embedded designers, because the nature of their applications, are much more focused, usually developing specific board-level solutions: choosing the processor and accompanying logic, developing the application, and debugging the hardware and software before committing to manufacture.

In the net-centric world, designers are going to have to learn a whole new methodology and get involved at the chip level in the design of their systems. Several factors are accelerating this trend. First, new process technologies make possible the fabrication of several million gates or transistors on a single integrated circuit. As a result, designs that originally required a board or a card of circuits to be fully functional can now be implemented on a single IC. Second, the requirements of cost, reliability and performance will require that systems designers move from board-level to chip-level designs.

Third, numerous manufacturers of alternative RISC based CPUs are doing their best to accelerate this trend. Compared to X86-type CISC processors which require anywhere from 1 to 4 million transistors or gates, most RISC processors require, at a maximum, no more than 100,000 to 200,000 transistors or gates. So, for an equivalent process capability, much more than just the CPU core can now be integrated onto a single chip. In addition to their standard processor designs, virtually every major CPU vendor today has developed a core strategy that allows them to build specific implementations targeted at not just the requirements of particular markets, but at particular customers.

For example, ST Microelectronics Inc. in its ST20 family of cores has developed a series of compatible microprocessor cores that enables the user to match the core to the system requirements for performance, price and power dissipation. Complementing these cores is a library of supporting macrocells and modules including standard peripherals, I/O, and embedded memory. With a suite of available macrocells, tools, and technology, the developer can then specify the chip that best suits the end application. The high level of integration that this approach offers reduces the chip count and system cost. All the library components have been designed using a modular methodology, allowing them to exist as independent macrocells and to be reused in a wide number of IC implementations. In addition to macrocells for peripherals, I/O, and embedded memory, other macrocells and modules that conform to the library definition can be added at any time. Among the modules that would be critical in many net-centric computing applications are an arithmetic accelerator, designed for applications that require fast hardware support for arithmetic operations including a multiplier that provides an N by N integer multiply in three cycles; a single cycle barrel shifter; and a single cycle adder. The accelerator is optional and will be omitted from low-end implementations of the ST20 core. Conversely, the accelerator can easily be replaced by other specialist modules that accelerate specific classes of applications. One of these is an instruction preprocessor that independently fetches, sequences, and issues instructions for the ST20 core. The instruction pre-processor is optimized for each ST20 core configuration and contains three main modules: instruction fetch unit and buffer, instruction sequencer, and instruction issuer. In the works is a Java-specific preprocessor that would make the ST20 even more efficient at executing Java bytecode.

A key element in this approach is an on-chip peripheral interface bus that has been defined to allow each macrocell module to be independently specified and designed, and hence the internal operation of each macrocell module can be decoupled from the rest of the chip. This is key to the concept of modular design and is achieved by each module interfacing to the bus via a common interface and conforming to a specified protocol. The ST20 peripheral interface bus is, in fact, two buses: a memory bus used exclusively for accessing internal or external memory; and a peripheral bus used for sending control information between on-chip modules. Each bus has an associated controller, and each on-chip module interfaces to each bus via a memory port and a peripheral port, respectively. The ports accept requests and pass them to the bus controller for arbitration.

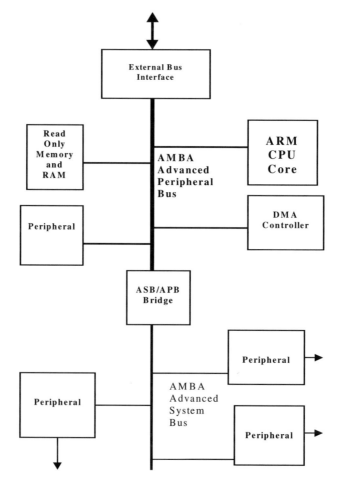

Figure 9.7 **Arm's AMBA.** Built into all Advanced RISC Machine CPUs is the Advanced Microcontroller Bus Architecture that allows designers to build their own customized versions by attaching predefined macrocells to the on-chip bus.

AMBAling Along

Taking the modular core design concept even farther is Advanced RISC Machines, which in virtually all of is CPU designs has incorporated a special Advanced Microcontroller Bus Architecture (see Figure 9.7). AMBA was developed with the support of the European Union Open Microprocessor Initiative (OMI) project and is being promoted as an open standard. The basic idea is that design time and development cost can be cut by re-using proven macrocells that conform to a common bus standard. It is based on a standard on-chip 32-bit system bus specification to which different microprocessor cores, memory I/O, and peripheral cells can be added. The design methodology extends from an initial hardware description language (HDL) circuit description to final test vectors. It includes an HDL behavioral model for each macrocell as well as on-chip test methodology. AMBA allows the system designer and the chip architect he is working with to quickly partition a design. AMBA's open bus architecture provides a mechanism for ARM and its partners - companies such as GEC Plessey, VLSI Technology inc., Texas Instruments, Digital Equipment Corporation, Samsung, Sharp, and Cirrus Logic - to pool development resources and respond quickly to market changes, important in net-centric computing where everyone marches to a different clock: the much more accelerated Internet Time.

Putting Macrocells To Work

Some segments of the net-centric computing market are moving rapidly in the direction of much more application specific designs. While network computers are still being designed using concepts familiar to the desktop environment, the shift to more highly integrated application-specific designs is already underway in Web-enabled set-top boxes. The next area to undergo the transformation is likely to be wireless personal digital assistants, telephones and handheld PCs, where a shift is underway from relatively low bandwidth 9600 bit/sec transmissions over standard analog connections to all-digital transmission where data rates are expected to reach 1 to 2 Mbits/sec.

The catalyst for this transition is the worldwide "third-generation" (3G) mobile communications standards being developed by the International Telecommunications Union, through its International Mobile Telecommunications 2000 (IMT-2000) initiative. The aim is to not only provide for global roaming between multiple countries and networks, but also for a smorgasbord of compute-intensive and signal processing-intensive services, such as full-motion video, Web browsing and videoconferencing, as well as the more pedestrian fax, e-mail access and paging services associated with second-generation mobile. A major challenge facing designers is how to get at least an order of magnitude more MIPS out of DSPs and CPUs without exceeding power consumption, space and cost constraints. There are two different types of processor requirements in IMT2000-based Web-enabled handheld terminals. In setting up the communications link the requirements increase dramatically along with the bandwidth increase. The traditional embedded processor that handled everything may be required to either do less than before—only the display, keyboard and scheduling—in a net-centric wireless mobile unit where applications are run on the server, or much more if it must run video and graphics applications.

In general, the architecture of IMT-2000 terminals will have one or more RISC microprocessor and-or microcontroller cores, one or more programmable DSP cores, and at least one type of specialized, dedicated hardware. Depending on the type of applications running on the CPU, it may need to deliver 100 MIPS or more, while estimates of DSP MIPS run anywhere from 300 to 3,000 and up, depending on additional hardware. Because of the varying requirements, designers of Web-enabled net-centric terminals based on IMT2000 will have to use the modular application-specific methodologies. For example, VLSI Technology expects to use a two-core controller/DSP system-on-a-chip architecture based on its Vector ARMThumb and Vector Oak DSP embeddable cores.

Siemens Microelectronics Inc. expects to use its 32-bit customizable TriCore architecture—which combines a microcontroller, microprocessor and DSP onto one device—for IMT-2000 handheld terminals. Most traditional cell phones combine two separate processors running in parallel, the host processor and the DSP. An architecture such as TriCore that combines multiple processors in one environment allows a flexible split between host processor and DSP tasks, so that a single device executes both control and signal processing instructions.

More Recent Developments

In many respects, the shift to net-centric computing is similar to the early days of the desktop personal computer. Then the fight was among a number of vendors of operating systems (Digital Research's CPM, Microsoft's DOS, and a number of other entries) as well as a number of different CPU vendors: Intel Corp. with its 8080 and 8086, Zilog Corp. with its Z80 and Z8000, National with a processor that had characteristics of both the Intel and Zilog CPUs, and Motorola with its M6800 and M68000, used by Apple Computer Corp. in the original Apple I/II and Macintosh computers, respectively .

With the entry of IBM into the fray, things changed and hardware and software standards emerged. With this standardization the market began its upward path to desktop dominance. And as pricing came down, software (languages and applications) was developed to take advantage of this new computing architecture. Now, exactly the reverse is occurring. Standards for languages, applications and operating systems are fighting for dominance. But it is the software that is driving the hardware evolution.

If Sun is successful in making Java the dominant operating environment for net-centric computers, it will dictate that in other areas than the full blown desktop the processor architectures used to execute the Java based applications will have to meet specific requirements: they must be able to process stack-oriented instructions efficiently, must have features that make it easy to execute object-oriented code, and must be able to efficiently deal with the built-in features, such as "garbage collection" that have been included to guarantee correct execution of code.

Balanced against these considerations is the fact that unlike in the 1970s and 1980s, when computer makers were going after one consumer - the personal computer user at home and at work - now there are multiple markets, many different kinds of end users and multiple architectural requirements. The network computer is different from the Web-enabled set-top box device and both are significantly different from Internet-connected information appliances such as Internet-ready wireless and wired telephones. Not only are the hardware specifications different, but the way in which they access the Internet and use its resources varies.

Certainly, Sun's Java, Oracle's NC, and Wintel's Windows and NetPC will be significant factors. But the Internet has grown up relatively independent of the hardware used to access it and will continue to do so. So the wise engineer who is designing hardware or software for that environment should first determine what is changing with regards to the Internet and the Web and how this might change the direction of the design. Only secondarily should what Sun, Intel, Oracle or Microsoft are doing be included in the equation. They are important factors in the direction of the Internet, the World Wide Web, and net-centric computing. But they are but not the only ones, or even the most important ones.

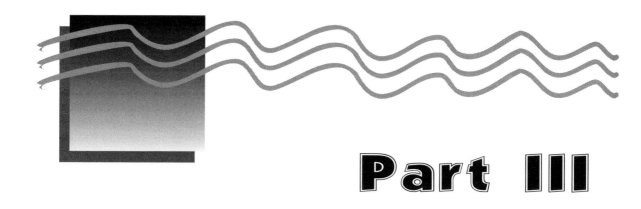

Part III

Listening for the Web Tone

Chapter 10

Building More Reliable Boxes

A major challenge, perhaps the one and only challenge, the one that will make or break the net-centric computer as a mainstream phenomenon, is reliability. To the average consumer, it is clear what reliability means. What he or she wants is the assurance one has with the public telephone system: the certainty that when someone picks up the phone there will be a dial-tone and when a number is dialed, the right person or company is at the other end.

In the net-centric computing environment this kind of reliability is emerging, but we are still very far away from the "Web tone" and all that this implies. For the average noncomputer user a device is reliable if it works. It is unreliable if it does not. The reasons and the sources of the problem don't count. But from a systems designer's point of view, a system's reliability can be affected by a number of factors: On the hardware side, a system's reliability is compromised by problems in the physical design of the boards internal to the box: the motherboard, the add-in boards, the physical routing between components on the boards, and between the motherboard and the add in boards. Hardware issues relating to reliability will be the focus of this chapter. In other chapters in Part IV, we will look at other sources of perceived unreliability on the part of the end user:

1. Failures and glitches because of unreliable code

2. Sources of communications breakdown between the various thin clients in a net-centric computing environment and the servers

3. Security problems which compromise the confidentiality of the files

4. Security problems that impact on the reliable operation of the programs on either the clients and servers

5. Network outages and slowdowns from data overloads and user's overloading the capabilities of the inter-network of communications that make up the internet, preventing access to services on the network

6. Catastrophic "hard" failures that crash an end user's application.

Building Reliable Boxes

To achieve the goal of reliable computing in a net-centric environment, it will be necessary to merge two disparate engineering disciplines. One was developed for open systems such as desktop computers. The other was developed for closed applications typical of embedded computing systems in industry and consumer electronics. But rather than x86-based desktop technology moving in and eventually taking over many applications and market segments that have traditionally been hard core deterministic, real time and embedded, it is the reverse that may true. This is especially the case as we move toward a net-centric computing world.

PC hardware and software technology have indeed established a beachhead in the embedded segment of the market. This is due to the economies of scale that allowed vendors to build components and sell them at a much lower cost that the traditionally much smaller volume embedded market allowed. But this has been true only to a point. Where the desktop has stopped, or slowed down, are on issues of determinism, real time, and, more to the point of this chapter, reliability and reliable computing.

In both the corporate and organizational world of Intranets and in the larger more consumer-oriented world of the Web and the Internet, it is not so much cost as it is reliability and the ability to perform reliable computing consistently that will determine how pervasive the net-centric computing paradigm will become.

The ability to achieve this goal will not depend on what engineers have learned about building desktop PCs, but rather on what they have learned about how to build reliable embedded industrial computers and controllers on the one hand, and what they have learned about building the reliable high integration circuitry embedded in most consumer electronics devices.

Compared to the "boxes" built by the telephone company or by the major electronics firms, the desktop computer that PC users have resigned themselves to would have resulted in bankruptcy for the firm in the consumer market. But sophisticated users have gotten used to the vagaries and are familiar with all the ways desktop PCs can hang up: mouse devices which freeze up; fax programs that interfere with the operation of data modem programs and vice versa; and graphics boards whose displays are corrupted as the higher temperatures of more advanced processors knock out sensitive analog circuits, resistors and capacitors.

Admittedly PC manufacturers and the companies who supply them have considerably improved the reliability of the average desktop system over the last 20 years. But, be honest. How many years did your VCR, TV or radio operate before it malfunctioned? Now think about your PC: how many months did you get of untroubled operation until a hardware problem of some sort developed? And how many weeks did your desktop operate without a software problem that crashed the entire system?

Users of net-centric computing devices by in large are an entirely different breed from the present PC user. The gap between the two is almost as wide as it is between the early adopters of black and white televisions, who got used to snow and horizontal and verticals scans that went out of synchronization, and present-day finicky viewers who accept nothing but the best quality viewing. Just as such consumers and users would not put up with an automobile whose wheels were constantly in danger of falling off or whose steering wheel constantly froze, the users of net-centric computing devices are consumers who will not put up with what the average PC user has come to accept as normal.

As hardware vendors and systems integrators move into the net-centric computing market, they are looking to a number of hardware strategies to make their systems as reliable, if not more so, than the telephone device and infrastructure that supports it.

Somewhat simplistically, there are three essential truths that engineers and systems developers will have to remember or relearn in order to build successful, reliable and cost effective net-centric computers for the office and home:

1.   A closed system is more reliable than an open system;

2.   The fewer the component parts, the more reliable a system is;

3.   Passive is better than active.

Closing Up The Net-Centric Computer

In one fundamental respect, makers of desktop PCs as well as their net-centric successors have learned a hard fact that designers of embedded systems have known for a long time: the more closed a system, the fewer the unknowns and the more reliable it is.

To the system administrator who must manage the use of desktop PCs, NCs, or NetPCs, within a large organization, the biggest unknown – and the biggest pain in the neck - is the average user: his level of expertise, her ability to read documentation accurately, his mechanical dexterity, and her ability to solve problems.

To eliminate the user from the equation, NetPC vendors have been the most active: Intel Corp. with its Wired For Management Initiative and Microsoft Corp. with its Zero Administration for Windows Initiative, both of which have been folded into the NetPC specification. Also incorporated into the NetPC specification are a number of concepts developed by the Desktop Management Task Force. This industry group consisting not only of Intel and Microsoft, but, of a wide range of NetPC and NC players, as well as net-centric infrastructure companies: Apple Computer, AST Research, Compaq, Dell, Digital Equipment Corp., Hewlett Packard, IBM, Novell, Sunsoft, Symantec, and Synoptics Communications, among others.

At the core are the requirements that each PC or NetPC (1) be locked or sealed physically so that the end user cannot fiddle with the system and (2) have no internal expansion slots available to the end user.

The starting point of the Intel Initiative is "plug and play" — a capability it has specified as a requirement of the NetPC. In addition to the ability to allow hardware to automatically install itself on a compliant system, the Wired For Management initiative specifies unique IDs for each installed device and add-on. Each system would have to be uniquely identifiable by a machine readable ID tag.

Controlling Software Usage

Microsoft's Zero Administration initiative has five key components: auto-configuration of system software; automatic desktop; automatic system state storage; central administration and lockdown; and application flexibility.

First, let's look at the auto-configuration of system software. In this concept, when a NetPC is booted up initially, the aim is to have the operating system automatically update itself from a server without user interaction. Complementing this is the automatic desktop feature that is also built into the operating system. It allows the user to access all the applications that he or she needs or is permitted to have, automatically installing them when they are called up by the user. The third element, automatic system state storage, is a mechanism by which all data created or entered on the thin-client NetPC is automatically stored, or reflected (copied), to a server. This is done to ensure that the user will be able to access not only his applications, but also the datal, no matter which terminal or NetPC is used to log on.

With the fourth component, central administration and lockdown, all aspects of a NetPC's configuration are under the control of a central administrator over the network. This control ranges from lockdown of a particular NetPC, or refusing a particular user access, to granting the user even more flexibility, allowing him or her to add various hardware or software options to the NetPC locally.

To allow the user as much freedom as before, the Microsoft initiative also specifies applications flexibility, which in the view of the Redmond, Washington-based company means that the NetPC can operate in either of two modes. The first is as a traditional "fat client" using the well-understood smart terminal or Windows terminal configuration. In this mode, full applications programs would be accessible to the user either locally or remotely, using protocols such as those developed by Citrix. The other is as a "thin client" similar to the NC, using Java-based or Java-like Web-style applications or applets that are downloaded as needed by a user.

Beyond Microsoft

Even this level of centralized control still gives the user too much leeway in the eyes of many systems administrators. Above and beyond the Intel and Microsoft initiatives, the Desktop Management Task Force has defined a desktop management interface(see Figure 10.1) and a management information format to more tightly control what the user has access to on the desktop. The desktop management interface consists of three layers: a group of manageable components at the bottom, a service interface in the middle, and the management application layer at the top.

Management Applications

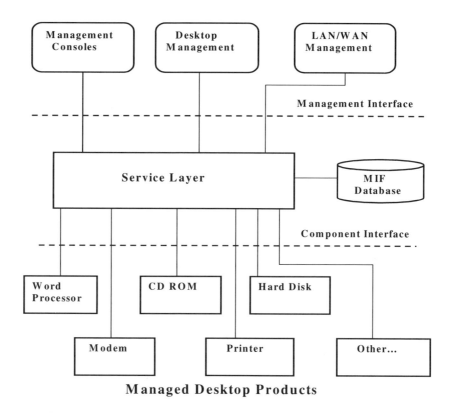

Managed Desktop Products

Figure 10.1 **The Human Element.** The Desktop Management Task Force initiative is aimed at incorporating features into the personal computer that increase its reliability by limiting the user's accessibility to hardware and controlling the use of software.

At the lowest level, components are divided into five groups: software applications; operating systems; hardware products (motherboards, add-in cards, mass storage devices, displays, and the like); peripherals designed to attach to a system by some external means; and system hardware, which includes things such as microprocessors, memory, and other board-level components. To give system administrators precise control over these various components, the task force has defined a special language, the Management Interface Format. MIF incorporates a well-defined grammar and syntax. This is used to describe the various components that are included in a special MIF file for each component. At the time of installation in the system, a file describing a specific component is then incorporated into a MIF database. The middle or service layer is a program that resides somewhere on the network that collects information from components, manages the data in the MIF database and passes this information to the upper management application layer consisting of such things as a management console, desktop management applications, and a LAN(local area network) management application.

Whether such a complex and, some would say, top-heavy, management structure will be successful, after users have become used to two decades of completely unstructured freedom, is questionable. But it indicates that the industry is waking up to the need for absolutely rock solid reliability if net-centric computing is to be successful. Much depends on the industry's ability to walk the fine line between the control that is necessary for reliability and the freedom that has made the desktop PC the world's dominant computing paradigm.

Opening up with USB

One such effort in the direction of freedom and away from control is the universal serial bus (USB). Closing up the NC and NetPC from the end user solves a number of reliability and cost of ownership problems, but it raises new ones. Not all users of computers are the same, and they have varying needs for peripheral support. In terms of input, most people are comfortable with a keyboard, but there are a variety of other types of input devices, each requiring slightly differing physical requirements. Storage of hard disk or floppies may also vary from user to user and over time. A network computer user who is working on a high-priority, high-security project may feel that storage on the intranet server is not secure enough and may want to save everything to a disk and keep it with him or her. Alternatively, the system administrator may want to have more options in how to configure a network computer for the lowest cost, the highest performance, or the best cost-performance ratio. Depending on hard drive and DRAM costs at the time, a system administrator may find that a network computer with 32 or 48 Mbytes of DRAM is too expensive and opt for a smaller amount of DRAM and install more mass storage disk drives, using the operating system's virtual memory capability to give the user the illusion of larger DRAM space.

While such techniques as Intel's "plug and play" have made installation of boards and other peripherals much easier and more error free, giving the user access to the inner workings of the computer raises the chances of damage and problems: the user may try to insert the board inappropriately – backward, loosely, or by forcing it in. A screw may drop into the wrong location on the motherboard. The user may try to force a PCI card into an ISA slot, or vice versa.

To keep the network computer as closed as possible, but give back some flexibility, builders of network computers can take advantage of the Universal Serial Bus to add a variety of peripheral functions external to the box. First developed by Intel Corp. as a way to move the PC industry away from dependence on the original ISA bus for peripheral devices, the USB in its original specification was a four wire serial bus (two differential signal lines, power and ground) that run at either 1.5 or 12 Mbits/sec. per second, roughly the performance of the original 8-bit-wide parallel ISA bus. A big advantage of the USB over the ISA is that it was designed to be plug and play and hot pluggable from the start. These latter two "idiot-proofing" features will be important in many net-centric computing applications. The first means that new peripherals and features can be added to a basic system automatically without intervention from the user of the net-centric device.

USB is a single master bus; that is, the host CPU in the desktop, NetPC or NC is always the master and all requests are initiated by the host. While similar in some respects to the way the SCSI (small computer system interface) was used as an external peripheral and mass storage bus on the Macintosh, USB is much more sophisticated. Rather than just a one dimensional linear bus onto which many devices can be linked, USB is a tiered star bus, where the lowest level is a device, with hubs used to create additional tiers and where devices only connect to hubs. By the end of 1997, there were about 10 million USB-enabled desktop PCs. The introduction of feature-limited NetPCs and NCs will no doubt accelerate its use, as system administrators employ it as a way to give users more flexibility without sacrificing the lower cost of ownership advantages of NCs and NetPCs.

By mid 1998, an even faster alternative to USB had begun to be implemented: the IEEE 1394 or Firewire connector. Although the NetPC must be locked to ensure reliability, Intel, Microsoft and Compaq have together started the Device Bay Initiative, which would use Firewire to let users attach a peripheral box into which devices can be easily plugged. Any devices that are plugged into the Bay must meet the NetPC's ID requirements, so network administrators can determine what has been added and still maintain control of system configurations at each location.

Eliminating System Complexity

A truism of engineering design is that the simpler a system is, the fewer the moving parts or constituent elements, the more reliable it is: whether it is a bridge, an automobile, or an electronics device, such as a computer.

The improved reliability of many of the consumer electronics devices with which we are all familiar - VCRs, radios, televisions, stereos, cameras and phones - is all based on this one fundamental truth. Manufacturers of these devices have discovered that systems built with discrete circuits for discrete functions are much less reliable than systems with higher levels of integration in which all the discrete functions are contained in one or a few integrated circuits. And a system built with 10 integrated circuits is more reliable than one built with 100. It should be no surprise that the overriding goal of any reliability-conscious designer is to reduce component count to its absolute minimum.

In engineering parlance, the reliability of an assembled network of components tends to be inversely proportional to the number of connections amongst components: a printed circuit board with 100 components on it may have upward to of 1,000 or so connections amongst them. Reducing the number of components to, say 10 and the number of connections to 100, for example, then increases the reliability by tenfold as well. This is because when a large number of individual components are assembled on a printed circuit boards there is a chance of loose contacts or bad soldering of wires, and it is more difficult to detect problems after assembly.

However, in the case of an integrated circuit containing tens or hundreds of thousands of transistors or gates, there are fewer external components to connect because the connections are a part of the package. So the reliability is higher. While the connections within the ICs could be faulty, the chance for encountering such faulty connections is much lower than for external faulty connections.

According to various estimates, the present reliability levels of current LSI, VLSI and ULSI components, including most second-, third- and fourth-generation microprocessors such as the Pentium, Pentium Pro, PowerPC and Alpha — are about 99.5% per 1,000 hours or so at a 90 % confidence level and 70° C. In other words, in the standard office or home environment at humanly comfortable temperatures, you can be 90% sure at any point in time that there is only a five out of 10,000 chance that the microprocessor will fail over the next month and a half if you run it constantly.

Moreover, the reliability of any LSI or VLSI part improves after testing for "infant mortality" in the first 100 or 1,000 hours of operation, a common procedure before any integrated circuit leaves the factory. Also, as the manufacturer of the integrated circuit gains more experience and on-chip failure mechanisms are identified and eliminated, reliability further improves.

Consumer electronics companies such as Fujitsu, Hitachi, Mitsubishi and NEC have used the improved reliability of highly complex integrated circuits embedded in their products to establish dominance in a wide variety of entertainment and communications electronics: stereos, TV sets, portable TVs, portable CD players, and VCRs among them. Many of these same companies are moving to establish a foothold in consumer information appliances by using the same strategy in building Internet telephones and Web-enabled settop boxes.

Even though integration levels have moved from the hundreds of thousands of transistors to millions of transistors on a single IC, it is expected that reliability levels will be at least as high as they were in the mid-80s, when integration levels were still in the 100,000-gate level.

With a 100,000-gate microprocessor such as the 8086, the work horse of the original desktop PCs, reliability estimates indicate that 90% of these CPUs would still be operating after 1,000 years, assuming operation at 70° C. Anyone who has bothered to dig out that old PC from the garage or storage and test out the components will find that 10 years or so later most of the CPUs are still functional. If anything has failed in the system it is probably the mechanical components such as the disk drives, or the printed circuit board itself.

At current 1,000,000-plus gate levels for processors such as the Pentium Pro, PowerPC, and Alpha, such reliability estimates are expected to hold. But with the new generation of "system-on-a-chip" designs that such high levels of integration make possible, it is likely that system reliability will improve significantly as more and more of the external peripherals and functions are integrated on chip.

Additional reliability issues arise because most desktop computers, including PCs, use an active-backplane approach, in which all the components necessary for the operation of the electronic device are active on the same motherboard as the buses and signal lines that move data and signals amongst the various components.

This presents designers with a dilemma. On the one hand, increased integration improves system reliability and reduces costs. But, on the other hand, beyond a certain point, increased reliability decreases flexibility at the system level. The more highly integrated the system, the less able it is to be adapted to a wide range of applications and configurations.

From Active to Passive

While using the passive is a "no-no" grammatically, it may be the only way to go if builders of not only traditional desktop PCs but also of the newer connected NetPCs and NCs expect to achieve the kinds of reliability on which corporations are beginning to insist, but do so without using flexibility.

While going to higher levels of integration will have an impact on total system reliability the same way it has in consumer electronics, the NetPC and NC have a business model and physical design in common with the desktop PC that will limit the usefulness of such strategies.

The big advantage of the desktop PC traditionally has been that, because of its bus-oriented modular design it is essentially an "anything machine." The same basic system can be sold are an extraordinarily low price to an end user and adapted to a wide variety of applications and needs.

One of the main motivations of major corporations for moving to NCs and NetPCs is to reduce the cost of repair and the cost of unanticipated interactions between programs and systems. However, this does not mean that they are willing to give up the flexibility of adapting the basic unit through the incorporation of add-in cards and circuits.

The problem with this approach is that, unless systems makers shift away from the current emphasis on active backplane designs in which all the critical components share the same motherboard space with the signal lines and bus interconnections, they will still face issues of reliability. While many of the reliability kinks have been worked out of the original ISA (industry standard architecture) bus for the desktop PC over the past years, yielding reasonably solid motherboard designs, the industry has shifted, on the desktop at least, to a higher performance alternative, PCI. Because of inherent electrical characteristics and its much higher bandwidth, this bus requires much more attention to detail in order to ensure a reliable design.

In the passive-backplane approach to system design, a simple connector-only or passive backplane is substituted for the usual active-backplane motherboard. The active components normally placed on the motherboard are placed on a card or cards that plug into the passive backplane.

The concept is not new to the PC world. Indeed, in the mid-1980s when Compaq was battling with IBM for dominance of the desktop market, the Texas company introduced a modular, passive-backplane design. But because IBM had the market clout, the design was not widely accepted. Even earlier, at the very beginnings of the PC revolution, when Intel was still battling with Zilog for domination of the PC motherboard, there was the venerable S-100 bus, a passive-backplane standard that was the most common architecture for desktop computers. This was before IBM thought it could do better and designed the active-backplane ISA-based IBM PC, which became the industry standard.

And as designers have moved PC hardware and software into the embedded world for use in industrial computing, it is passive rather than active backplanes that have become the norm: passive ISA ; STD16 and 32, VME 16 and 32, Multibus, and, now, a variety of passive PCI variants including PC/104+ and CompactPCI..

This approach has a number of advantages as far as reliable computing is concerned. Because various functions are separated into a number of different boards, problems can be isolated much more easily. Mean time to repair is also significantly improved, since various functions such as the CPU, memory, and peripherals can be removed quickly and replaced. This feature would make the passive-backplane approach to designing net-centric computers very attractive to corporations who are looking for ways to reduce their total cost of ownership. As described in earlier chapters, a significant portion of the total cost of ownership has to do with the downtime, the time it takes to repair or upgrade a desktop system and bring in online again.

There are also a number of inherent advantages of the passive approach, above and beyond its modularity. While all card interconnect schemes, where the number of cards is a variable, suffer from problems such as impedance matching, the mismatch is much more easily controlled in a passive rather than active scheme. A passive backplane provides alternating signal and ground layers in which the signals exhibit better transient behavior. In active-backplane designs where flexible interconnects or signal paths are not controlled either by length, geometry or shielding, there are much wider swings in impedance and a greater chance that reflection spikes will cause unreliable operation. Signal skew can also be a problem since inductance and capacitance may vary widely between signal lines.

As of late 1997, despite the obvious drawbacks to the PCI active backplane approach used in most desktops, this paradigm had led to net-centric alternatives such as the NC and NetPC as well. The NetPC specification specifically requires version 2.0 of the original PCI standard, a 32-bit, 133-Mbyte/sec. with a three-card expansion bus capability. In the NC world, no specific bus structure or motherboard design approach is defined. This has been left up to systems designers. As with the choice of CPU and how it is implemented in a design, the choice of active versus passive and the specific type of internal bus architecture has led to a plethora of designs. Some companies, such as Wyse Technology, have developed their own proprietary passive backplane architectures and others have stuck with the original PCI and ISA implementations. Other companies, such as Philips Electronics, are looking closely at some of these alternatives, particularly in Web-enabled set-top box designs, for their hardiness and as a means of quickly upgrading their designs with readily available standardized add-in boards. In its MyWeb set-top box, Philips Consumer Electronics has opted for a sparsely populated motherboard that uses only a few highly integrated circuits for all of the key functions. Because the consumer environment is much closer to industrial applications in the kind of treatment electronics devices undergo, Philips incorporated a PC/104 bus and connector as an expansion interface for additional applications because it incorporates safety features that insure reliable connections no matter how much abuse the unit receives..

The reason companies are moving in this direction is simple: in addition to greater reliability an industrial quality connector such as the PC/104 or PC/104 Plus provides, lower costs come with the design that makes use of standardized components and boards. In the case of PCI, a number of bridge chips have emerged that connect many of the RISC chips used in the various network computer designs: Motorola's 68000-derived Coldfire, the PowerPC, the DEC Alpha, the DEC StrongArm, the MIPS CPUs and Intel's 80960.

It is more than likely that some form of passive-backplane version of PCI will emerge as a standard for all net-centric computers, if costs can be brought down. But systems designers will not have to reinvent the wheel. A number of passive-backplane versions of the PCI bus have been developed for use in industrial computers and embedded PC designs (see Table 10.1) that would be more than suitable, especially from a size and compactness point of view. Both are important considerations in most network computer designs. At least two alternatives come to mind: Compact PCI, and PC/104 Plus.

CompactPCI

When the outline of CompactPCI was originally brought to the PICMG (PCI Industrial Computers Manufacturer's Group), in early 1995, it was an organization of approximately 20 members with a specification for passive backplane PCI (slot card) systems, using the standard PCI card form factor. Today, bolstered by the adoption of CompactPCI, the organization is approaching 200 members, with executive members including leaders from the industrial, embedded, telecommunications, and computer industry, names like Digital, Sun, IBM, HP, Ziatech, Pro-Log, Lucent (AT&T), Dialogic and National Instruments. Even Compaq Computer has joined.

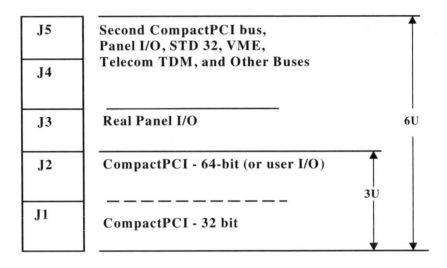

Figure 10.2 **CompactPCI.** To improve reliability and reduce downtime, the PCI Industrial Computer Manufacturer's Group has developed a passive backplane standard using the standard card form factor.

CompactPCI products come in two sizes, 3U and 6U, the different Eurocard heights referenced in the specification. These are the same form factors defined by the VMEbus,

long a standard bus in many industrial and embedded real-time computing applications. But the VME specification must spread the VME bus data lines across two connectors, which means that the smaller 3U cards cannot offer the bus performance of the 6U cards. Also, the size of the VME bus logic and the technology of the time made 3U cards less attractive.

CompactPCI is different, in that all the CompactPCI signals for 32-bit and 64-bit buses are accommodated within the connectors on a 3U card (see Figure 10.2). The fact that the bus interface (PCI) is built into the chips and that integration has progressed at such an astonishing rate means that 3U cards are likely to play a much larger part in CompactPCI systems than they did in VME.

Significantly for designers of net-centric clients, network computers and servers who are looking for a source of passive backplane add-in cards, most standard interface cards such as Ethernet, Serial, VGA, ATM, etc., are appearing in the 3U form factor. The small size of the boards is also a significant advantage in such applications. The backplane board can be of any size that fits the requirements of the net-centric application, as long as it can accommodate the extremely small 3.8 by 3.6 inch boards, which are one-quarter the size of standard long PCI add-in cards and half the size of the short PCI boards. As more and more 3U cards are made for CompactPCI, it is likely that they will not only

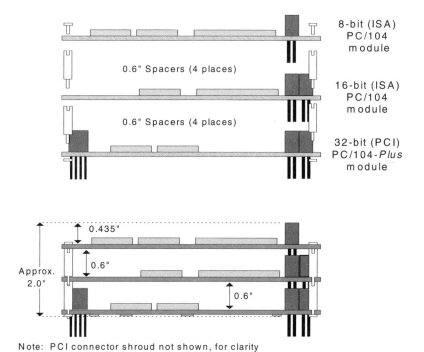

Note: PCI connector shroud not shown, for clarity

Figure 10.3 **PC/104+.** Building on the original ISA-compatible PC/104 specification, this compact board piggybacks PCI compatibility on top of the original 3.575 by 3.575 inch form factor. (Source: Ampro Computers, Inc.)

supplant the PMC (PCI Mezzanine Card) format as the most common PCI expansion mechanism for industrial and telecommunications systems, but could also be common in NCs, NetPCs, Web-enabled digital settop boxes, and even digital television systems.

CPU product offerings on CompactPCI now include Pentiums, dual Pentium Pros, PowerPCs, SPARC CPUs and digital signal processors in a variety of 3U and 6U form factors. The core peripheral products are also well represented, with Ethernet, SCSI, IDE etc., but the rest of the I/O world is still a little sparse because of the relative newness of the bus. This area should see dramatic growth over the next few years. A variety of products, ranging from ATM and Gigabit Ethernet interfaces to network servers with 24 by 7 (24 hours per day, seven days a week) availability are in the works from many manufacturers.

Because of the initial shortage of I/O cards, CompactPCI applications have so far appeared in two main areas: (1). applications needing a standard PC in a rugged format with little extra I/O and (2) large or specialized applications in which the original equipment manufacturer designs his or her own CompactPCI I/O. But as the standard proves itself in these areas and the need for a smaller form factor and higher reliability than standard active backplane designs can provide, makers of net-centric clients and servers should move rapidly to the CompactPCI standard, especially if volumes increase and the costs of the still expensive CompactPCI drop..

The goal of the original CompactPCI specification was to define a clear and unambiguous way of connecting up to eight cards in a Eurocard backplane, using standard PCI silicon and maintaining compatibility with the underlying PCI specification. Extensions for future developments, such as 64 bits and the mechanical provision for "hot swap" were included in the original specification.

The core of the specification was the 32/64-bit CompactPCI definition on the 3U form factor. The inclusion of the 6U form factor also opened up some interesting possibilities, namely what to do with all the extra pins now available in 6U. Most of the CompactPCI specification work underway within the PICMG involves these extra pins in the 6U format and the implementation of hot swap capability and 64-bit extensions. Before we discuss these future capabilities in more detail, lets review some pertinent details about the CompactPCI specification. The core CompactPCI specification defines the pin assignments of connectors J1 and J2 and leaves the others as User I/O. User I/O is further divided into Rear Panel I/O, where the signals are brought straight through the backplane to a rear connector and connected via cards or cable to external devices, and Bused I/O, where signals are routed across the backplane to other cards in the systems.

Cards designed to operate in conformance to the CompactPCI specification will be able to communicate via the PCI bus under all circumstances. The other pins can optionally be assigned for other specific applications, such as telecommunications or DSP buses. Keying is provided on the both the connector and the ejector handles to prevent incorrect insertion in systems. Thus, most standard CompactPCI peripheral cards will use only the J1 connector to implement the full 32 bit PCI, whether those cards are 3U or 6U format. System CPU cards will need both J1 and J2, even for 32 bits, because the system card is responsible for driving all the clock and arbitration signals, some of which originate from the J2 connector. Many 6U cards may implement J3 for rear panel I/O to provide an optional connection scheme for applications that do not want to cable from the front of the computer.

Sixty-four-bit PCI is already accommodated in the CompactPCI specification and was included in the extensive simulation work that was performed prior to the bus becoming an official specification. The only thing that CompactPCI needs to make 64-bit cards possible is actual implementation from PCI chip makers. By mid-1998 these designs hard started entering the market. It is worth noting that 64-bit peripheral cards can operate in systems with 32-bit CPUs. The dynamic bus-sizing characteristics of the CompactPCI bus make this possible, with the 64 bit cards "dropping back" to 32 bits when necessary. Of course, to talk in 64-bit lingo, these cards need other 64 bit cards to talk to, but this can occur without the data going through a 32 bit CPU.

PC/104 Plus

A very compact PC-derived bus that in essence rides on the back of other more traditional passive-backplane busses, the PC/104 Plus specification is an extension of the PC/104 Standard, a stackable passive-backplane board architecture that has found wide usage in industrial computers. Developed originally by Ampro Computer Corp., the PC/104 Plus connector, and it was developed by Samtec, a manufacturer of the original PC/104 (ISA) bus connectors. It is a new self-stacking, 120-pin high-density (2mm) connector, which incorporates a unique connector pin shroud that: (1) guides the male portion of the PCI connector as it mates with the female portion of the next connector in the stack; and (2) protects the PCI connector pins, which are slightly thinner and therefore somewhat more vulnerable than those of the ISA connector.

The resulting combination of PC/104 with PCI, called "PC/104-Plus," meets all the objectives listed above. In addition, extensive electronic simulations have been used to validate its reliability for both 33- and 66-MHz PCI bus chip sets. The basic mechanical dimensions of a PC/104-Plus module form factor match that of PC/104, so PC/104-Plus can be said to be "PC/104 form-factor compliant." The new self-stacking 120-pin high-density (2 mm) PCI bus connector fits between the standard PC/104 mounting holes along the edge opposite the regular PC/104 (ISA) bus, so it consumes minimal space (about 10%). In a typical application, both 8- and 16-bit PC/104 (ISA) modules can be stacked along with 32-bit PC/104-Plus (PCI) modules, provided that similar type modules are adjacent to each other.

Of a size that would make them attractive in many small footprint net-centric applications, a PC/104-Plus is even smaller than the CompactPCI configuration, measuring only 3.575 x 3.775 inches with a minimal 0.6 inch spacing between stacked modules. However, data throughput is 26 times that of the original PC/104 and equal to the standard PCI, 132 MBytes/sec. Five PC/104 Plus boards can be connected to the PCI bus, except that they are stackable, including the base CPU module. PC/104 Plus is essentially a repackaged version of PCI that is optimized for the unique requirements of embedded systems, where space is scarce and ruggedness is paramount, but it would find wide acceptance in many net-centric designs where the same issues are of concern..

Although PC/104-Plus modules provide connectors for both the 120-pin PCI bus and the 104-pin ISA bus, only the PCI bus actually connects to circuitry on such a module. The ISA bus connectors are only there to pass the ISA bus on for possible use by the next module in a stack. Over a dozen companies have already announced support or begun development programs based on PC/104-Plus. Products being developed by both Ampro and other companies include high performance single-board computers, full-motion video interfaces, high speed LANs and communications interfaces (100BaseT, USB, IEEE-1394, etc.), and PCI bus adapters and bridges.

Which Way?

What are the chances that a passive-backplane version of the PCI bus will become a standard in net-centric computing? On the face of it, it would seem that the active backplane PCI approach, backed by Intel and adopted by virtually all desktop computer makers would have momentum going for it in NetPCs and connected PCs. Because of Intel's stranglehold on the computer market, the active backplane PCI approach is at least initially the architecture of choice for some network computers. This is despite the fact that other alternatives would be more suitable from a reliability point of view. Because of their much more stringent space and cost constraints, set-top boxes and wired-wireless Web-connected information appliances may be the only holdouts, simply because the form factors involved are not compatible with most implementations of the active backplane approach.

COMPARISON OF PCI FORM-FACTORS

	Desktop PCI	Passive Backplane PCI	PMC	CompactPCI	CardBus	Small-PCI	PC/104-*Plus*
Dimensions (in.)	Long: 12.3 x 3.9 Short: 6.9 x 3.9	12.3 x 3.9	5.9 x 2.9	6.3 x 3.9	3.4 x 2.1	3.4 x 2.1	3.8 x 3.6
Area (sq. in.)	Long: 48 Short: 24	48	17	25	7	7	13
Bus Connector	Edge-Card	Edge-Card	Pin & Socket	Pin & Socket	Pin & Socket	Pin & Socket	Pin & Socket
Includes ISA bus	No	Yes	No	No	No	No	Yes
Installation Plane	Perpendicular	Perpendicular	Parallel	Perpendicular	Parallel	Parallel	Parallel
Expands Without Additional Slots	No	No	No	No	No	No	Yes (self-stacking)
Positive retention	No	No	Yes	Yes	No	No	Yes
Standards body	PCI-SIG	PICMG	IEEE	PICMG	PCMCIA	PCI-SIG	PC/104
Primary Application Area	Desktop: motherboard expansion	Industrial: backplane expansion	Industrial: VME mezzanine	Industrial: backplane expansion	Laptop: end user additions	Laptop: factory options	Embedded: SBC expansion

Table 10.1 **Comparison of PCI Form Factors** (Source: Ampro Computers, Inc.)

It would not be the first time that a less than ideal solution dominated the marketplace. The computer industry is replete with examples where a technically superior solution has lost out to one that compromised on technical purity in favor of getting to market first. IBM's first mainframes used the less efficient, but familiar, punch card for inputting statistical data and beat competing firms who opted for the more technically elegant and ultimately superior, keyboard entry approach. Intel's original 8080 microprocessor established a lead position in personal computers which the company has never lost, despite the fact that it was more of a calculator chip more appropriate for embedded control. It won out over more elegant and easier to program alternatives such as Motorola's 6800 microprocessor, even though the latter was more computer-like in its architecture.

Despite the dismal history of more technically superior solutions in the computer industry, such may not be the case as far as the passive backplane alternatives. Several factors may make it possible for passive alternatives such as Compact PCI to win out over active backplane PCI. First, server vendors have been using it or something very close to it in their new designs, taking advantage of its modularity and increased reliability. Second, passive backplanes such as CompactPCI are favored by many of the major network router and telecommunications switch companies, on whose communications backbone the Internet rides, and who are in the forefront as far as Internet appliances and other consumer Web products are concerned.

Indeed, network-telecom companies and the hardware vendors who supply them have been in the forefront of adapting CompactPCI to the requirements of net-centric computing. Of all the features commonly required in networking and telecom, all but expandability and hot swap are fully resolved by the current CompactPCI specification. PICMG is presently working on three extensions to the base line CompactPCI specification that will allow a system designer to fully implement a telecommunication system. Since the original CompactPCI specification was developed for standard computer applications, it did not include all of the features required by network and telecom vendors. There are a total of eight subcommittees attached to the PICMG CompactPCI effort, two of which are especially important to telecommunication users: Hot Swap and PCI-to PCI Bridges. Both of these are important in the design of reliable network-centric computers.

In addition to the hardware requirements of hot swap, work remains on implementing the software that will support hot swap. Both OS's and BIOS's must be modified to support this. There is however, good news on this front. Compaq Computers and Microsoft have been working on a "Hot Plug" implementation that will allow the standard PCI Local Bus to support hot swap. This technology will be targeted for CompactPCI hot swap implementations.

A second area of interest to network-telecom companies is the PCI-to-PCI Bridge (PPB).. This extension defines a standard way of implementing PPB technology on the CompactPCI bus. The use of a standard PPB interface, such as Digital's 21150, or the use of a smart interface, such as Intel's i960RP, is compatible with this proposed modification to the Compact PCI specification..

Given the new developments, CompactPCI can be considered a viable solution for use in in a wide range of telecom and networking applications, and eventually in a variety of net-centric computer designs.

If the average consumer is to willingly buy into the net-centric "information superhighway" vision he or she will do so not only because the cost of the system and its cost of ownership are sufficiently low, but also because it is reliable. Can you imagine that the average automobile driver would put up with what the average desktop computer user does? Ford did not win the automobile wars of the late 19th and early 20th centuries merely because he found a way to mass produce low- cost affordable autos. It was because he produced RELIABLE low-cost autos. The first users of automobiles - after early experiences with autos in which wheels fell off, headlights went out, steering wheels froze up, and radiators blew up - quite naturally went to the first maker of autos who gave them a product that did not do these things. That is why reliability, as well as low cost, will determine who will become the Ford of the information superhighway. If it is not passive PCI, CompactPCI, PCI/104 Plus, or one of a number of other similar alternatives, it will be something that looks like it, tastes like it, smells like it and feels like it.

Chapter 11

Securing Your Network Connections

A critical determinant in the success of net-centric computing as a mainstream paradigm will be reliability. The user must be able to assume with the certainty of almost absolute truth that the connection he or she has with the Internet or intranet is secure from (1) persons who want critical information, such as credit card numbers, which may be on the messages going back and forth; (2) hackers who are attempting to break into a system or into the server or servers that are being used; (3) computer viruses written by malicious souls and unleashed for a variety of reasons on the Internet; and (4) bad code, or poorly written programs, that when downloaded to the client, have unforeseen interactions with other programs or program segments.

In the simple black and white world view of the non-PC-savvy consumer, all of these problems are indicative of a "broken" net-centric computing device. Some of these issues have already been faced in the precursors to today's network computers and Internet appliances: dumb terminals and diskless workstations. One of the most troubling aspects of the Java and ActiveX paradigms, their use of the downloading of programs from the network server to the client, is not a new concept.

While enormously attractive, network loading of system software raises two important issues: security and dynamic binding. Anyone involved in the management of a network of any size has learned one unbreakable rule of thumb: system software should only be loaded across a network if there is a way to be certain that the software can be trusted — despite the fact that such software comes from a remote server across a network path that could be long and complicated. This guarantee can be assured only if appropriate security measures are in place. Absolutely rock solid security, however, will require that at the very least the software component that enforces security policies be present when other components are loaded.

Trusting Your Software

As managers of in-house LANs and intranets have learned from bitter experience, unless there is certainty about who created the software, it can not be trusted under any circumstances. Misplaced trust has serious consequences, the very least of which is corrupted OS code. Nearly everything a terminal or diskless workstation on an intranet or LAN involves data that are likely to touch the network, compounding the problem geometrically.

On the Internet, the problem is even more acute. Particularly vulnerable to attack are consumer Internet appliances that boot from a network. The link that that carries the code feed will do it either through some sort of wireless connection or through miles of unsecured cable. But because the stakes and/or payoff in attacking consumer electronics devices (currently limited to games and the like) have traditionally been very low, this route has typically not been problematic.

The emergence of net-centric computers raises the stakes considerably. Soon, network-connected smart consumer electronics may be involved in banking, gambling, shopping, home control and transmission of other sensitive information. This information is generally transmitted encrypted and then decrypted at the point of reception - the Internet appliance. If an attacker can find a way to replace the software in the appliance with a form of Trojan horse virus, for example, he or she can access data in its decrypted form and may even gain access to the keys.

A smart Internet appliance that will handle sensitive data must use a secure operating system, with system software allowed to be modified only the strongest of security precautions in place. The easy way to meet this latter requirement is never to change the system software - period. Because security precautions tend to be expensive and inconvenient, this easy solution may be the right one. But in the era of the downloadable Java or ActiveX applet, it may not be realistic.

Safeguarding The Network

The reason that the easy solution will be the last one to be employed in the age of net-centric computing is that network loading of system software provides a host of broad scale business benefits. Phone companies, cable companies, and utility companies like the concept because it gives them a way to avoid extensive maintenance on the equipment they own. Additionally, they like the fact that users themselves would be able to install upgrades that benefit the utility, further reducing their costs .

If there are no serious bugs in the software, failures won't force software updates. On the other hand, no degree of "correctness" will eliminate the need for upgrades. Upgrades may not always be triggered by new features that the end user would happily purchase. The utility company will want upgrades that benefit them. The network or telecom provider might want to be able to download to the user device a new protocol that decreases network load.

How to provide such built-in system protection? The key is an operating system with good security that will keep user state code from doing any harm. The greater the degree to which updates can be contained in user state, the more effective the system can be protected from damage. While this is sufficient for the operating system and the things it protects, but it may not protect the user's application programs or data files. Also, because user-state code probably sees a lot of proprietary data , it is open to exposure through the use of such things as log-in Trojan horses. This is an old trick that could be applied to any system with a procedure resembling a log-in. For this to work, the log-in program is replaced with a program that does everything log in does, and sends the user's log-in information to a remote database.

Even more than in a closed network environment such as a local area network or an intranet, system software and important utilities delivered to clients on systems with access to the wide-open Internet must come from trusted sources. But in the absence of a practical way to secure the network and its servers, how can this be done?

Engineers and developers need a way to know that a software module is "good." One alternative that has been widely implemented is the application of public-private key encryption similar to that used for digital signatures. Each good module needs to be signed by a trusted entity, so that the Internet-connected information appliance knows who to trust. This procedure is usually performed in two parts: the originating computer, either a server or a net-centric device, verifies that modules can be trusted, and a certifying agent checks the modules and stamps them with an authentication code. Making use of the public-private key methodology, this code would include a cryptographic checksum of the software in question encrypted with the agent's private key. If the device decrypts the checksum in the signature and finds it equal to the actual checksum, then the module is as trustworthy as the agent. The public key of at least one verification agent is required. This scheme protects sensitive modules from time the verification agent stamps them until they are checked at the net-centric device. Problems introduced by a compromised server or a compromised network will all cause the signature to differ from the one created by the originating module.

Dynamic System Configuration

Another issue associated with the downloading of system software is that of dynamic system configuration in a network computing environment, especially on clients with constrained resources and where the load on the system varies and is hard to predict. Unless the network computer or Internet appliance has DRAM to spare, unused software needs to be purged. Furthermore, when a system does not need to support a device, the support software should not be present. In the case of system utilities, this requirement is taken for granted. But device drivers, file managers and other system services that can be installed and uninstalled at any time are not universally supported. Preinstalling everything is only a viable alternative under two conditions: where there is sufficient memory to accommodate the installation and where configuration of the system can be predicted accurately.

Today's dynamically linked net-centric systems are much different from the situation in the mid-1980s when nearly all operating systems were statically linked. Any change in the system configuration required a system programmer to link (sysgen) a new operating system. Since the early 1990s, operating systems have been moving increasingly toward dynamic linking. Now one can find loadable device drivers and installable file systems in mainstream operating systems such as UNIX, OS/2 and Windows. And while Sun's Java has gotten a lot of credit for the concept, as we have learned in earlier chapters dynamically loadable and installable system software is an old idea. Many real-time operating systems have employed the technique for many years. For example, Microware's OS-9 has dynamically linked all system services for nearly 20 years, and QNX has used a dynamic structure since its inception.

The principle of dynamic linking is based on indirection in that components of the operating system are not allowed to call one another directly. In some architectures, these components most often use messages to call one another. Many modular RTOS kernel architectures use indirect calls. No matter what the details, there is usually a level of indirection that lets the system components be added, deleted, or replaced . A dynamic system requires software to use indirect calls and use - with caution - persistent pointers to memory controlled by another component. If a file manager were to make a copy of the pointer to a device driver, and keep it beyond a single operation, it could find that the device driver has been replaced - and this pointer could be pointing to memory that contains something else entirely, such as a Trojan Horse or other kind of computer virus.

In many embedded systems, a technique is used that might find application on net-centric devices if the problem of viruses become endemic .That is to store critical systems software components in read-only-memory that can only be read and cannot be written to and modified. .This runs counter to the idea of dynamically downloading not only applets and applications but also system software updates. For this technique to work, the operating system kernel needs a way to indicate that the components are out of date without modifying the components. This means that the kernel has got to maintain a directory of software components, usually stored in DRAM so it can track and dynamically update changes to attributes of ROMed modules.

Kernel support for adding and removing software components may be enough to support dynamic reconfiguration. But run-time upgrades might involve components that the system cannot remove, even momentarily. For example, it might be necessary to upgrade some of the software handling the communication link used for the upgrade. Since software modules are seldom hot swappable, the kernel will need some sort of mechanism to manage the delicate transition from the old software to the new. A non-secure network can be used to load software into a secure machine if the machine uses digital signature technology to check the authenticity of the software. This approach has limits. It only allows system to load all their system software at boot time. Also, it is only practical for static applications and for only those applications that can easily be rebooted with each new application.

Viruses and net-centric Computers

Such tried and true techniques have worked well enough in the relatively closed world of LANs and Intranets. But in the more open environment of the Internet and World Wide Web, these not enough. As the Internet and World Wide Web have become more public and accessible, the problem of computer viruses, in particular, has become endemic. And it will probably get worse as the normal mode of operation for a computer is in a networked environment. By some estimates, more than 10,000 different viruses have been identified, and during any one year at least one million computers a year are infected.

Generally, there are two ways to classify viruses, by what they attack and by the way they behave when they try to conceal themselves. In the first category are six types of viruses: file infectors, companion and cluster viruses, boot-sector viruses, macro viruses, and stealth viruses, all of which can be further classified by their mode of operation, either as direct action or resident. A direct-action virus works by selection of a program or programs to infect each time the program that contains the virus is executed. The resident virus is a bit smarter. It hides itself in memory the first time an infected program is executed and then remains in memory to infect other programs. Meanwhile, it constantly checks logical conditions and monitors hardware and software interrupts to spot conditions that might reveal its existence.

About 85% of all known viruses are file infectors, which depend on hiding in an application such as a database generator, a spreadsheet or a game program. When an infected application is executed, the virus code contained inside installs itself in the system DRAM so that it can copy itself into applications that might follow. The user does not know it is there, because as soon as it has found a comfortable home it returns control to the application that acted as a carrier, which, ironically, in most cases remains uninfected.

Distant relations to the file infector are the companion and cluster viruses. Rather than modify an existing executable file, a companion file takes advantage of the fact that in the Windows/DOS environment, at least, the command interpreter always executes a COM file before an EXE file. It creates a hidden file with a copy of itself, giving it the same name as the chosen EXE file, but with a COM extension. As a result, the virus is always executed, rather than the intended program. When it has finished its destructive tasks, it exits and invokes the original file, returning things to normal. The cluster virus modifies directory table entries enabling a virus to be loaded and executed before the one the user wants. Only the directory entry, and not the program itself, is altered.

A boot-sector virus, while more serious, generally accounts for no more than 5% of the known strains. It downloads to a system in a manner similar to the file infector, but makes its home in a very specific, but critical portion of the system, the boot-sector of a floppy, removable or fixed hard drive disk. Because the boot-sector contains the program code for loading of a computer's operating system, a virus located there can be loaded into memory, re-infecting the system each time the system is restarted.

Another category that is growing more prevalent is the macro virus, which operates independently of either the operating system or an application program. It usually infects files that contain data, rather than executable programs. But these files are of a particular type, containing subroutines or scripts that the user can implement to automate a sequence of actions. The types of files that are infected are those generated by spreadsheets, database generators and word processors. While it has traditionally represented only about 10% of all viruses found, this type is occurring more and more often. More than 1,000 have been identified, a nd that number is growing. Stealth viruses are among the newest type of virus and can act in a manner similar to all the above, but have the additional ability to avoid detection from anti-virus scanners by simply overwriting the original program, or adding a new cluster to the FAT (file allocation table) chain, replacing only a portion of the original code. More proactively, some attack and disable the anti-virus software before it does its job while others are designed to avoid infecting anti-virus programs, which would reveal their existence. In all these cases, in addition to the code that allows them to replicate, these viruses also contain code for whatever the originator wants to accomplish, from the relatively benign display of a message or an image to the more malicious kind, which destroy files, applications and even the operating system itself.

As the nature of the computing platform, its operating system and the applications that it runs has evolved, the occurrence of these viruses appears to be cyclic. As changes have occurred in how the operating system handles files and protects the boot sectors of a drive, the occurrence of file and boot sector viruses is dropping, while macro viruses are on the increase. The increasing use of Java and ActiveX to download not only data but software and executable files can be expected to increase the prevalence of all three types, but selectively. In particular, in a networked environment, stealth viruses may be the most serious, moving across dozens of systems without being detected. In network computers and terminals, which do not use either floppies or hard drives, boot-sector viruses will not cause problems until they get to a system, such as a server or a network-connected PC.. But such systems could act as carriers, passing along the virus and making it more difficult to trace back to the source. However, since network computers and terminals depend on the servers for execution of significant parts of their program, such viruses could be more of a problem for servers, where an infection in one system could affect thousands of users.

The Download Conundrum

The introduction of dynamically downloadable Java and ActiveX onto virtually every computing platform introduces a new range of complexity to the problem of detecting viruses. To start with, there are a number of destructive mechanisms unique to the Java environment. Then there are useful features of Java that can be turned to the advantage of the virus makers, making a serious problem even more so. In Java, viruses and hostile applets are divided into two categories: attack and malicious applets. Attack applets are similar to traditional viruses in that the focus of their efforts is other applets and the files that they generate by corrupting data on the hard disk; revealing private data on the hard disk to others; and rendering resident applications and applets inoperable.

More subtle, less direct, and harder to pin down is the second category of Java virus: the malicious applet. Such virus applets do not actively attack other applets and files, but create situations that make it hard for a computer user to conduct business. There are three classes of malicious attack: denial of service, invasion of privacy, and simple annoyance. Denial of service applets do such things as use up all available CPU by looping continuously on a particular operation, by allocating all of the available DRAM, or using up all the available screen space.

Not only can Java initiate viruses and act as carriers, but the very features that make Java so powerful also make it much more dangerous as a vehicle for transmitting destructive programs. First, its platform independence means that a virus or vicious applet is not limited to one type of processor or operating system. Now it can migrate from computer to computer, from a Windows to an Apple to a Macintosh to a UNIX, and if it is polymorphic, it will be able to mutate with each infection adapting to the requirements of each environment. Second, the dynamic downloadability of Java applets as well as ActiveX applications changes dramatically the rules of the game, both in the way viruses are propagated and in the speed at which they might propagate throughout the public Internet or the private Intranet.

While the Sun/Netscape security model effectively deals with some of the more destructive kinds of viruses by simply not permitting applets any access to or control of critical system resources such as the hard drive, the dynamic downloadability of ActiveX brings a new level of complexity. An ActiveX control is nothing more than a Windows DLL that is executed by a browser as a complement to the downloaded applet or application. In effect, an ActiveX control is just another virus, benevolent though it may be. But such dynamically downloadable "viruses" can also be malevolent as well. And unlike downloadable Java applets which are restricted to Java and to the security model that it uses, the same is not true of ActiveX controls. An ActiveX virus can be written in Java, C, C++ and Visual Basic, and the Microsoft security model does not limit its access to critical systems components. Writing an ActiveX control that would, say, reformat a hard disk is something that requires only a few hours work.

A Change in Time Scale

Traditionally, computer viruses have spread in a manner remarkably similar to biological viruses, from person to person, via diskette or email; or within groups in close contact as part of a business or industry. The emergence of dynamically downloaded applets has changed the dynamics of the problem. What has changed is the time scale. In the past the speed with which computer viruses spread was dependent on the ability of one person to interact with another. So, once a virus has been introduced, it took days, weeks and months for it to proliferate to the degree that it might turn into a problem sufficiently serious to affect a large number of users or an entire system.

Dynamically downloaded ActiveX and Java applets change the nature of viruses significantly. Other than at the initial introduction into the network, they are no longer under human control and no longer depend on human actions to propagate. If the evolution of Java and ActiveX as a means for downloading executable programs from companies to customers continues, soon software agents may routinely be sent automatically by a server to perform specific actions automatically.

No longer dependent on humans or their time scale and mechanisms of interaction, malicious applets and destructive viruses will now be free to spread at rates that are orders of magnitude faster than they are now. Given that the typical Web surfer uses his or her browser to access anywhere from 4 to 20 Web sites per hour, a virus could be downloaded from one infected site to a client, do its destruction or plant a copy, mutate and attach itself to a browser mechanism, and move to another server and another group of users within hours or minutes.

So far, such problems from destructive applets have been few and far between for two reasons. First, Java is relatively new and people are still learning what it can do, so it takes a relatively sophisticated programmer to create such destructive programs. Second, the Java community has the experience of virus hunters in the more general environment on how to deal with the types of viruses that have been created. In the last two to three years that Java has been available, many of the possible holes in the Java security mechanism that would let these kinds of destructive agents in have been plugged up. But crackers are tirelessly inventive and a new variation could occur at any time. Given the rate at which a Java or ActiveX applet propagates throughout the network, it only takes one rare success for a virus to cause severe problems on the Internet.

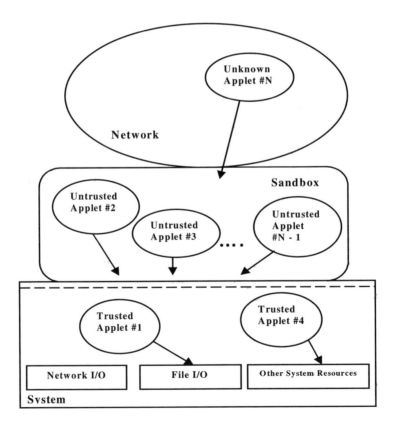

Figure 11.1 **Java's Sandbox.** To protect system software and applications, downloaded applets are confined to a sandbox where they can be monitored and are allowed access once it is determined that they are well behaved.

Java versus ActiveX Security

While many of these problems appear on their way to solutions, it is important for the system designer to remember that the new net-centric technologies are scarcely out of their embryonic stage, and all babies make mistakes, stumble and fall numerous times before securely standing alone.

In the area of security, the systems designer building an Internet access device and the infrastructure surrounding it has a choice between three basic paradigms. One was proposed by Sun for use with Java applets and quickly adopted by Netscape, and another is used by Microsoft for both its ActiveX and Java components. A third approach, developed by IBM, uses a human immune system mode, and takes a network-wide preemptive "immunization" approach.

The Netscape/Sun model uses what developers call a "sandbox" paradigm (see Figure 11.1). In this approach, component code for an application is safeguarded by the operating system in a safe location where it can be easily watched. Java applets, good or bad, are confined there and not allowed access to many system services until it is determined that it is not a rogue application which might surreptitiously reformat a hard drive or attach

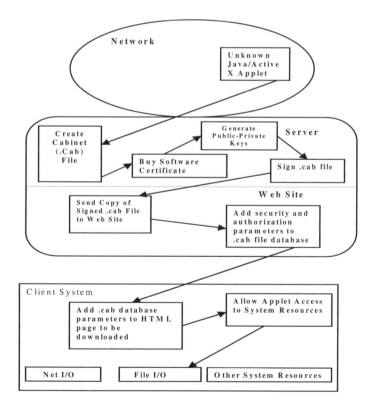

Figure 11.2 **Microsoft's Traffic Cop**. In this scheme, downloaded applets are allowed complete access to a system's resources if they satisfy the Authenticode requirements with the appropriately approved digital signature.

viruses to executable files found there. While this insures that only wel- behaved code can gain access to the client system's software, it also restricts the components from doing many services they might otherwise perform.

With its ActiveX technology, Microsoft - predictably - thinks it has a better idea: the "traffic cop" mode (see Figure 11.2). It takes advantage of the fact that an ActiveX component is just another DLL that has full access to the operating system services. In this model, a component is allowed to operate anywhere, as long as it has the right certification and then only under close supervision by the system. If the component is a dangerous applet or a virus, its activity is reported back to the operating system.

Java Security

The problem with Java is that ordinary applets are themselves "viruses" in the traditional sense of the word: independent, self-executing programs that use communications links and protocols to enter a system. Rather than wreak havoc, Java applets are designed to execute specific programs on the client, either at the request of the user, or under the direction of the Webmaster or Web page designer whose pages are being downloaded.

The sandbox security model that Netscape-Sun provides has both strengths and weak-nesses. This model provides the most flexibility because it allows any applet into the sandbox. But the price is performance, because of the additional overhead of software to maintain the sandbox and limits on the activity of the applets to a specific set of programs and files and no others.

Java, as a language, provides the most secure development environment of all of the object-oriented languages. In addition to such features as strong typing and multiple threads, it requires object-oriented classes that can only be accessed through object handles. Rather than actual pointers as in C++, which can result in erroneous or non-existent pointers, only symbolic pointers are used. By prohibiting pointer arithmetic, Java reduces substantially such security breaches as object spoofing, dangling references, renegade pointers, and memory corruption. The result of this is that systems are not as vulnerable to poorly written code or code that attempts to access local resources through operating system vulnerabilities.

Java's provision for automatic garbage collection, which dynamically frees up memory no longer needed by the program, limits the ability of hackers to enter backdoors in the application, making the data all the more secure. Because it is strongly typed, Java can't access the values of uninitialized local variables, which means that all variables must be declared, preventing their types from being changed or recast to another type, a common way for hackers to Trojan-horse their way into a system. In addition to the built-in secu-rity features of Java itself, the Netscape-Sun model incorporates three additional lines of defense: the byte code verifier, the class loader and the security manager. If the model is to work, each of these elements must do its job properly.

In circumstances where there is a suspicion that the Java run-time compiler has been spoofed and is not trustworthy, or where it is suspected that the applet is not following the Java language rules, the byte-code verifier is used to put each applet class through a series of tests. First, the byte-code verifier looks for bad applet behavior, which is usually exhibited as object encapsulation rule violations, pointer forging and stack overflows. The verifier then proceeds through a battery of tests. First, it determines whether or not class files are in the correct format. Next, it makes sure that the byte codes adhere to structural limits. Third, it assures that parameters, arguments and return types are correctly used. Fourth, it determines whether stack overflow and underflow are occurring. Fifth, it makes sure that the various access rules - public, private, and protected - are followed. Finally, it checks to see if register accesses and stores are valid.

The second line of defense is the class loader, whose function is to divide and isolate Java classes into small, secure, well-mannered, and well-guarded groups. It does this by separating classes. Applets from trusted realms, such as the local desktop or server, are not allowed any interaction with those that come from untrusted realms, such as the public Internet. Also, to prevent untrusted classes or applets from spoofing trusted ones, the class loader does not allow classes to call methods across realms. Even applets in the same realm are not allowed to communicate directly with each other, but must instead go through a public interface. Talk about tough! Finally, Java employs the Security Manager to act as a jailer, controlling the methods that are to be called and restricting methods that attempt file or network I/O access.

Despite these inherent safeguards, Java is a relatively new language and still rife with security leaks and holes. As of mid-1997, eight to ten security problems had been found in various implementations that would have allowed hackers to breach the security of a system. Many of these have been covered in detail in Java Security by Gary McGraw and Edward Felton (John Wiley & Sons, New York, 1996).

While the specific holes have been fixed, it is likely that in the future breaches of Java's security will use similar techniques, especially as programmers and Java hackers (Jackers?) get more experience with the language.

ActiveX

In Microsoft's traffic-cop model, there are apparently no constraints on the applications that run, but there is still significant overhead in the form of a variety of security tools.

At the center of the Microsoft model is the Authenticode technology, which implements code signing, providing digital signatures that allow a user to positively identify himself on the net as a user, a software developer, a Web site, or a Web server. Authenticode ensures accountability and authenticity for software components being used or sent across the Internet, verifying that the software has not been tampered with and identifying the developer of the software. Also available with the ActiveX software developer's kit are a number of new tools and APIs. CryptoAPI, for example, is an extensible architecture that allows developers to build applications that take advantage of system-level certificate management and cryptography. The WinVerityTrust API provides a set of services that enable applications to sign and verify software components using digital signatures. The Certificate Server is a WindowsNT-based server for managing the issuance, revocation and renewal of digital signatures.

A number of other schemes are being investigated by Microsoft. One is HTML signing, which requires a specific Web page to be open before an ActiveX component is run. Another is a set of proxy server/ firewall improvements that contain databases of names and characteristics of unfriendly components. Also being investigated is the formation of an industry-wide clearing house for Java applets and ActiveX controls.

There are a number of problems with the Microsoft model, not the least of which that it is still up to the end user to leave the security features in place on the Internet Explorer browser. Although the security settings are set to ON as the default, the user can still turn them off. According to a number of industry observers, there are other problems because the model does not take into account some common features of viruses. For example, most viruses are "silent running" when they enter the system, remaining benign for some time after entry before activating. So what is to stop a malicious ActiveX component from installing a virus that activates hours later after the user leaves the Web, but before the next virus scan? There also does not seem to be any safeguard against another common feature of some viruses: the ability to spawn a new process that wipes out all evidence of the original one.

The IBM Model

The actions of computer and network viruses, in both their impact and method of solution, seem to have analogies to the way things are done in the human body's immune system. Taking this analogy to its logical extreme are researchers at IBM's Thomas J Watson Research Center. Their proposed solution has two parts. First, they have developed a self-learning neural network algorithm to recognize viruses by scanning for tiny segments of a program, usually no more than 3 to 5 bytes in length. Taking what it has learned, this algorithm then turns around and analyzes a suspicious file or program by looking for occurrences of these segments. This is done on the theory that while a software file or program might have one or two of these fragments naturally, the occurrence of numerous instances one type or one each of several different types is indicative of a virus. This part of the solution is already available commercially as part of IBM's AntiVirus software.

As of early 1998, the second and more far-reaching portion of the solution was still in the research prototype stage: a network-wide immune system capable of fighting computer viral infections within minutes of their detection (see Figure 11.3). Each computer on the network uses the antiviral algorithm as well as a set of heuristically-derived characteristics of system and program behavior. Any changes in the system that do not match the predetermined set of characteristics or detection of a program or file that contains one or

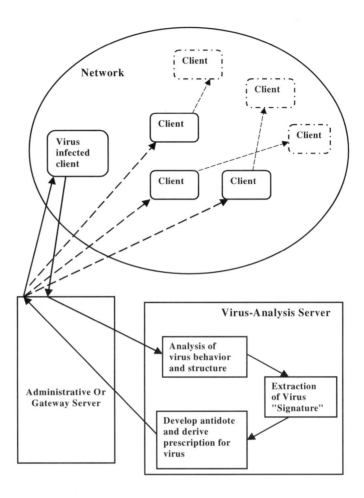

Figure 11.3 **IBM's Virus Killer.** To match the speed with which computer viruses could spread through the Internet IBM has developed a network-wide immune system capable of detecting and neutralizing infections within minutes of detection.

more of these suspicious segments triggers a monitoring program that immediately makes a copy and sends it over the network to a special intermediary server.

Once the special intermediary server receives the suspicious program or segment, it immediately forwards it to another server that provides an analysis and test environment. In the scheme that the IBM researchers have developed, they take advantage of one characteristic that all viruses have: to replicate successfully they typically attach themselves to programs that are used often and are ubiquitous, existing on numerous computers.

Essentially, what is done is that specially-designed decoy programs are released on the test server that mimic a potentially likely carrier by executing, writing to, and copying often. Once a virus is lured into attaching itself to the decoys, other programs in the test server extract viral signatures, analyze them and develop an antidote. The analysis-test server then sends this information and the antidote back to the requesting intermediary server or to the requesting computer, where it is then stored in a permanent database of cures for viruses. Each infected computer is then directed to locate and remove all of the instances of the virus. Using the same mechanisms by which viruses replicate themselves quickly and proliferate themselves throughout a network, the antidote is quickly spread to all other systems in the network.

There Is No Universal Solution—Yet

It should be increasingly apparent to the engineer designing a system for use on the Internet that, unlike the homogenized world of the desktop PC, the security model employed in the heterogeneous world of net-centric computing depends very much on the specifics of the design.

Both the Sun-Netscape and Microsoft models assume clients with reasonable system resources. The Microsoft model is based on the assumption that the client is either a PC or something very much like it: a NetPC or some other variation, with a large operating system, large programs, a sophisticated file system, disk drives, and I/O peripherals. It also assumes a relatively sophisticated operating system, as well as dynamic memory requirements in the 16- to 32-MB range, far beyond that available for most net-centric computers such as NCs or information appliances.

In some cases, the choice of security scheme will be a no-brainer, dictated by hardware choices. The Microsoft security scheme is an integral part of its browser and in future versions of its operating system. If a system designer chooses to base his or her Internet appliance on one of the X86 CPUs or to use Microsoft's WinCE OS on other architectures, it makes sense to use the Authenticode methodology, if system constraints allow it. If the choice of system CPU is one of the other CPU architectures or one not supported by Win CE, then the use of the Sun-Netscape model makes the most sense.

On purely technical terms, the system designer needs to determine which methodology best suits the environment and the end user. While Microsoft's traffic-cop model is a relatively new implementation as far as the desktop world is concerned, it is based on principles that have been used widely through out the workstation and server market. However, this model, where the suspected malicious or dangerous applet is tracked and its source identified, but not prevented from executing, has a serious flaw. It assumes a relatively sophisticated user who has come to expect occasional downtime, disrupted operation and corrupted files. It also reflects the desires of the IS manager who not only wants to prevent breaches of the system by hackers, but also wants to know the source so that who ever is responsible is punished for violating the system.

However, in the new world of net-centric computing where information appliances, Web-enabled set-top boxes, smart phones and Internet access terminals are mass consumer products, one has got to assume a user whose view of reliability is black and white—it works or it doesn't. He does not care who or what the source of the problem was. All he cares about is that it does not occur again. In this kind of environment, it may be that the "prison guard" model used by Netscape-Sun is the most appropriate.

No matter which approach becomes popular, or is proven to be the most effective, the IBM network-wide instant-response approach is a necessary, even vital , addition. For if the key to the success of the Internet and World Wide Web is reliability, finding a way to deal with the problem of viruses quickly before they proliferate is a requirement that should be non-negotiable.

CORBA as an Antidote Broker

One possible way to proliferate the IBM virus hunting and killing scheme throughout the Internet may be the Object Management Group and its 700 to 1,000 member companies, using CORBA as the mechanism of dispersal. Unlike Java or DCOM-Active X, which evolved in response to market demands and to which modifications and enhancements are still being made to improve security, the Common Object Request Broker Architecture is a well defined standard for interoperability amongst systems on a network. It would be logical to assume that it has some well-thought-out solutions to the same problems. It does, but CORBA itself has a number of security issues that complicate matters. The CORBA governing body, the Object Management Group, is addressing these issues and as of this writing, is considering a number of proposals. And the IBM scheme has been added to that priority list by a number of companies participating.

In some ways the security problems facing distributed net-centric computing systems depending on the CORBA model are even more complex. This is because distributed objects face all of the security problems of present client-server and time-shared systems as well as a few that are distinctly their own.

In traditional client-server systems, a systems designer cannot assume that any of the client operating systems on the network will protect the server resources from unauthorized access. Nor is it realistic to assume that once the client machines and the servers are trustworthy the network will be also. This is because the network itself can be highly accessible and open to break-ins. So systems designers need to keep in mind the unique aspects of a system based on distributed objects and the ways in which security can be broken.

Let us count the ways. First, unlike traditional client-server systems, where architectures contain "class"-based servers that are only servers and clients that are only clients, the distributed object environment is much more democratic: an object can be both client and server. This is important from a security point of view because the general assumption is that servers can be trusted and clients can not. So in a distributed object based net-centric environment where it is not always possible to tell who is a client and who is a server, the big question is: "who can you trust?"

Second, distributed objects are dynamic; that is, they get created and are then destroyed once they are not being used. While this is great as far as flexibility is concerned, it is a security headache of major proportions. Related to this is the fact that distributed objects are in a constant state of evolution, and you seldom can see beyond the first couple of layers to determine how it has changed. One time a remote user accessing a server might see a particular configuration of objects and the subclass of objects that make up that "shell" object. Another time, the shell might be the same and the subobjects might be as well, but the relationships may have changed. Finally, because of the way objects can be encapsulated and contained in other objects, there is no way to determine if the contained objects or their relationships have changed.

As of this writing, the OMG was considering a number of alternatives to solve the security issues associated with distributed objects. Generally, these proposals fall into three areas - audit services, authentication, and authorization – all of which are used to some degree or other in present systems.

In a system with audit services, every discrete event is logged and noted as to who or what initiated the event, where it came from, and where it is going Also audited are attempted log ons, and which servers, clients or objects are being used. Just knowing audit services exist in a system are often enough to discourage users from tampering with servers using their own log on. But tampering using someone else's log on is another issue. Authentication is the CORBA mechanism for preventing this. But in a distributed object environment it requires much more than simply using passwords which has been the time-honored approach used first in early time-shared systems.

What is being proposed to OMG is to encrypt the password as well, a technique already used in some proprietary systems and within the U.S. government. But this requires additional levels of infrastructure. Specifically, it needs trusted third-party authenticators whose responsibility it will be to manage the secret keys and to which objects must prove that they are who they claim to be. Once identified and assigned an authenticated ID a client can access any server object from anywhere. The interesting thing about this approach to authentication is that while the password may still be stored on the client system, where it is available to hackers to find, the authentication ID is not. It changes each time the user logs on and proves to the system who he is, and is not available to anyone else who wants to surreptitiously access the system. Finally, there is authorization, which complements authentication. Here, special server objects are responsible for certifying which operations the clients are allowed to perform. This is done by creating access control lists for each computer resource, either hardware or software. These lists contain names of users permitted to access the system and the type of operations that they are allowed to perform and on what resources.

Convergence

While both Microsoft and Sun-Netscape are still stubbornly protecting their fiefdoms as far as object models are concerned, there is some convergence in security models. The latest releases of the Java SDK, for example, incorporated a signed applet scheme that uses something like the Microsoft-ActiveX authentication model. As with the Microsoft model, it will let users employ these signatures to determine whether an applet from a particular source should be trusted with additional file and network access privileges.

On its side, Microsoft has incorporated features into Authenticode that allow it to be used to sign Java applets. But it requires an either-or decision. Right now, it is not possible to make use of the advantages of both security models. If the user decides to sign an applet with Authenticode, it can execute within Microsoft's IE browser without being restricted by the Java sandbox. It can perform all the same functions as a traditional ActiveX control, such as write to hard drive or open a communications session.

The cybernetic immune-system approach that IBM has developed offers considerable promise, since it theoretically could immunize all computers on an intranet or the Internet very quickly against a new emerging virus. In the prototype, it takes the virus test-analysis server only 5 minutes or less to identify an intruder and produce an antidote. While the IBM solution uses its own virus-hunting algorithm, the immune-system approach seems to be algorithm independent and could be adapted for use with any or all of the various virus algorithm schemes now on the market or with any of the security models proposed by Microsoft, Sun, Netscape and the OMG..

Given new methods of network-mediated transport that are largely out of the control of humans and allow the transmission of destructive code in hours and minutes, it is critical that some mechanism be developed for quickly identifying the infection and coming up with a method of immunization.. The trick will be to develop a mechanism to proliferate the cybernetic immunization scheme through out the Internet. Conceivably, it could evolve the way immunizations now occur and how diseases are slowly stamped out. In this scenario, duplicates of the IBM prototype could be copied and used to clean up the thousands of intranets and the larger information services such as America OnLine and the Microsoft Network. Eventually, over time, the entire World Wide Web could be inoculated. Alternatively, the same mechanism could be proliferated by installing it on all the various Internet Service Providers, using the existing access servers as the intermediaries, but linked to other test-analysis servers specially designed for virus detection and immunization.

Whatever the approach, it will require industry-wide and Internet-wide cooperation to adopt such a scheme. In the past, Internet providers have been very good about coming up with standards of various sorts. The big questions are (1) does IBM still have sufficient marketing clout to make it happen? (2) if not, can Microsoft, Sun and Microsoft be convinced to cooperate, or will each go their own ways with their own solutions? And, (3) if the previous two alternatives are not viable, can the combined clout of OMG's 700 to 1,000 member companies make it happen?

Chapter 12

Building Better I/O

As any regular user of the Internet and World Wide Web is aware, often connections are not made, are cut off, or proceed at a much slower rate than is normally possible with a modem or network connection. While the reasons for such communications brown-outs, gray-outs, and blackouts may sometimes be due to an inherent hardware or software failure, more often than not there is simply a traffic jam on the "information superhighway."

This is because the superhighway is not so super, not yet, anyway. There is a lot of talk about and some work on upgrading these info-highways. One idea is to simply increase the rate at which data can move between servers, routers, switches, desktop systems, network computer, set-top boxes, and Internet appliances using technologies such as Asynchronous Transfer Mode and Gigabit/sec Ethernet. The real-world analogy to this is moving the speed limit higher or widening an existing two or three lanes to five or six or more along some of the more heavily used thoroughfares.

However, it will be a while before ATM and the Gigabit/second networks become common. Even when they do, it is very likely that bottlenecks will still occur. This is because the traffic will increase and the "vehicles" will get larger. Instead of text files, graphics files and mixtures of both, future Web sites will be multimedia-rich with lots of audio and video. The present relatively modest, mostly text email messages will be replaced or supplemented by video emails; multimedia narrow- and broad-casts using a variety of push technologies; and video teleconferencing. Both in terms of numbers and in the way the net-centric paradigm has been conceived, Internet appliances and network computers will place an ever-increasing load on these servers. First, there are the numbers. One measure of the pervasiveness of the Internet-connected computer is in the number of computer users who are now and will be connected to the WWW and Internet.

According to one of the many research firms analyzing present and future Internet usage and its impact on the PC - International Data Corporation - the number of PCs capable of Internet access is growing worldwide from nearly 60 million in 1996 to 265 million in 2000. Nonetheless, IDC projects that shipments of non-PC Internet appliances will undergo explosive growth over the next five years, with a 1996 through 2000 compound average growth rate (CAGR) of 151.6% in the United States and 176.4% worldwide. By the year 2000, non-PC Internet appliances will make up 20% of the total U.S. net-centric computing market and 21.7% of the total market worldwide.

The hidden "gotcha" that has not been raised in many discussions of network-connected computing devices is the load that this new paradigm places on the workhorse servers out there on the intranets and the Internet. By some estimates, considerable memory and processing power will have to be set aside on the server to handle remote users. Right now, an application on a local PC requires anywhere from 2 to 4 Mbytes per session for most desktop applications to as much as 16 Mbytes per user for heavy-duty client-server or database applications. Now, multiply this by the number of users one can expect in this new computing environment, which can range from a normal Internet installation with 20 to 100 simultaneous users to more widespread Internet applications where anywhere from 100 to 10,000 simultaneous users can access a site. This simple calculation well illustrates the kind of input-output (I/O) loads servers, routers, and switches on the network back-bone will have to absorb if new approaches are not devised.

The situation is not unlike that facing many a suburban or urban community that got along with two-lane dirt roads but then increased those to three- and four-lane asphalted roads or freeways. With the influx of tourists and new residents due to the improved roads, more cars and bigger cars were the result, and congestion remained. A few cities or megacities faced with the same problem have found that building more lanes and widen-ing the highways only exacerbates the problem. In the physical world, traffic planners have found that the best way to boost the efficiency of the existing infrastructure is to improve the capacity of the on-ramps and off-ramps, enhance the capabilities of the traffic lights at intersections, and build more sophisticated interchanges by which traffic can move from one destination to another smoothly and quickly.

Engineers and systems designers planning systems for use on the information super-highway are going to have to think about improved and more secure performance in the Internet in similar terms. They will not be able to depend on the fact that using techniques such as ATM and 1 gigabit/sec Ethernet in the long run will be enough.

Improving the Traffic Lights

According to pessimists, the state of health of the Internet and World Wide Web as an information superhighway is reaching a critical stage, bordering on total breakdown. In the eyes of optimists, the health of the Internet, while approaching serious to critical, is not irreversible and is amenable to a number of fixes that will bring it back to an acceptable level of performance.

Both are looking at the same sets of numbers, but making different interpretations. Despite their disagreements, both agree that the Internet has grown faster than anyone predicted a few years ago. To improve the existing not-so-super information highway system, a number of changes have taken place or are in the process of being implemented to improve the Internet protocols and the way they control data moving across the matrix.

One basic problem with the Internet is the underlying protocol, TCP/IP, and the delay time incurred when it is used to make a connection. In the familiar public telephone switching network environment, the average user has gotten used to virtually instantaneous response. In fact, the systems that manage the connections have been designed to keep latency times to no more than 10 to 20 ms. On the Internet, the strategy has been exactly the opposite. The TCP portion of TCP/IP sets a limit on the amount of data that the user can send to the network and introduces delay as well. This is done to reduce congestion, much in the same way traffic and highway planners put stoplights on freeway on-ramps to control traffic flow and congestion. This transmission delay is usually on the order of several hundred milliseconds, but during high traffic periods can approach a second or more.

To improve the real-time response of the Internet without making it more congested, Internet designers and organizations have proposed and implemented a number of new standards and taken a second look at some old ones: UDP, RSVP, ST2, T/TCP, and RTP. They are also already in the process of rethinking the mechanisms by which they route information around this heterogeneous mix of interconnected networks we call the Internet.

One significant modification recently approved by the IETF (Internet Engineering Task Force), and adopted by most major router vendors is the Resource Reservation Protocol, or RSVP. A system set up under RSVP is somewhat like setting aside special lanes during the rush hour for bus traffic or autos with more than one person inside. Under RSVP, a sending system, such as a desktop, a network computer, or a server, can request a specific quality of service by sending reservation messages to the host server or desktop from which data are being sent or requested. RSVP then reserves the appropriate network resources. When a desktop or network computer requests a specific quality of service for the data stream being sent, the host system that is to transmit or retransmit the data uses RSVP to deliver this request to each router on the network path to be used. RSVP also allows the transmitting host to maintain control of the routers along the path providing the service.

Similar to RSVP is the IETF's Internet Stream Protocol (ST2), an experimental resource reservation protocol which is currently in its second revision as of late 1997. Like RSVP, it also allows the host to set aside specific routes through the Internet. But, in addition, it allows scheduling of enough bandwidth for a given use. ST2 does this by creating a point-to-point or point-to- multipoint stream for data packets as part of the transmission setup phase, during which routers are selected and resources on the Internet are reserved.

Rather than improve service for all messages, as RSVP and ST2 do, other approaches, such as the IETF's transactional TCP, look to enhance the throughput of only certain types of Internet traffic. It does this by using one packet to carry both data and control information, thus reducing the overhead latency involved in setting up a session between a server and a client system.Still another alternative is the IETF's Real Time Protocol (RTP), which is designed to work in conjunction with the tried and true UDP. It enhances UDP and boosts real time performance by using the well-known technique of forward error correction to reconstruct a transmission in case a small amount of data is dropped. Particularly useful for any kind of multimedia transmission (audio, video, video conferencing) RTP provides flow control and synchronizes the timing of transmissions between the sending and receiving systems.

Finally, the major Internet router companies such as Bay Networks, 3Com, and Ipsilon Networks, are rethinking the way they design their systems. Traditionally, routers on the Internet guarantee data reliability by reading each data packet and deciding each time the route that each should take through the network. While ensuring reliability, such schemes make data transmission slow. On the public telephone networks, switching systems are used in which only the first few packets in a data stream are looked at. Based on this the switcher then determines a single path by which all the packets should go. While this speeds up data transmission, it makes for a less robust environment. For most voice conversations, in which analog signals are converted to digital data and back again, enough redundancy is built in that this is not noticeable. But for data sent over the public switching networks there is a marked reduction in reliability.

Boosting Server I/O Efficiency

Where the previous techniques deal with the issue of traffic congestion by dealing with the way the flow is controlled, the alternative method is to control and improve the efficiency of the on and off ramps at sources and destinations on the information super-highway. This can be done by addressing and improving the I/O capabilities of the servers that feed data to and accept data from the network. While there has been an enormous increase in the number of servers on the Internet, the majority of the newest ones are derived from 32-bit designs built around Intel's 32-bit Pentium Pro architecture, which is designed to support up to multiple Pentium Pros in a symmetric multiprocessing configuration. While capable theoretically of supporting up to 16 Pentium Pros in a symmetric multiprocessing cluster this is only of marginal aid in helping such systems in handling the I/O processing chores imposed by the Internet.

Server performance can be measured in many ways, including host CPU clock speeds and MIPS, but the bottom line is that applications must run as fast as possible and move data into and out of the system quickly. The problem is that I/O processing steals cycles from the host processor and reduces the amount of application-processing bandwidth available to servers. If the I/O requirements are not high, the impact is minimal. However, servers demand high-speed I/O, and this results in an unbalanced system, with system performance suffering.

A variety of software modifications to the underlying OS and applications software can increase I/O performance by 25% to 75%. Substantial improvement in server and desktop performance can be achieved by using a technique called logical volume management, widely employed on mainframe systems. Logical volume management allows physical drives to be accessed as a virtual device, enabling physical storage segments to appear as a single address space. The capability provides for increased system performance, by allowing a system to be tuned for I/O performance throughput.

There are a number of ways to measure performance. One measure is the number of instructions per second that a CPU can execute. Another is the performance of a system running some known workload or benchmark. A third measure is the number of users that a system can support with an acceptable response time. In the world of net-centric computing the best and most critical benchmarks will be those that effectively measure the I/O service rate. The more expensive architectures typically support greater I/O parallelism, resulting in more I/O operations per second. Often the only performance difference between a more expensive and a less expensive system is the I/O service rate.

Software Aids to Better I/O

Before we discuss some of the techniques employed in increasing I/O performance, it is important to understand some of the basic properties of I/O, mass storage and controllers as they relate to moving data into and out of a system, and on and off the disk drives.

To start with, disks operate on requests serially. That is, a disk completes servicing the first request before it begins servicing the second request, Therefore, a single disk does not provide any opportunity for parallelism, This does not mean that you cannot improve the I/O performance of a single disk system, but your opportunities for improving performance are greatly limited when compared to that offered by multiple-disk systems. As a general rule, you will achieve greater performance when configuring a system with two disks of size 1/2 N than a single disk of size N. While two smaller disks together will almost always cost slightly more than a single larger disk, the difference in price is small when compared to the frequently dramatic improvement in system performance. If spending a few hundred dollars more by using multiple, smaller drives significantly increases the performance of a $10,000 system, it should be considered a bargain. Parallelism is the reason why multiple disks improve performance: two disks can typically service two requests about as fast as a single disk can service one. But why stop at two disks? Three disks will generally offer better performance than two, and so on. As a general rule, the more disks that you can configure, the better will be the performance, especially in I/O constrained systems.

No matter how many drives or controllers there are in a server, however, little or no performance benefit will result if the system is only accessing data on a single drive. The key to exploiting the potential parallelism on a system is effective data placement. For example, in a multiple drive system, suppose that the user's file systems and databases are divided equally amongst the drives. Also suppose that the data accesses are distributed more or less equally across all of the user data. If that is the case, the result is a well-balanced high performance configuration that exploits the parallelism inherent in the system's architecture. But in the real world this is almost never the case. It is rare that all of the file systems and databases in a system receive an equal amount of access. Typically, most of the disk access will go to a single file system or database. What comes into play is still another expression of the 80/20 rule familiar to most engineers. In this case, 80% of the accesses will be performed on 20 % of the data. If the 20% of the data that is receiving the bulk of the disk accesses is located on a single drive then the I/O service rate of that drive will likely be limiting the I/O service rate of the entire system. The other disk drives will be idle most of the time and will not be contributing to the effective I/O service rate of the system. Another factor to be considered is that even if over the course of time all the file systems or databases are used an equal amount, when viewed through a short time window, I/O operations typically tend to be clustered in fairly small areas, such as a small set of files or a portion of a database. Over time, this area of heavy access may move between files or regions of a database, creating the illusion that the I/O load is well distributed. The result of this clustering is that at any given moment most of the I/O is typically being serviced by a single disk, leaving the other drives largely idle.

So how does one go about placing and moving data on drives in units smaller than a single file system or database? And doing so in such a way as to achieve the most effective data placement? It is done by using logical volume management, a technique for file management that applies the principles of virtual memory to disk space. In a system with virtual disk management, databases and file systems are created on virtual rather than physical disks. The virtual disks are mapped by the logical volume management layer of the operating system to physical disks. While a virtual disk appears to have a logically contiguous address space, it may actually be made up of physically discontinuous storage segments from multiple disk drives. Logical volume management is the mechanism by which parallelism can be applied to optimal distribution of storage across multiple disk drives. Some logical volume manager products support the movement of storage segments between disk drives even while the system is in use. During operation, it is possible to identify data that are causing a bottleneck, split it in into multiple segments, and move each of the segments to another disk, achieving much better system throughput without disrupting the operation of the system.

A number of techniques for software-based I/O improvement have been developed which can be enormously useful in improving the performance of servers on the Internet. If one or more file systems or databases on a single disk are receiving the bulk of the accesses, performance can almost always be improved by moving a portion of the data to another drive. Using logical volume management, this can be accomplished without any disruption to system availability. A good understanding of 'hot spots" and knowing which data regions require movement can be accomplished with any number of volume manager products that provide an I/O analysis capability. This feature provides a clear indication of current disk performance with respect to the levels of read and write activity to regions of each disk. For example, areas of the disk receiving heavy access are identified as high activity regions which may require movement to another device.

If a disk were receiving a disproportionate share of the I/O requests, volume management allows for offloading one or more hot spots to another drive. The ability to move only a portion of a file system or database to another drive is an important feature of volume management. It allows the system designer to avoid the situation where moving an entire file system between disks would result in merely moving the access bottleneck from one disk to the other. As mentioned above, the placement of data can have a tremendous impact on the effective parallelism of a system. If there are two regions of simultaneous access on the same disk then it is likely that the average access time will be above average due to frequent disk head movement between the two regions. A basic performance optimization strategy is to attempt to limit each drive to a single region of heavy access, thus significantly shortening the average access time for that drive. Using volume management, this can be achieved by either moving one region of heavy access so that it is adjacent to the other region on the same disk or by moving one of the regions to another disk. Of course, involving a second previously idle disk will provide the additional benefit of spreading the I/O across more disks, thus further exploiting the potential for I/O parallelism.

Mirroring and data stripping are other techniques engineers have used to increase the I/O handling capabilities of their net-centric systems. In mirroring, a virtual disk is made up of one or more storage segments that have been combined. Each virtual disk, or logical volume, has at least one copy of all the data in the virtual disk. If more than one copy is attached to a virtual disk then that virtual disk is said to be mirrored. Each copy of a mirrored virtual disk contains identical data. Normally used to provide protection against physical media failures, it can also increase read performance because . read accesses to a volume can be spread across multiple copies and hence multiple disks. With data striping, the I/O load is distributed across multiple pieces of a virtual disk located on separate physical disks and the virtual disk address interleaved across each of the storage segments that make up the virtual disk, reducing both read and write time. .

Mirroring and striping are widely used on servers with Redundant Array of Independent Disks (RAID). Striping is known as RAID-0, while mirroring is known as RAID-1. RAID-3, RAID-4, and RAID-5 describe extensions to striping, where parity data is maintained for each stripe. The purpose of these parity data is to allow for the regeneration of the data on a failed disk following a drive failure. The benefit of this approach is that less redundant disk space is required to achieve improved data availability than in the case of mirroring. RAID-3 defines an array where the data for each I/O is spread across each of the drives in the stripe. This technique can enhance read/write throughput for a single-user application, especially when the disks are attached to multiple controllers so that data transfer can be overlapped. However, as each disk is involved in servicing each and every I/O, it does not allow for parallel operations and will degrade the performance of a multiple-user environment. RAID-4 and RAID-5 support independent access to the data on each drive and are intended for a multiuser or multiserver server environment. RAID-4 differs from RAID-5 in that all the parity is placed on a single drive in RAID-4, while the parity is rotated across each associated drive in the stripe for RAID-5.

Off-Loading the I/O Processing

But system software fixes and clever use of RAID-based mass storage will not allow mainstream 32-bit servers to scale up to the 3x, 5x or 10x performance boosts in I/O capability dictated by the demands of the Internet. So a variety of other techniques have been employed, mostly making use of the symmetric multiprocessing capabilities of CPUs such as the PentiumPro. While servers can grow through symmetric multiprocessing

(SMP), but scalability is usually limited to a few processors before performance bottlenecks are encountered. Beyond the SMP limit, servers can be replicated, but this usually requires replicating the entire database. Total replication can be prohibitively expensive, because even the infrequently accessed data must be replicated on all systems.

The obvious conclusion is that the server architecture needs to allow for shared access from any processor to any disk. But a simple shared-disk architecture is still not ideal, because it forces all traffic to pass through main memory on the way to and from the disks and communications lines. In many cases, this necessity to pass through main memory could be avoided if there were a direct path allowing I/O-to-I/O transfers. Looking at current usage and the anticipated demand over the next few years, various server makers such as Sun, Silicon Graphics, Tandem, and Compaq, among others, have incorporated a variety of mechanisms to offload the main processor from the chore of handling the I/O transactions.

Sun, for example, in its UltraSPARC-based Ultraserver family has applied an array of hardware and software fixes to solve or alleviate the I/O latency problem. On the hardware side, in addition to 64-bit wide I/O channels and faster disk I/O, the servers have boosted memory and I/O bandwidth by two to three times to 600 Mbytes/sec. sustained and 1.6 Gbytes/sec., peak. This has been done by replacing the conventional three-tiered bus hierarchy with a three-port crossbar switch that acts as a bridge between the processor, memory and I/O buses. The crossbar switch's decoupled memory port allows data transfers to occur in parallel with other operations and permits burst transfers to operate on four bytes of data per slice. Also incorporated into the CPU hardware are features that accelerate certain operations in the operating system that have to deal with I/O. One is a block load-store function, which allows the operating system to directly write a page of memory without disturbing the cache memories. Instead of overwriting the cache to swap pages in and out, which reduces the system's ability to respond quickly to I/O operations, the OS leaves the cache untouched, executing 64-byte loads and storing directly into main memory.

Accelerating load and store in hardware benefits the system I/O by enhancing crucial network operations. Within network layers, data can be moved as much as seven times before it reaches its final destination. With hardware acceleration, block movement of data is much faster and clean. It also simplifies sharing applications and data among multiple users on a network, since copies of applications or data structures can be generated quickly, even while others are using them, which is useful in many workgroup and collaborative client-server applications. Another improvement is the use of split data and instruction buses: 144-bit-wide data and 48 bit wide instruction. With two buses, contentions between data and instructions are eliminated as are dead cycles for switching between the two. The chances of a system crash, locking up while accessing email on a network is greatly reduced. Further, because the buses are packet-switched, the processor can accept multiple requests in parallel.

Even more sophisticated in its approach to the I/O bandwidth and latency problem is Tandem Computers Inc. (Cupertino, California), now a subsidiary of Compaq, which has developed a system area network (SAN). In this approach, Tandem engineers use an array of low-cost, high-speed router chips (see Figure 12.1) to construct a point-to-point packet-switched interconnected mesh inside the server itself. Each SAN router has six bidirectional ports, so designers can arrange the server elements into any number of different connection topologies - meshes, trees and hypercubes, among others. The particular choice depends on the requirements of the server in a specific application. The routers rapidly switch data among the various I/O devices (see Figure 12.2) and compute elements: between CPUs, between the CPU and the I/O devices and, between I/O devices.

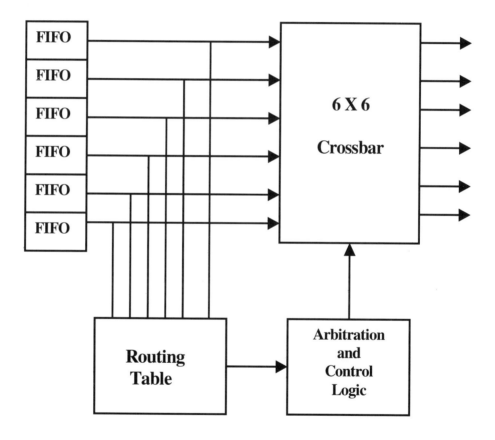

Figure 12.1 **System Area Network**. At the core of Tandem approach to boosting server I/O efficiency is a router circuit with six bidirectional ports that allows point-to-point connections using a number of different connection schemes.

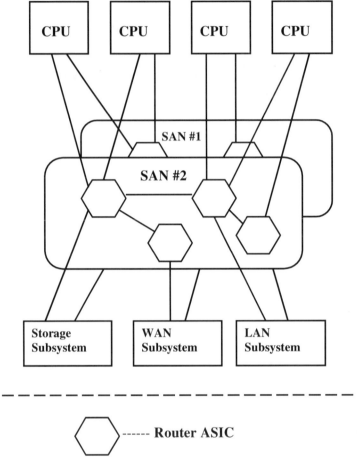

Figure 12.2 **Any-to-Any**. Using special router circuits, a system area network allows creation of any-to-any links by means of a number of different schemes: byte-, serial, point-to-point, and full duplex.

Internet servers in the future must be designed as clusters if they are to keep up with the rising demands for increased bandwidth and functionality. SAN technology can solve the key problems in building this type of cluster-scalability to many nodes, reliability, and efficient protocols to reduce CPU consumption. The growing awareness of the need for scalable servers has already prompted Compaq, Dell, NEC and others to design future server clusters around the Tandem's implementation of the SAN. A key feature of the SANs the ability to create direct "any-to-any" links by using byte-serial, point-to-point, full-duplex connections between node interfaces and the six-port crossbar router ASICs. The routers can be cascaded to build extremely large configurations, with hardware latencies of just 300 nsec. per router stage. The low latency is obtained by the use of a technique called wormhole routing, which begins to route a packet as soon as the destination address appears in the first few bytes of the packet header. The ability to perform both reads and writes allows a programming model similar to a standard I/O bus. For instance, the processor can read and write registers on an I/O controller or read and write remote memory locations without the need to execute CPU cycles on the remote node. The read capability is also used extensively in the message system to reduce the number of interrupts and context switches.

In the Tandem implementation of a SAN, called ServerNet, request packets include a 4-byte I/O address that is mapped into the physical address at each destination node. When a packet is sent to a CPU, this I/O address is looked up in an address validation and translation table, which translates it into the physical page number within that node. The address validation table also includes access rights to make sure that the requesting node was allowed to either read or write the requested range of addresses. The translation stage also improves performance by providing a hardware-implemented scatter-gather mechanism. Each packet includes a variable length data field, with a maximum per packet of 64 bytes in the current generation. If a network allows very long packets, the maximum latency in the network is also very long, because short packets can be stuck behind a large number of long packets.

Because all the data into and out of the server move directly among these elements and not over a central bus, the SAN approach is much more efficient. Data paths are shorter, and the routers can find alternative pathways, with the result that there is less chance of I/O bottlenecks during heavy loads. To accommodate I/O devices with different speeds and buffer sizes, the SAN is designed to pull (read) as well as push (write) data, allowing it to support many active I/O devices and channels simultaneously. Because the routers have multiple interconnections to the CPU(s) and memory, a fully connected ServerNet SAN with 4,680 routers can deliver a maximum switching bandwidth of 410 GB/sec.compared to about 1 to4 Gbytes/sec for present server and superserver configurations.

Adding I/O Processors

For the mid- to low-end of the server market that is emerging to service the needs of the Internet and the many Intranets using Pentiums-Pentium Pro hardware and Windows NT, such esoteric and sophisticated solutions as Tandem's SAN are not cost-effective. So there has long been a search for more scaleable and lower cost solutions. Initially it was thought that the symmetric multiprocessor capability of the Pentium/Pentium Pro CPUs would provide a mechanism for off-loading or at least reducing the amount of computer resources that the main processor had to devote to the I/O chores. The problem is that the Pentium/Pentium Pro CPUs are designed to work in only a symmetric multiprocessing environment that can handle anywhere from 4 to 16 clustered processors simultaneously. While it is an advantage in general-purpose computing because it allows computing chores to be shared, SMP does not allow a single processor in the array to do just I/O chores. Each CPU has equal-access I/O devices and main memory. All the CPUs cooperate in executing a signal stream of code, and all run the same operating system and application code concurrently.

Processors such as Intel's Pentium or Pentium Pro were designed with high performance processing in mind, not I/O, and have such features as 14 to16 stage pipelines, out-of-order execution, and transaction-oriented buses to improve performance. These fea-

tures of a server's host processor significantly improve application processing performance but require higher- bandwidth I/O to keep the high speed host processor fed with
data. New network and storage protocols, such as Fast Ethernet, ATM, and Fibre
Channel, can provide high-bandwidth connectivity. However, increasing the I/O bandwidth by an order of magnitude (such as the switch from 10/100 Ethernet to Gigabit
Ethernet or ATM) causes an order of magnitude increase in the number of low-level I/O
interrupts that are sent to the host. Interrupts drastically slow down the pipelined
processing which was designed to improve performance in the first place. To compound
this problem, the internal I/O architecture of the PC or server has not changed much, even
though technology advances have caused dramatic improvements in host processor
performance and network protocol bandwidth capabilities. I/O architectures are still
"flat" in that the host CPU must mediate the demands of all I/O subsystems and interface
cards. Heavily-loaded CPUs may still spend up to 30% of their cycles on overhead
functions such as peripheral status polling and data-block transfers.

The key to solving this problem is in its definition: application processing and I/O
processing have very different processing characteristics. The solution is to balance the
system by using different processors for each. Implement one or more high performance
Pentium or Pentium Pro processors for application processing and implement an intelligent I/O processor, which excels at interrupt processing and data movement, for the I/O
processing.

One highly effective way that an I/O processor can be used to maximize system efficiency
is to minimize the number of asynchronous interrupts that the host CPU must process.
Asynchronous interrupts reduce host efficiency, since servicing an interrupt forces the
CPU to stop what it's doing, save the complete machine state, and begin executing from
a different location. Upon completion, the service routine must reload the original state,
reconfirm process access rights, and resume execution wherever earlier processing left
off. Interrupt processing also reduces cache efficiency as service routine instructions
may flush instructions already present in the cache. An intelligent I/O processor can
ensure that an entire command sequence or network transfer protocol is completed with
no host intervention. But requiring the host CPU to process even one interrupt per
command sequence is still suboptimal I/O processors can further reduce the host interrupt load by building a list of command-completion messages within shared memory.
When new messages are placed into a formerly empty list, the I/O processor interrupts
the host CPU. As the host unloads command-completion messages from the head of the
list, the I/O processor can continue appending new completion messages to the tail. The
host can thus defray the overhead of a single interrupt across multiple completion messages. More importantly, this scheme seems to scale well. The more completion messages posted by the I/O processor to the host, the larger the number of transactions
processed by the host per interrupt.

In such an environment it is difficult to assign one or more processors the majority of the
I/O tasks since the operating systems in use, such as Windows NT, typically assign tasks
on a priority queuing basis. This makes it difficult to predict how many computing re-

sources will be devoted to the I/O. In a typical server that operates in both an intranet and Internet environment, there can be as many as three or four communications cards, and I/O processing can eat up as much as 25% to 35% of the main CPU's resources. This might be manageable if the I/O drain was predictable. But in today's complex networked environment, such predictability is a dream. If you set up the priority queue to handle I/O requests first, there are times when the main CPU is devoting a major share of its processing to that, reducing the response time or computer resources available to normal compute chores. If you design the priority scheme so that main processor chores are handled first, very little bandwidth is available for the I/O.

Building Asynchronous Server Clusters

The way that many server makers have chosen to solve the I/O problem is to modify the traditional asymmetric model for server clusters slightly, by adopting architectures that are slightly less symmetric or completely asymmetric. Whereas in a symmetric architecture all CPUs are equal and any processor is available for any task, in an asymmetric design each CPU is dedicated to a specific task. At one end of the spectrum are solutions such as the pioneering Flex/MP from Compaq Computer Corp.which was computationally symmetric, but allowed the processor that originally booted the system to be devoted to servicing I/O interrupts. Typical of solutions at the other end of the spectrum is the totally asymmetric approach taken by NetFrame Systems Inc. which uses a single dedicated Intel CPU, most recently a Pentium or Pentium Pro, to run Netware or LAN Manager from main memory. The main CPU bus, usually PCI or EISA, was connected via special bus controllers to two to four 4 I/O channels, each of which supported two dedicated peripheral processor boards. Each board was able to handle up to 12.5 MBytes/sec., for a total of 100 MBytes/sec. of total I/O bandwidth. Increasingly, however, server vendors are taking the middle ground, mixing symmetric processor clusters with dedicated I/O processors. One example of this is the Powerframe architecture from Tricord Systems, Inc. with a seven slot backplane that accepts up to four symmetric Pentium/PentiumPro boards and up to six asymmetric I/O boards.

Moving To a Common I/O Structure

The move toward the use of intelligent I/O solutions in servers has gained impetus with the development of a new proposed standard, called I_2O (or more colloquially I2O) for Intelligent I/O architecture, that would eliminate the need to write unique code for each combination of OS, system, and peripheral board hardware. The cost of such code development has been one of the chief reasons that intelligent I/O in general has not been adopted widely in the low- to mid-range server market. This proposed model splits the driver architecture between OS specific and hardware dependent modules (see Figure 12.3), which communicate with one another via a hardware and software independent message passing protocol. Among the companies that have been active in the development of the I2O specification are software vendors such as Microsoft, Novell, Integrated

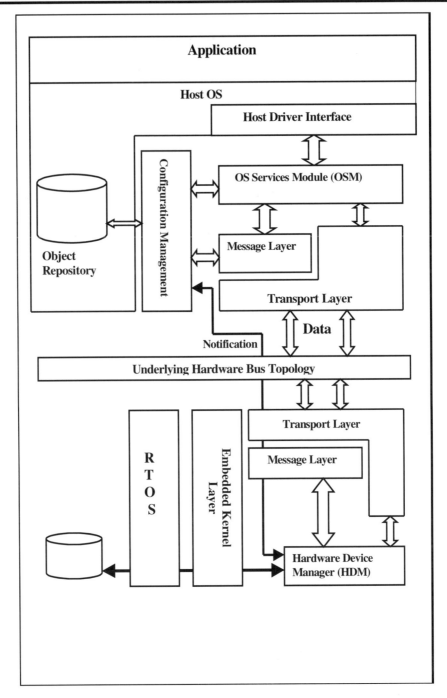

Figure 12.3 **Intelligent I/O.** By using a standardized model which splits the driver architecture between OS specific and hardware dependent modules, the new I_2O specification
eliminate the need to write unique code for each system.

Systems, and WindRiver Systems . and major server-desktop hardware vendors such as Compaq, Hewlett Packard, Intel, and NetFrame Systems.

The main objective of the I2O specification is to provide a standardized I/O subsystem architecture that is independent of both the I/O device being controlled and the host operating system. This independence is achieved by logically separating the driver segment that is responsible for managing the device from the specific implementation details for the operating system that it serves. There are three key areas on which the I2O standard focuses: the messaging layer, device class definitions, and standard interface definitions. Typical device drivers today include both OS-specific code and hardware-specific code in a single module. The I2O specification inserts a messaging layer between the OS-specific portion of code and the hardware-specific portion of code. This messaging layer splits the single driver of today into two separate modules: one for the OS and one for the hardware. The messaging layer provides the communication service between the OS and hardware modules. The only interaction one module has with another is through this messaging layer. To partition the OS-specific and hardware-specific portions of the drivers efficiently, interactions between the two must be clearly defined. For this reason, I2O devices are categorized into classes. Each I2O class has its own set of messages and protocols for interacting with the system OS and other I/O modules. Therefore, any hardware implementation of a specific class would appear identical to the host. The I2O Revision 1.0 standard specifies message classes for each of the following devices: block storage devices, such as hard disk drives and CD-ROM drives; sequential storage devices, such as tape drives; LAN ports, and Ethernet or Token Ring controllers; and SCSI ports and devices. True industry advancement using the I2O specification will depend on including newer technologies (such as Fibre Channel, RAID, and ATM) in later revisions of the specification.

In addition to the message layer interface, I2O also defines an embedded hardware interface within the intelligent I/O subsystem consisting of shell and core interfaces. The shell is another name for the messaging layer and provides the interface between an I/O subsystem, the host, and any other I2O subsystem (peer). It defines the subsystem behavior in relation to the host. The shell consists of a register-level hardware interface, a set of message definitions, and a protocol for exchanging these messages. The shell provides both OS and I/O subsystem independence. Internal to the I/O subsystem is the core interface. The core provides the communication between a device driver and the I/O platform. It describes the I/O platform environment as it relates to the device driver. Like the shell, the core interface provides an operating environment for device drivers that is both OS and I/O platform independent. The shell and core interfaces are layered, with each layer isolating or hiding the nature of a particular characteristic, such as bus type, processor, instruction set, or device type. These two independent interfaces allow the I2O specification to isolate the host processor, operating system, and I/O subsystem. Thus, the I2O specification provides an open standard to be used across multiple operating systems and devices. In addition to the near term benefit of reduced driver development, a number of long term benefits are theoretically possible with I2O standardization including complete decoupling (isolation) of the operating system and I/O subsystem; increased error isolation and robustness; higher performance I/O subsystems and improved bus efficiency; all of which are important, if not critical, to the success of net-centric computing.

Building an I_2O *Proceessor*

A good example of the performance edge that can be achieved in typical server applications is the use of the i960RP, from Intel Corp., one of the first hardware vendors to dedicate silicon to the I2O approach. It combines onto a single chip (see Figure 12.4) all the elements essential for high-bandwidth I/O: a complete 32-bit i960 JF processor; , a PCI 2.1-compliant PCI-to-PCI bridge; an I2O messaging unit; three-channel DMA; memory controller, and a variety of other peripherals interfaces.

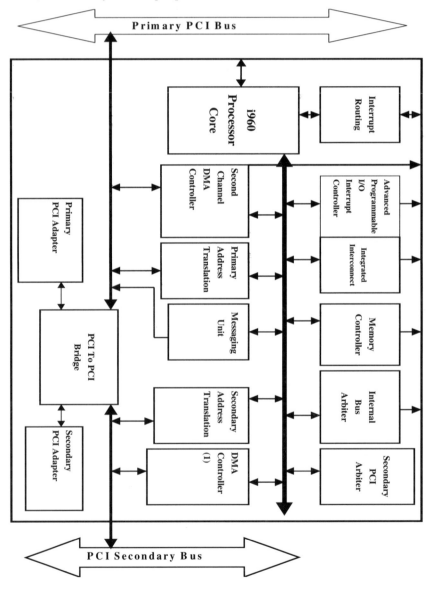

Figure 12.4 **I/O Processor**. Intel has targeted the I_2O market with a processor family, the RX series, which surrounds 960 core with support logic for PCI-to-PCI bridge, I_2O messaging unit, three channel DMA, and a memory controller.

The i960 RP processor addresses two key server requirements: bus expansion and high performance I/O processing. The RP's PCI-to-PCI bridge provides the server with expansion capability, enabling a secondary PCI bus for PCI peripherals. However, the RP also adds a new dimension to PCI bus expansion - intelligent processing capabilities. This enables the server to balance applications processing and I/O processing across different processors, resulting in higher performance systems. This architecture enables flexible systems with a variety of intelligent I/O implementation options. The options include (1) the traditional intelligent adapter residing on the primary PCI bus; (2) the i960 RP processor resident on the server motherboard and non-intelligent PCI adapters on the RP's secondary PCI bus; and (3) full server motherboard integration. The intelligent adapter provides not only intelligent I/O performance benefits, but also enables multi-port capability using the RP's secondary PCI bus. SCSI and LAN controllers can be implemented on a single card, with data transferred directly from the disk out to the network under the i960 RP processor control. This accelerates I/O throughput by keeping data congestion isolated to the I/O subsystem. The motherboard-PCI adapter model provides the server with the flexibility to enable off-the-shelf dumb PCI cards to perform at intelligent I/O levels, energizing the dumb PCI adapter with software executing on the i960 RP processor. The last alternative, the server motherboard integration model, incorporates all of the pieces of the I/O subsystem on the motherboard, allowing a designer to use PCI components already on the motherboard, such as SCSI, to provide very low cost intelligent I/O for mature functions such as RAID.

A critical system performance factor that the designer must consider is PCI and host bus utilization, including: host processor bus utilization, contention for host memory, and unnecessary LAN PCI traffic. The I/O processor improves the system utilization of tiered buses (host bus and PCI bus) by performing I/O tasks intelligently at a level closer to the I/O interface. It also uses shared system resources more efficiently. For example, each time the host processor performs a context switch for an interrupt, the code must be fetched from the host memory, a use of the processor bus that is costly. By the same token, the contention for the host memory affects the network throughput. Since the host processor builds the individual packets in this memory, the amount of contention for this bus increases. In addition, most PCI systems use a CPU core logic chip set to access the PCI bus and these accesses can burden the chip set with unnecessary PCI transfer requests. The access latency through the PCI bus can vary depending on the traffic loads required by the other PCI agents. Also, as the LAN controllers transfer the data packets across the PCI bus, the overhead for each packet is transferred unnecessarily. By eliminating unnecessary LAN traffic, (in particular, small control packets) and buffering and packing receive data, the PCI bus is used more efficiently. This takes advantage of PCI burst performance and avoids small transfers which drag down the performance of the PCI bus by increasing the number of wait states and arbitration cycles, reducing PCI bus concurrency and throughput.

Existing intelligent I/O designs provide all the intelligent I/O performance benefits but are limited by the I/O processor bus data paths. The i960 RP processor's architecture has optimized these data paths and the system designer can structure software to take advantage of this and dramatically improve I/O performance. Current intelligent I/O architecture schemes place peripherals and memory on the I/O processor's local bus, connecting to the host's PCI bus via a PCI bridge. This imposes the requirement for data traffic to be transferred between peripherals and local memory before being sent up through the host PCI bus to the host processor. The i960 RP processor changes this model by including a PCI-to-PCI bridge on board. The processor enables data traffic to be transferred directly through the bridge path, bypassing the local memory. For example, in network adapters, data transmits can be streamed through the bridge, interleaving the data packets with headers from local memory, thus eliminating a memory copy. In SCSI adapters, disk reads, which consume 70% of disk traffic, can be streamed through the bridge, also eliminating one memory copy. Taking advantage of this data path architecture provides a 30% to 40% increase in I/O throughput.

The Future of Intelligent I/O

After a slow start, the I2O standard is making its way into the mainstream of server designs. By the end of 1998, most of the major server vendors had introduced systems with I2O integrated onto the motherboard. Also, most of the major server OS vendors - Microsoft on NT, Santa Cruz Operations on UNIX, and Novell on its implementation of UNIX - and at least one RTOS vendor, Wind River Systems, Inc, had committed to the standard and made the necessary modifications to their operating systems.

Also, at least two other I2O-specific processor implementations have come to market, one for the PowerPC and another built around the ARM CPU architecture. The PowerPC implementation is the IOP 480 from PLX Technologies, Inc. Built around a PowerPC core it contains the same data pipe architecture used in the i960, a PCI-to-local bridge, and DMA and memory controllers. The company has also developed the PLX PCI 9080, an I2O-ready PCI-to-local bus interface chip which acts as an I2O accelerator when used with the i960, or as a generic interface device that will allow designers to develop I2O subsystems built around virtually any microprocessor.

Finally, as of mid-1998, Intel Corp. had introduced its second generation of I2O processors, the i960RM and i960RN, which offers a 4X to 5X performance improvement over the i960RP, enough to solve many of the Internet's I/O-bandwidth bottlenecks well into the 21st century. The difference between the two processors is in the external PCI bus width each supports: the RN has 64-bit, 33 MHz primary and secondary external PCI bus interfaces and the RM is designed for 32-bit, 33 MHz PCI buses. Based on a design developed in collaboration with major server OEMs, the new architecture retains the basic 32-bit i960 as the core, but the logic that surrounds it has been totally revamped: fatter internal buses, deeper queues, multiple read and write queues, hardware acceleration of key I2O functions normally done in software or externally, and support for 66 MHz synchronous DRAMs.

Not to be ignored in this barrage of architectural enhancements is the fact that the new chips based on this architecture will represent a shift in direction away from generic solutions to I/O problems and toward a more application-specific strategy. The primary mechanism for this customization is an on-chip Application Accelerator Unit (AAU) that effectively increases the I/O processor's bandwidth four-fold.

Chapter 13

Adding 64-bit Muscle to the Internet

While the vast majority of new additions to the server population on the Internet have been built around 32-bit CPU architectures such as the Pentium Pro, there is a growing trend, at least in the backbone, to systems based on 64-bit architectures. Similarly, while the existing installed base of Internet and intranet bridges, routers and switches is still built around 32-bit processors, system designers are finding that one "no-brainer" approach to beefing up things overall is also a shift to 64-bits.

In the not too distant past, when 64-bit RISC processors and operating systems began to appear in the general computing marketplace, the consensus was that it was not likely that such designs would find their way into the mainstream much before the end of the 90s. Also, the common view was that if 64-bit processors were going to have a place it would be at the very high end of the market. And rather than networking, it was thought that the applications most likely to need 64-bit CPUs were scientific and technical projects requiring the precision, word size, throughput and terabytes of direct memory access that such architectures make possible.

The Internet and its growing popularity and the consequent demands on it now and in the future are changing all that. Stretching the Internet to a computational limit that only a shift to 64-bits can handle is the convergence of a number of key events. These include: (1) the increasing numbers of users; (2) the emergence of network computers and other net-centric devices; (3) the move toward network-centric computing in which applications run primarily on servers; and (4) the development of multimedia-based connected PCs.

Bigger pipelines based on new technologies such as Gigabit Ethernet, ATM, cable modems, ADSL and ISDN will not be enough. Without improvements in the server, router and switcher "nuts and bolts" that hold the worldwide system together, the Internet cannot keep up with the demands placed on it.

The biggest beneficiaries of this trend have been the handful of CPU vendors who have pushed the state of the art in CPU design to 64-bits: Digital Equipment Corp., HAL Computer Systems, Hewlett Packard, Silicon Graphics, and Sun Microsystems. As of the mid-1998, Intel Corp. had yet to enter this club. It has, however, revealed more details on "Merced" and plans to be in volume production by the end of 1999.

One indication of the power that 64-bit architectures bring to Internet and Web applications is the use of DEC's 64-bit Alpha CPU in its much heralded Alta Vista Web search engine. It consists of two 64-bit Alphaserver 8000s with 10 processors each and an additional eight with one or two processors, as well as a Very Large Memory (VLM64) architecture that gives each processor 16 G/bytes of directly accessible memory. With this kind of capability Alta Vista can search 40 million Web pages in a fraction of a second.

Another good measure of the growing acceptance of the need for 64-bit servers as a solution to the Internet's computational bottlenecks is the fact that Netscape Communications Corp., with an estimated 80 million hits a day, was one of the earliest purchasers of DEC's Alphaserver 8000 series.

The Burdens of Serving

General-purpose 32-bit servers, which constitute the vast majority of systems on the Internet, face a number of challenges which demand that they be upgraded and strengthened. One burden lies in the fact that the Internet is not designed to allow efficient use of either servers or clients. Rather, all technical enhancements to date have been aimed at providing a transparent and platform-independent way of transmitting, storing and viewing data.

For example, a common chore that the World Wide Web performs, and that places a considerable burden on the servers, is forms processing, a staple in most Web pages. Inordinately complex and inefficient, the protocols require that the client contact the server twice and the server do two entirely different things: first it must request that the form that is returned from the server contain everything that the browser needs to know about how to submit the form while the second request to the server contains information extracted from the browser that is necessary in order to answer the query.

The entire process would be much simpler if the Web script simply waited for the user to fill out the form and then continue from there. This would eliminate the overhead of the second connection to the server, reduce the amount of information sent between the client and server and eliminate the need for a separate form and script. The reason that this and other procedures are done this way is because the hypertext transport protocol is what is called a "stateless protocol" that is designed to eliminate dependencies on any particular architecture and is guaranteed to work in any OS or CPU environment.

Numerous ways have been tried to reduce the load this "one connection per request" scheme places on the Internet. One of the most common and straightforward schemes is caching duplicate copies of pages at special server sites across the Internet. Just as it works on a desktop computer such cache servers contain the most recently and most commonly accessed data of particular primary server data. The impact on reducing loads on servers can be dramatic, not only reducing overall traffic but also allowing system wide load balancing of such activities throughout the Internet. It is faster to fetch a document from a local server than from one far away. And if each document is fetched from the main server just once and then reused, the load on the relay and the wide-area network is vastly reduced. The load on the originating server is reduced because it has only to deal with the requests from the caching server, and the load on the caching server is lower because it has to deal with only a local-area network, an intranet or a smaller segment of the Internet.

Why 64-bit Servers?

With all these advantages, it is not surprising that cache servers are now being widely implemented on the Web. But the Web is also growing at a much faster rate, not only in terms of browsers and clients, but in terms of low- to medium-performance servers based on 32-bit OSes such as Windows NT and 32-bit architectures such as the Pentium Pro. The load has increased to the point that even cache servers based on 32-bit technology are running out of steam. So a new generation of dedicated cache servers based on 64-bit CPUs, such as DEC's Alpha and the Silicon Graphic's MIPS 4000 emerged. Network Appliance Inc.'s NetCache, for example, is based on a 64-bit Alpha CPU and a proprietary RTOS, which provides a four-fold performance improvement over software-only solutions running on standard 32-bit NT-based general purpose servers. Another player is Cobalt Microserver, Inc. whose CacheQube is built around a 64-bit MIPS RISC and customized version of the public domain Linux operating system which has been optimized for real time caching transactions.

But the pressure for more capabilities out of servers is still on For example, there is the wider spread use of distributed-object-based computing, based on CORBA and Java, on the Web. With the increased load, the 32-bit address "handles" now available on most servers may not be enough to deal with the overwhelming number of objects that will have to be retrieved from across the net. Another problem arises with the much promoted ObjectWeb-based net-centric computing model. With the majority of the computing shifted away from the clients and onto the servers, the amount of memory space directly available on servers will require a shift to 64-bits. According to preliminary estimates, the traditional approach to net-centricity using thin clients and Windows-based terminals alone will tax server technology in a way that it has never been before. As we learned Chapter 8, a single session using the traditional Citrix protocol to connect terminals and network computers to servers would require anywhere from 2 to 16 Mbytes per client per server per Internet session. Certainly, the use of Java and more modular applications using that language would reduce the memory load by balancing things between the servers and clients, but it would increase the I/O burden substantially. A 64-bit server architecture with two terabytes of directly accessible DRAM may be the only way to solve both problems.

Not even factored in here is the continued expansion of the Internet and the intranets in terms of available bandwidth, numbers of users, and type of usage. Also, network-connected multimedia desktop computers will test not only the bandwidth limits but the capabilities of the servers, as will the sheer numbers of users who are likely to access the Internet on more multimedia-modest Internet appliances. If the solution to some of the Internet's performance problems still rests with the widespread proliferation of cache servers, the heavy usage of full motion video - which even in compressed form requires tens of megabytes of disk space for a single minute - will also require the terabytes of directly accessible memory that 64-bit architectures provide.

The 64-bit address space available on a new generation of 64-bit architectures such as the Alpha 21264 from Digital Equipment, UltraSparc from Sun, the Sparc V9 from Hal Computer Systems, the PA8000 from Hewlett Packard, and the 64-bit PowerPC from IBM Corp., will certainly ease this burden, but much depends on how quickly the costs of the systems come down in the future.

Sixty-four Bits Means More Than Performance

While the improvement in the capabilities of the Internet and the many large intranets alone would be enough to move to 64-bit based servers, there are numerous other reasons for making the transition. Specifically, the emergence of the World Wide Web has created a huge demand for systems that utilize all aspects of the operating system, (e.g., fast process creation, large and fast file systems and databases, fast I/O, and fast networking). And as more and more machines get on the Web, the demand will far surpass anything a 32-bit based server can deliver. In addition to the greater demands, current Web servers have a number of performance-inhibiting problems that continue to crop up: limited or slow I/O; limited address space; firewall penetration; and a number of CGI related issues. While CPU performance has been increasing rapidly, I/O speeds have not. Although the increase use of intelligent I/O will ease the pressure, there will be an ever accelerating demand for even more I/O performance

Also, the Common Gateway Interface (CGI), which manages the way in which servers extract information from a client and distribute it to programs on the server, has some serious limitations. These have primarily to do with the high overhead of spawning a new process for each incoming request, which causes severe degradation in scalability and throughput. While server APIs address this limitation to some degree, they cause the extensions to be linked together. This can cause serious bottlenecks, because one extension can get in the way of another, slowing down its ability to execute an operation. In addition, work has to be done to make the extensions into a "library" that can be linked to the server. These extensions may require the designer to consider multithreading (adding locks to access shared data). But in a 32-bit environment this can be difficult, if not impossible, for some arcane applications.

Firewall implementations also have severe performance flaws. Ideally, firewall software is implemented as a separate user process (called the firewall proxy) that filters incoming packets. However, this causes a huge performance penalty because of the time needed to context switch into the user process and copy the data back and forth between the firewall proxy and the kernel. Several vendors have attempted to put this code into the kernel but this has substantially increased the size of the operating system.. By moving to 64-bit architectures, the current programming paradigm of a single process supporting multiple threads making system calls to the kernel for services is substantially eased, because there is no need to context switch and copy back and forth. It is also a relatively minor matter to put the code into the kernel, because the increase in size is minor compared to the large virtual address spaces available with 64-bit systems. It is enough to map large objects in an enterprise. For example, if each application used 1 terabyte (TB) of virtual address space and assuming a quarter of the 64-bit address space is available for mapping, there is more than enough room to 4 million 1-TByte applications simultaneously.

High availability (including failover and fast restart) of Web services is going to be crucial in the future. In a 64-bit environment where addresses are unique, either an address can be mapped by distributed shared memory (DSM) or not. If it cannot, another machine can supply the data. Support for caching and running complex objects (beyond Common Gateway Interface scripts), using ActiveX (from Microsoft) or CORBA (defined by OMG) links is becoming commonplace. So, network-wide caching of objects is going to become crucial. With the luxury of 64-bit architectures and memory addressability, objects such as Web pages and Java applets can be retrieved from across the net from any node and forwarded to the requester, without any of the current back and forth communications traffic that is currently accepted as necessary overhead. . Only objects with 64-bit handles are able to provide the features that allow such applications. A 32-bit address handle may not be enough because of the sheer number of objects in the net. Java applets may also be cached and addressed with a network-wide address in the 64-bit address space.

Sixty-four bit servers will allow the implementation of special applications (like the Web server) that are permanently mapped in a fixed place in the large address space. Using this new paradigm, this fixed address space can be accessed from user programs without an address space switch. This kind of stable and contiguous address space survives crashes better, not just locally, but across the network. This is achieved through the use of distributed shared memory (DSM). Sixty-four-bit enabled DSM allows an address space to be transparently accessed across the network. Permanent mapping substantially diminishes the context switching and data copying overhead. Firewall proxies and database servers attached to Web servers will substantially gain from this feature.

Sixty-four bits will also lead to new ways to create arrays of processes that are multiple copies of the same process with little context switching, creation , and destruction overhead. In a 64-bit address space, each of these array elements can be located in a unique address space. The arrays do not require multithreading. As a result, problems such as those relating to Common Gateway Interface (CGI) scripts will be substantially diminished.

Processors Adapt To Network Requirements

Just as the implementation of 64-bit processors in servers will have a fundamental impact on how the Internet operates, the reverse is true also. Manufacturers of next generation 64-bit processors, such as Sun Microsystems, are making additional modifications to their architectures to make them better at network operations. Tthe foremost requirement that a system developer must look for in the construction of an intranet or a Internet server is a processor capable of delivering enough throughput to keep pace with the large applications and complex data flowing over the network. Fast Ethernet and the widening acceptance of Asynchronous Transfer Mode (ATM) is transforming the network into a virtual superhighway of high-speed information.

Processors that interact with the network, whether in servers or routers/servers must be able to absorb the flow without creating a bottleneck. They must be able to move data quickly among the different layers of the network and to support large-bandwidth applications and movement of information. To handle such demands, many 64-bit CPU companies are adapting their architectures to make them even more net-centric in their features and capabilities. A good example of this kind of modification is Sun Microsystems' UltraSparc (see Figure 13.1), Some of these improvements, such as the UltraPort Architecture and modifications to the CPU hardware to accelerate block-store operations, are among the network enhancements that have been moved from the company's 32-bit designs to 64-bits. In addition, the architecture incorporates enhancements that take advantage of its 64-bit design to further improve network performance.

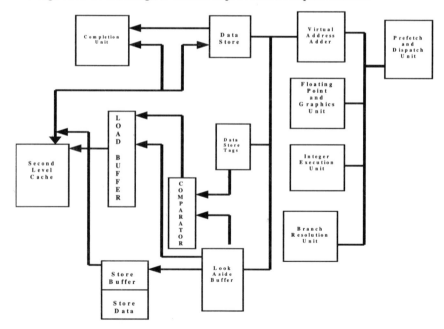

Figure 13.1 **Sparcing up Servers.** To boost throughput on servers, Sun's Ultrasparc makes extensive use of 64- and 128-bit data buses and hardware to accelerate block/store operations, which are critical in networking.

A network processor must support an inherently multitasking environment with multiple users. Support for multitasking memory provides large-bandwidth memory throughput for executing tasks in parallel. To handle this, the UltraSPARC includes a direct interface for large- bandwidth multitasking communications with cache coherency. As for networked multimedia and other data intensive applications, the only realistic way to deliver the throughput is to share compute tasks across multiple processors. Multithreading is the most efficient way to implement multiprocessing. With multithreading, single and multiple applications are parsed into separate functions or threads that are executed on different processors either within a single computer or scattered through the network. On UltraSPARC these kinds of operations are accelerated through the use of faster context switching, along with an atomic instruction that promotes tighter synchronization of threads for compare and swap operations. Together they result in faster execution of multithreaded network and end-user applications. For example, a multithreaded ATM interconnect will operate much more efficiently because each layer can be threaded out to different processors. This tighter synchronization also translates into better support for multithreaded operating systems.

Security, too, is a burning issue for networking. Cryptography needs to be supported on all levels of a network to ensure the protected flow of information. The UltraSparc's visual instruction set (VIS), developed to primarily handle multimedia operations, also supports flexible data encryption with filtering capabilities, For example, UltraSPARC can multiply a set of four 16-bit words simultaneously with an 8-bit filter, quickly encrypting large streams of data. The variety of integer operations and their ability to work on several integer words in parallel give the cryptographer a wealth of tools to fashion a robust encryption scheme. Data integrity is another crucial area in net-centric computing Data cannot be allowed to be corrupted as it flows over the network, or the network is no longer a viable means of communication. Check summing with signature generation is one class of algorithms used to examine incoming data for integrity. The processor adds up the data that come in and looks at the overflow, or signature, and if it matches the signature that was sent, the data have not been corrupted. Algorithms for check summing with signature generation move through vast amounts of data. These kinds of operations are also accelerated using the Visual Instruction Set. For example, included in VIS is a check-summing algorithm than can add two 16-bit words in parallel using 64-bit registers. This allows for a large overflow area, 16 bits for each of the two streams, ensuring accurate signature generation even for large amounts of data. Creating two check sums in parallel significantly streamlines data integrity operations at all levels of a network.

Routers and Switches Need Help, Too

Many network applications, especially routers and switches, impose a unique set of requirements not found in desktop. For example, the handling of unaligned, big and little endian data types; cost limitations on memory subsystems; and fast, robust interrupt handling are typical system parameters. Also increasing the load on network routers and switches are the support of critical code and interrupt routines, the occurrence of frequent variable profiles only discernible at compile time, and the need to efficiently handle deterministic branch trajectories and other optimizations that result in higher performance.

To solve such problems, the CPU configuration of choice in many new networking systems is 64-bit CPUs. Among the companies servicing this need are Motorola, IBM, Hewlett Packard, Digital Equipment, Sun and a variety of 64-bit MIPS vendors, including Integrated Device Technology, NEC and Quantum Effect Devices. To understand what this shift will do to the performance of routers and switches, it is necessary to review some of the basic performance parameters of such systems. There are three key parameters that a systems designer needs to be keep in mind when selecting processors for use in the routers and switches in a network: interrupt handling, context switching and total switch time.

In networking and data communications applications, system peripherals use interrupts to signal the processor that an event has occurred that needs attention. Two types of interrupts are of concern in this environment: those that can be handled rapidly without causing a context switch and those that will generate a context switch. To handle interrupts efficiently, processors normally have mechanisms to save key system registers internally, when they occur. Certain processors save the entire machine state regardless of the type of interrupt, causing unnecessary degradation in performance. One aspect of interrupt handling in a networking context that is particularly important is the processor's interrupt latency. In other words, if you have a time-critical task, one key item that must be determined is the length of time from when the interrupt occurred to the point where the processor was serviced. The time this takes depends on how long it takes the processor to recognize the interrupt and fetch the interrupt handler. In many 32-bit processors, particularly those designed originally for desktop and workstation applications, this time depends on the instruction stream being executed. Many block- oriented or multicycle instructions, common in such applications, are not interruptible, making the interrupt latency difficult to determine.

Another critical parameter is context switching time. A context switch can be either synchronous or asynchronous. The first occurs when a task completes normally or when a task suspends itself while waiting on some event. The task calls the operating system kernel, which in turn suspends the task and switches to another. An asynchronous task switch occurs when an interrupt causes a call to the kernel, forcing the currently executing task to be suspended and an interrupt handling task to be started. The main difference between the two is the amount of context that must be saved before the new task can execute. Asynchronous tasks switches, common in networking and data communications applications, generally require more context to be saved than synchronous switches. Just as important as the specific context switching time in networking applications is the processor's total switch time, which is the length of time it takes for it to go from doing real work in one task to doing real work in another task. This time is affected by the time spent pulling data off the network and into registers by a new task so it can begin executing.

Boosting 32bit CPU Performance

Recognizing these constraints, a number of second-generation 32-bit CPU designs have emerged that have been optimized for networking applications. The i960Hx RISC processor from Intel Corp., for example, retains the core of other members of the family, including the basic pipeline and superscalar organization of the Cx series chips. But to this Cx core were added instructions, larger caches and features specifically for networking applications..

For example, a "quick data invalidate by region" feature was added, enabling the efficient parsing of volatile packet data by allowing designers to invalidate a specified cachable data memory region independently, while preserving the remainder of the cached data. Second, to optimize the processor performance despite the limited bandwidth imposed by the low-cost memory subsystems mandated by the market, Intel designers increased the amount of on-chip memory. The i960 Hx processor includes 16 kbytes of instruction cache in a four-way set-associative organization, which maximizes the efficiency of the large cache by increasing its effective hit rate. Also, each way of the cache (each 4 Kbytes) can be incrementally loaded and locked with critical code or interrupt routines. This gives network application designers the ability to guarantee that 4, 8, 12, or even 16 kbytes of key code will be resident on chip, completely avoiding off-chip fetch latency. The unlocked portion of the cache continues to function as a 4- way set associative cache. This flexible strategy allows critical code to be locked without forfeiting processor performance on unlocked code. The data cache contains 8 Kbytes, also organized in a four-way set associative configuration. The data cache has several features for manipulating cached data. New instructions allow the invalidation of a line (four words) of cached data at a specified address for pinpoint control. The quick-invalidate feature, described earlier, allows efficient management of transient data structures, such as network data packets. Cached data in specified memory regions can be invalidated independently while leaving other cached data intact. Both of these precise invalidation mechanisms reduce bus traffic by preventing the need to refetch useful data in the cache. Also included on the chip are 2 Kbytes of data RAM, which systems designers can use to set up the data RAM for frequently used variables. Alternatively, portions of the data RAM can be used as additional register-frame storage for up to 15 on-chip frames. Both techniques significantly reduce bus traffic by providing on-chip storage of frequently referenced data structures.

To further optimize the utilization of the limited bus bandwidth, the i960 HX processor eliminates bus traffic associated with over aggressive instruction fetching and data loading. Aggressive instruction prefetching actually reduces overall performance due to the small basic block size (number of instructions between branches) and limited bus bandwidth of network switches and routers. The i960Hx uses an instruction fetch policy that specifies a fetch of either two or four words, depending on the instruction pointer (IP) alignment. Thus, the processor typically saves the time to fetch two instruction words that are not needed when the IP falls in the latter half of a quadword. Similarly, the allocation policy for the data cache utilizes only word accesses (matching the valid bit granularity, rather than the traditional approach of retrieving a full cache line, which is a quadword. This approach reduces unnecessary bus traffic generally caused by the wide distribution of networking data, or low data locality.

Networking applications require manipulation of packet and frame data that are arbitrarily byte aligned. The processor must handle unaligned short-word (16-bit) and word (32-bit) accesses with minimal overhead penalty, and the processor must readily accommodate both big and little endian data types. So, all the co-processors on the i960Hx (the bus controller and the data cache, data RAM, and address generation unit) were redesigned to handle all unaligned short-word and word accesses entirely in hardware instead of in microcoded routines, which stall the processor core. The handling of unaligned data accesses in hardware allows the processor to continue program execution with no slowing for unaligned accesses. The processor breaks multiword unaligned accesses into word accesses to match the unaligned hardware granularity. This improves handling unaligned data as much as four times, which becomes significant with fractional-speed buses. Since the i960 processor architecture does a full-context switch (saves all local registers) on interrupts, the most time consuming component of interrupt latency is performing a frame spill to external memory. A frame spill is necessary when the on-chip storage for register frames is full. To reduce the frame spill latency, the chip provides reserved register frame space for high-priority interrupt handling.

The N-Squared Path Problem

Despite such enhancements, the 32-bit CPUs used in the vast majority of the installed base of routers and switches may not be enough to handle the increased traffic requirements of today's intranets and the Internet in general, especially with the emergence of the new generation of hybrid switchers and routers. One thing that going to 64-bit CPUs gives the makers of these network devices that 32-bit architectures with all the enhancements described earlier cannot is the direct accessibility to terabytes of very fast DRAM. To understand why this one feature is so important to makers of the next generation of switchers, routers, and new hybrid combinations of both, it is only necessary to look at what they call the N squared path, or more correctly the N x [N-1], problem.

Suppose that there are N nodes in a network and each node needs to know how to reach every other node. To do this, the network would have to "know" about N x [N-1] paths. Routers do not have to know about each and every individual path to each and every end point in the network. But in the new router-switcher combinations they need to know about paths as close to the edge of the network as possible. The result is a geometric growth in the number of "virtual circuits" that it is necessary for each switcher-router combination must handle. Running in an Intranet with 1,000 or 2,000 routes is something that is manageable, just barely, in the present 32-bit environment. But in an Internet context, with as many as 50,000 connected sites, each with their own set of nodes, the situation faced is entirely different. Not only is it difficult, if not impossible, to set up and tear down that many connections quickly, but the high number of routes in such a network can be problematic, since it would require keeping track of, mapping and switching a huge number of VCs, or virtual connections. And with the Internet growing so rapidly, with a four fold to five fold increase to as many as 250,000 to 300,000 routes, the mapping and switching job will quickly become impossible with the present generation of 32-bit processor technology.

Using the greater addressability and direct memory access of 64-bit processors, makes the N-squared problem much more amenable to solution, because it would not be necessary to switch between the main memory's DRAM and the disk-based mass storage for new addresses and routes. All of the routes on the Internet now and in the future can now be maintained in DRAM, allowing the instantaneous matching and tagging of literally hundreds of thousands of IP addresses, rather than suffer the delays that would be involved when accessing the hard drive's much larger virtual memory for additional locations and routes.

The larger addressability and use of terabytes of directly accessible DRAM alone are making it possible for switcher-router manufacturers to look a number of new configurations. One pproach is a route-aggregation system called multipoint to point, or MTP. MTP builds a tree data structure within each switch, doing away with the need for VCs from each endpoint to all others. By setting up special VCs to act as trunks for all of the sites on a switch, the problem is simplified from one of geometric to one of linear growth. Where 64-bit processing comes into play is that while this approach addresses the growth issue, it does not address the need to create and tear down many virtual connections very rapidly. Essentially, 64-bit designs are used to "brute force" the problem, using them to build permanent MTP VCs from each switch to every other switch. To handle Internet Protocol routing, software is located on each switch and router so that VC translation tables can maintained on each interface card.

At alternate approach is the so-called tag-switch solution favored by the Internet Engineering Task Force's Multiprotocol Label Switching working group. Essentially a protocol-independent methodology, in tag-switching, routes are aggregated or assigned to tags with varying granularity. In this approach, a single tag could represent either one application to application flow, or hundreds, effectively reducing the complexity within the switcher/router environment.

CPUs To Match Network Requirements

Similar to what happened in 32-bit designs, companies targeting state of the art router-switchers, such as NEC and QED, have added major enhancements to their second generation 64-bit MIPS CPUs. For example, In its new RM7000 CPU (see Figure 13.2), QED has incorporated features that boost raw performance by almost ten times, improve the effective bandwidth and enhance interrupt handling capabilities, all of which directly or indirectly affect network performance.

First, the RM7000's clock rate was raised from its original 200 MHz to 300 MHz.. Further improvements include such things as a dual-integer, superscalar design where any two integer or integer/floating point instructions can be executed within a single clock cycle. With networking in mind, the RM7000 contains considerable modifications in the way it does caching. Because of delay issues relating to the move to 0.25 micrometer process technology, it is not possible to simply add more Level 1 (L1) cache to the chip and maintain overall performance. Beyond a certain L1 cache size, the actual chip performance degrades due to the longer metal signal delays across larger and larger memory arrays. As an optimum solution, the RM7000 incorporates a 16-KB instruction and 16-KB L1 data cache size, each four-way set-associative. Beyond that , the RM7000 incorporates a 256-KB L2 cache, normally off-chip in many CPU designs, significantly improving cache hit rates and reducing memory latency.

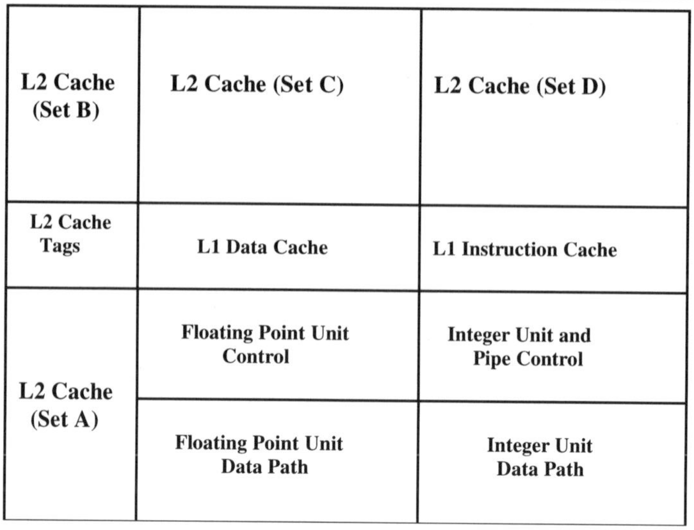

L2 Cache (Set B)	**L2 Cache (Set C)**	**L2 Cache (Set D)**
L2 Cache Tags	**L1 Data Cache**	**L1 Instruction Cache**
L2 Cache (Set A)	**Floating Point Unit Control**	**Integer Unit and Pipe Control**
	Floating Point Unit Data Path	**Integer Unit Data Path**

Figure 13.2 **Building up Routers**. In its RM7000 MIPS CPU, QED has added features such as a 64-bit arithmetic logic unit and data paths and on-chip Level 1 and 2 data cache to accelerate I/O processing in routers and switches.

Several specific features have been added to further enhance the memory bandwidth performance for on-chip and off-chip caches, which is important in handling the I/O traffic in many network applications: non-blocking caches, cache locking, and cache allocation. The nonblocking cache feature is of particular value to the network systems designer. Non-blocking means that a miss in the cache does not necessarily stall the processor pipeline by allowing the CPU to continue to execute other instructions that do not use the data referenced from the cache. Nonblocking can effectively double system performance when interleaving execution threads.

An extension of the system interface protocol to support a pendant bus was included to further enhance the nonblocking feature. The pendant bus allows two outstanding reads on the bus at one time - meaning that the interface can support a second read request to the memory system while the first read is pending. In addition, QED designed the interface such that the outstanding reads can be returned out of order for execution. This is especially valuable to the system designer as the first read may go to a slow I/O device and the second to L3 cache. This pipelining technique helps create a higher-performance memory system and increases system efficiency up to a factor of two.

To maximize overall system performance in networking applications, the RM7000 contains features that give systems engineers more control of how specific data get put into caches. For example, in a network router, packet header data are used for a short time relative to routing tables which is more long-lived. With on-chip L1 and L2 caches, the most efficient allocation is to put the packet header data into the single-cycle L1 memory, bypassing the slower on-chip L2 and even slower off-chip L3 caches. In many memory systems, the same data is automatically entered into all three caches: L1, L2, and L3, for future use. To meet the cache allocation requirement, a new coherency attribute scheme was implemented. The new coherency attribute allows the bypassing of the L2 and L3 caches. In networking applications, this allows packet header data to come into the primary cache for only a short period and then be expelled without bumping routing tables out of higher-level caches. In a sense, using the new cache coherency attributes can make the L2 cache look like a 256-KB on-chip SRAM for instruction and data storage. This approach also saves the systems engineer the overhead cycles needed to specifically invalidate such data from the higher-level caches, in addition to the primary.

Also increased were the number of interrupts available and conversion from software to on-chip hardware management. As a result, each interrupt has its own vector that can be freely assigned among the priority levels. This not only improves system interrupt latency, but also provides the necessary control of the CPU in complex networking systems where many communications channels are competing for common resources.

64-bits Enhance Network Reliability

In addition to taking advantage of the higher performance and improved interrupt capabilities, a number of switch and router manufacturers are using 64-bit designs to improve network reliability and uptime, essential if the typical net-centric computer user is to come to depend on the "Web tone" with the same sense of certainty that telephone users associate with the ubiquitous dial tone.

While higher performance 64-bit CPUs improve network reliability simply by making it possible to handle peak network loads without going into overload or shutdown conditions, there are other, less direct, ways such processors contribute to network reliability. To date, most of the networking equipment and systems designed for traditional networking and the enterprise have not required the high reliability that systems like WAN-access switches demand. For instance, a few hours of repair time on a malfunctioning LAN in an enterprise have generally been acceptable. But not so in an Internet networking application like a WAN-access switch. Here, fault-tolerant subsystems based on powerful, 64-bit microprocessors are crucial for maintaining the 99.99% availability of the Internet service to the subscriber.

Network reliability and high availability start with the CPU architecture as a major contributor to self-monitoring diagnostics. For example, a high-speed, yet simplified five-stage pipeline, such as that used in NEC 's VR4200, make it easy to flush out the rest of the pipeline and restart the CPU when a subsystem experiences a failure. Hence, when the CPU goes through its reset, it brings the system back up much faster and downtime is minimized. Also, the extra code cycles that a high performance 64-bit processor makes available give the system engineer many extra spare CPU cycles with which to support high levels of reliability and high availability. For instance, those extra cycles could be used to run code for checking and monitoring system health within the WAN access switch. These spare CPU cycles can also be used for communication between software modules and a fault-tolerant application manager, often included in advanced router-switcher designs. It provides fault detection, notification, isolation, and service restoration for failures in such system areas as system control, forwarding modules, power supply, clock card, management bus and others. It also maintains "keep alives" with all software components to ensure that system software is operating correctly. These are are periodic messages sent from one software module to another to verify that each module is alive and functioning correctly. Where a 32-bit design might have to go offline, or at least operate at less than optimum performance, a router-switcher based on a 64-bit processor has sufficient extra cycles to handle both in real-time, with no sacrifices in performance.

Without such structures, an Internet router-switcher can have memory leaks which, in turn, lead to rebooting a system every so often. Memory management to alleviate the long-standing problem of memory leaks requires a large amount of extra CPU cycles, but adds considerable reliability.

Beyond 64-Bits

As bandwidth and usage increase, will 64-bit servers, switches and routers be enough? Certainly, 128-bit CPUs are one possibility. But the overall impact on software, especially the operating system, and concerns about retaining compatibility with existing 32- and 64-bit applications would seem to rule that out.

The most likely prospect, and the one that would be the most evolutionary, in that it would allow most existing software to run unchanged, is a move from the present 200- to 500-MHz designs to processors that allow 1-GHz performance and beyond. This is likely to occur much more quickly than anyone anticipates. It is no longer a matter of when the Intels and the HPs of the world determine that it is in their interest to introduce 1-GHz designs. It is the Internet and World Wide Web that are setting the agenda and it says : performance, performance, performance. As soon as possible.

Photo 13.1 **Gigahertz CPU.** By making all data and instruction paths 64 bits wide and using copper rather than aluminum for signal paths in its PowerPC CPUs, IBM has pushed clock speeds in excess of 1,000 MHz. (Source: International Business Machines Corp.)

The big question for CPU manufacturers is how to get to 1 GHz without substantially increasing costs. It has been assumed that to get an improvement of this sort would require radical changes in process technology, such as substituting much more conductive copper for the aluminum now used for interconnect in most integrated circuits. However, recent work by researchers at IBM indicates what with some innovative circuit designs, it is possible to get to the 1 GHz target economically with existing silicon processing technologies. At the 1998 International Solid State Circuits Conference, IBM announced a version of its 64-bit PowerPC architecture that runs at 1.1 GHz (see Photo 13.1).

The RISC design contains one million transistors and was developed using IBM's existing 0.25-micron CMOS 6X technology. The microarchitecture, circuits and testing techniques resulting from this project when applied to microprocessors using IBM's recently introduced CMOS 7S "copper chip" technology, would boost clock rates another 25 percent to about 1.25 GHz. Among the architectural innovations used to achieve the 1 GHz clock rate were a multifunctional execution unit, which combines addition and rotation operations into a single circuit, and an innovative cache design that combines the address calculation with the array access function. Also playing an important role is the use of a dynamic circuit approach that greatly reduced the number of stages through which signals must propagate. The design also uses innovative clocking methods that further reduce the chip's cycle time and overcome the challenge of generating and distributing a timing signal - or clock - with a high degree of precision.

Such designs are still in the early development stages at companies such as DEC, HP, Intel and MIPS Computer. If IBM can maintain its momentum and move this design to market quickly enough, it could gain an edge in the market for servers, bridges and routers. Such 64-bit, 1 GHz designs would provide the CPU power to match the Gigabit per second bandwidths that are anticipated for connecting the Internet backbone and would provide more than enough headroom for the World Wide Web to grow.

Chapter 14

Monitoring and Safeguarding the Network

Once the hardware designer has completed the design of the net-centric computer and the software engineer has debugged and deployed the code, the system may be operating according to specifications. But the job of the systems designer in ensuring that a reliable consumer-ready net-centric computer is ready for use is far from done.

What will also be necessary is the development and installation of a test, monitoring and maintenance infrastructure to ensure continuous, fail-safe bug-free operation. This is easier said than done, for as computer scientist and cybernetist Marvin Minsky has pointed out in *The Society of Mind* (Simon and Schuster, 1986) as the complexity of a system increases, the interaction between elements in that system, no matter how bug free, can result in inevitable unforeseen interactions: some times failure, but more often, actions that were not planned for in the original design.

This is a phenomenon with which most designers and operators of complex systems have long been familiar. In addition to the anticipated failures that do occur due to congestion; bad or poorly written code in a router; bridge, or switch; or even malicious computer viruses, there are the breakdowns that result from unanticipated interactions between programs in a system and between nodes on a network. And the move to net-centric computing introduces another set of uncertainties and sources of bugs and errors. In the world of relatively isolated desktop computers and many embedded applications, the environment in which the systems designer is working and where the applications run is a closed system and all of the potential sources of error can be defined very specifically. With the network aspect added in, there is a shift to a more open and less predictable environment, with many variables and sources of error that might bring a system down.

In the Internet environment, with literally millions of servers and clients connected by hundreds of thousands of processors embedded in switches, bridges, and routers, the problem is even worse. Here a problem in one node could bring down an entire network or, worse yet, be propagated throughout, affecting millions of users. Still another variable is the introduction of object-oriented languages and operating systems such as Sun's Java, which represent a double-edged sword. Certainly, as an object-oriented language there are a lot of features, such as garbage collection, that have been added to ensure correct and bugless code. But a key feature of Java is the concept of dynamic downloading of applications from a server to a client. How is the software developer to protect against errors and bugs being introduced, accidentally or intentionally, during this loading process?

What is necessary is a renewed emphasis on writing absolutely rock solid code that has been thoroughly tested and simulated. Traditional embedded systems developers have in general had to be much more focused on this than those in the desktop environment. But now, rather than just being a another factor to take into consideration when evaluating the cost of development, it must be among the top considerations, with everything else flowing from that. What needs to be kept in mind is that in the new mass market environment of the Internet (i.e., settop boxes, information appliances and network computers) the cost of errors could far exceed the savings from reduced development costs. The cost of fixing those errors in the field could also wipe out any profit gained from getting to market sooner with a not well-tested product. It could also kill a company's reputation.

Beyond this, there must come an acceptance that no matter how stringent the code development and testing process, errors will creep in and unforeseen interactions will occur. What is necessary beyond the emphasis on correct coding practices is to build in mechanisms to first test the software thoroughly and completely. Once released into clients, servers, routers, and switches on the Internet, it is then necessary to isolate and contain such errors when they occur; identify them, and, finally, correct them as quickly as possible. Given that complexity is its own source of errors in the form of unanticipated interactions between elements, it may be necessary to take techniques used during the development process and in development, such as trace analysis and instrumented software, and use them as mechanisms by which to constantly monitor and search for problems on this complex matrix of networks we call the Internet. Ultimately, what will be absolutely essential, if the information superhighway is to become the bedrock of the world's economy, is to come up with mechanisms for software fault tolerance that are as well thought out and as robust as procedures for hardware fault tolerance.

Test, Test and More Test

In general, there is a conscious effort on the part of most engineers and designers to adhere as strictly as possible to good coding practices. But this is not enough if there is not a corporate-wide effort in the same direction. What is necessary in the high reliability environment of net-centric computing is that applications developers adopt many of the same sophisticated procedures and techniques that the suppliers of their operating systems and development tool suites employ. And the builders of the operating systems and software tools will have to adopt test strategies that in the past have mainly used in only mission-critical and safety-critical applications. But this is not easy, nor inexpensive. What it comes down to is how much a company thinks it is going to cost if a serious bug gets through the development process and into the final product compared to what might be saved in development costs or profit gained in faster time to market.

A major attraction of net-centric computing is the pervasiveness of the applications and the opportunity that this provides. But as network computing devices become more pervasive and accessible to an ever increasing numbers of users, more and more applications will become critical to everyday life, whether literally or figuratively, requiring a considerably higher degree of reliability. This is true not only of the programs that reside on or are downloaded to the net-centric computing device, but to the servers, routers, and switches that provide the nuts and bolts of the intranets and the Internet. It is inevitable, then, that the industry will have to change and produce software developed to a higher standard, moving away from the largely ad-hoc development and testing procedures employed today. What will be necessary are strict, formalized methods to produce certified software similar to those used in many safety- and mission-critical applications.

If the computer and Internet communities do not come up with these more stringent standards, they will be imposed by outside influences, particularly the government. Because as net-centric computing devices become more ubiquitous and a common place way of moving information, the Internet infrastructure will become as important as bridges, highways, and roadways are for vehicles. Just as standards for highways, roadways and bridges have become common place as our economy became more dependent upon them, standards for the equivalent on the information superhighway will become commonplace in the future. And if the industry does not take a role in defining the standards for reliable code these standards will be enforced upon it, just as they are for mission critical and safety critical applications in airplanes, air traffic management and medical instrumentation, among others.

Hardest hit will be developers of software for the desktop, as they transition their products to an increasingly net-centric environment and desktop programmers as they try to adapt their methods to net-centric computing on the Internet. The cost of buggy software is already high. According to a study by the Standish Group while the annual investment in software is $250 billion, half of that is devoted to the costs associated with defective software. According to a study by Bender Associates the cost of a defect to a moderate to large organization is $20,000 to $40,000. But in a net-centric environment, where computing is done in groups on a network and where defects on a single client or server are magnified due to their effect on other clients and servers in a network, such failure rates are not acceptable any longer. A good example of the pervasive nature of defective software is that Microsoft Corp.'s Windows 95, its flagship software product, shipped with an estimated 10,000 defects according to a report in Infoworld in July, 1995.

Memory Leakage and Illegal Operations

The problems quantified in such studies are the result of bad code; that is, the software equivalent of the wrong bolt used with the wrong nut, the wrong material used in the construction of a building, or an engine that was not put together in the correct way. There are also other problems that are less obvious and take a lot more analysis and testing and have to do with, more than anything, badly written and sloppy code and construction. The equivalent in the physical world might be in the use of 2 inch by 4 foot instead of 3-by-8 boards in the frame of a house. The house may function as it is supposed to and remain sturdy for many years, but under the right circumstances the wall or roof may cave in because it was not constructed correctly..

Similarly, a program, be it an application or an operating system, works, but not very well, and under the right set of circumstances could crash. Typical of this sort of problem is memory leakage, which is endemic to the desktop environment, where relatively well tested software components are intermixed with others that are less so, and where the interactions between components are not rigorously tested and run to ground. The most common manifestation of this kind of problem is the user's system freezing up and an error message popping up saying "out of memory," "illegal operation," or "insufficient system resources." What has happened is that during the day, every time the user has accessed a tool or program, run it, and then closed it down, rather than freeing up 100% of the memory it was using it, frees up only 80%. Another program may only release 75% of the memory space that it occupied. Eventually the PC just runs out of memory space, or it over the course of the day, degrades in performance.

Under Windows 3.1 and Windows 95 there is a rather inelegant and sloppy way to solve this problem: reboot by hitting the control-alt-delete key. Memory leak control is particularly important in net-centric applications involving the use of browsers, which continuously allocate and free up memory for images and HTML. Microsoft has worked hard to eliminate such problems, but not in Windows 95, probably because the average computer user has gotten used to dealing with such situations and there is no concerted market demand that Microsoft deal with the problem.

In Windows CE 2.0, which Microsoft has targeted at net-centric applications, the operating system does clean out the memory of allocated programs when an application terminates. Moreover Win CE 2.0's memory management can track memory leaks to the process that is the source of the problem. Also development packages are supplied with Win CE 2.0 to help developers find memory leaks. But the desktop world, and the software derived from it, such as Windows CE 2.0, still has a long way to go. The embedded software world has a lot to offer developers of net-centric systems. For, unlike Microsoft, which leaves it up to the developer to find and fix such problems, most real-time OSes, such as OS-9000, VRTX and VxWorks, keep track of all memory resources that the kernel allocates and deallocates from processes. Upon termination, these real time kernels return any memory consumed to the system pool whether termination is fatal or normal.

But aside from taking more care in the choice of languages and operating systems, using better compilers, and employing better and more integrated development environments at the front end, the best way for a the designer of a net-centric computer system to avoid such problems is better testing and code coverage. The tools and methodologies are available; they just are not being used properly or enough.

Increasing Test Coverage

A not uncommon experience for a software developer using the serious code coverage procedures typical of avionics and medical instrumentation equipment for the first time is to find that only 30 to 40 percent of the code has actually been tested. What this means is that 60 to 70 percent of the code is passed on to the end users with bugs or with code that interacts with other code, the hardware or the operator in unforeseen ways.

This kind of skimpy code coverage occurs when the engineer or developer makes a common mistake in assuming that simply increasing the amount of time devoted to code testing has any relation to the amount of code that is covered. One European company ran a suite of tests for 8 hours and found that only 30% of the code had been tested. So they ran a second 48-hour suite of tests over a weekend and found that only got them up to 37%. The problem is that most systems developers, particularly from the desktop world, typically have a very narrow view of testing, looking only at the way a system is supposed to work. What they do not test is all the other conditions the system may undergo. In an embedded application, such as the controller in a microwave oven, an engineer would test to see if the sequence of lights or messages on the display were correct and assume that the system was operating correctly. What is not tested is whether there are conditions under which other sequences might occur. Like an algebraic equation with more than one solution, a typical system has a range of behaviors only one of which is correct, or, at least, which fits the requirements of the application.

A system may be operating correctly for 90% of the time, but when one looks at the underlying collection of behaviors that results in that operation, there may be more than one way in which a sequence of operations could be arranged to reach the desired result. To do a proper job of code testing, it is necessary to find out all the possible sequences, which are allowable, functionally, and then how often the system is running correctly.

The procedures developed by governments and standards bodies for mission- and life-critical applications are extremely detailed and comprehensive. At the very grossest level, there is what is called "statement coverage," where every statement in a program has been invoked at least once to determine if it is initiating the right response. In an aircraft, this is about the level that would be required to test something to be used for entertainment, say, a set of earphones or a seatback display. Beyond this are at least two additional levels of coverage: decision coverage and modified decision coverage.

In decision coverage, every point of entry and exit in a program is invoked at least once and every decision has taken on all possible outcomes at least once. Beyond that, it must be determined that every control statement and every branch point in the program have taken all possible outcomes at least once. Also, every Boolean expression in the program that contains a variable has been evaluated to both a true and a false result. Finally, in those systems requiring modified condition decision coverage the designer is required to know what impact each of these variations has on the correct functioning of the system. At this level it is necessary to know all the possible outcomes of every decision of the program, all the outcomes of every subroutine, every statement, and every expression.

Quite obviously the level of detail involved in such requirements is only affordable in the most mission- and -life critical applications. But the engineer and systems designer involved in developing net-centric systems and applications should at least have a general understanding of the details of the operation of his or her program or application. With this kind of understanding, an engineer can knowledgeably make decisions about the level of coverage required. Only then can the engineer assess the level of coverage (30%, 40%, or 80% of the code tested), the costs involved, and the time that will be required. These issues then must be balanced against time to market and the levels of reliability that a company or end user might accept. Without this level of understanding, the system developer is making such decisions in the dark.

Broadening Test Coverage

Beyond depth of software test coverage, systems designers have got to expand the breadth of test coverage and move beyond just simple functional "black-box" testing. Functional tests are at the heart of traditional validation and verification processes and usually receive the most attention. But they are not, and should not be, the only kind of test considered. Depending on the net-centric application, there are a number of other kinds of tests that the prudent designer should perform: error tests, free-form tests and white-box testing.

In error testing, the developer must ensure that the hardware and/or software handle all error conditions properly. It is also necessary to determine whether there are modes for "graceful failure" incorporated into a system that at are not catastrophic or have the appearance of catastrophe. What is catastrophic is often dependent on the end user. To a typical PC user, the system freezing up or an error message is not necessarily catastrophic. It just means that it is necessary to reboot. To users of Internet appliances and Web-enabled settop boxes, such operations are catastrophic and mechanisms need to be included to allow for graceful failure and recovery.

White-box testing is more comprehensive that traditional black-box testing. In the latter, user level inputs are entered and the operation of the program or device is evaluated without considering the inner workings of the device. What occurs in between is usually inferred from the results. However, such inferences can be mistaken. With white-box testing, it is possible to look inside the device and create tests to find weaknesses in the system's internal logic. This kind of testing allows the designer to examine communication protocols, data structures, and timing relationships.

Free-form testing has usually been at the bottom of the test hierarchy. Since it is most useful where relatively technically naïve users are involved, it will be a very important test methodology for the system developer to consider. This kind of test defies formalization because it is not done according to product specifications. Rather, it is a process by which users with different knowledge levels use the device in unexpected and unconventional ways in an attempt to invoke a failure. Because intuition, hunches, persistence, and luck all play a role in this kind of testing, it is often the hardest of all. Despite this, in many net-centric applications, where the typical user is a technically naïve consumer, it may be the most important.

Up to now, the level of software and hardware testing that net-centric products have had to undergo was relatively superficial. But this is changing, for two reasons. Governmental bodies and standards groups are apt to get involved as these systems become more and more common and ubiquitous. Second, the level of acceptance of the end user is less flexible and tolerant. Someone used to a TV set or a VCR has a very simple view of reliability: it either works or it doesn't. It is black or white, with no areas of gray.

Java-Imposed Test Requirements

The introduction of Java-based applications and systems may change fundamentally the depth and breadth of testing a net-centric system needs. From a code and software reliability point of view, the Java language, with all of its built-in mechanisms to prevent the programmer from writing bad code is probably the best thing that could have happened to programming, especially in the increasingly net-centric computing environment.

However, Java does nothing about writing lousy code a program that is not a good citizen in terms of its interaction with other programs. And when this is combined with Java's ability to generate applets and applications that can be delivered across a network and downloaded dynamically this introduces a level of complexity into the problem of ensuring the correct operation of a net-centric device's software. Unlike the past, where a system or software program was expected to operate in a relatively static manner over time, Java-enabled products can now mutate in the field. In a typical network computer or Internet appliance, the software may be changing constantly as new applets are downloaded. This introduces complications vis-à-vis system reliability on at least two levels. First, unless there is a detailed understanding of all the operating modes of a device or system, how will it be possible to predict what will happen when a new applet is introduced or a JavaBeans-ActiveX component is replaced in a system's software? Second, even if the base system is well understood, it will be necessary to test the applet in a wide range of systems and under a wide range of conditions to understand how it works and does not work.

One possible way to solve the latter problem, or at least place some limits on it, is to adapt the security mechanisms that have evolved for the control of malicious applets and viruses (see Chapter 11). Bad code and insufficiently tested code used in a downloaded applet or application can be just as dangerous as one written deliberately to cause damage. In the Netscape approach the jail paradigm is used in which an applet is allowed into a system, but is limited in terms of the resources that it can access. The more well understood the operation of the downloaded applet and the conditions under which it interacts with the system, the greater the access. The alternative approach is Microsoft's traffic cop methodology in which all applets must apply for a license and satisfy a minimum set of rules. Beyond that, they are allowed unlimited access to a system's resources. If there is a problem, the "license number" is noted and the source of the application is interrogated or blocked out of the system. When either or both of these are combined with the IBM real-time network-based virus-hunting methodology, the combination that emerges is a possible way to proactively monitor a network or the entire Internet for not only malicious code, but faulty, error-ridden and badly conceived software.

For any of these approaches to work as a means of insuring the reliable operation of not only each net-centric device but an entire network three things have to happen: (1) there needs to be an industry-wide approach to the problem of assessing software reliability; (2) some central location where applets can be tested needs to be established; and (3) a minimum commonly agreed-to set of software quality requirements that both the receiving system and the downloaded applet must satisfy must be developed.

Containment And Isolation

No matter how much effort at insuring correct operation of the software is built into the front end of the process, it is a fact of life that code errors will slip through and bad design decisions and assumptions will affect the performance of the software. Especially in the network-linked environment, the open environment is conducive to external events out on the Internet could cause unpredictable effects in the clients, in the servers and in the routers and switches. With this fact very much on their minds, a growing number of developers are looking to solutions in hardware that may aid them in containing and isolating such problems.

One solution has been there all the time in some CPUs, such as Intel's CISC-based X86 architectures and its RISC-based i960 family. It is a technique that is being re-integrated back into many of the RISC CPUs that are now targeted at the net-centric market. This is the use of memory protection mechanisms, on-chip in the case of the X86 and the i960 family and resident in an external memory management unit in the case of RISC designs. The role of such hardware is to assist in the detection of software errors by: (1) Placing boundaries on the range of memory available for each memory access; (2) Assigning privilege levels to each software element, with validation of the appropriate privilege level provided before each access; (3) Assuring that the error detection and correction software runs whenever an error occurs; and, (4) Providing information about where the error occurred. There is nothing radical or based on some previously unknown technology here. These are standard techniques that have been used for years in traditional multiuser operating systems on mainframes and minicomputers.

One reason the X86 architecture is still used in some segments of the telecommunications industry and in many network computers, even though other architectures far outstrip it in performance and interrupt response, has been its protected mode features. The wealth of development tools and the familiarity amongst many designers made the X86 the lowest-cost alternative. But it is the memory protection features of the architecture that allowed designers to isolate the system programs from the user programs and various segments of a user program from one another. If a program or subroutine had a coding error, or reacted in an unexpected way in actual operation, the bug did not impact on the rest of the system programming.

Intel designers added protected mode capability to X86 processors in the early 1980s in order to simplify porting UNIX to PC systems. The designers added significant hardware capability to support a multithreaded, multiuser capability, first to the 80286 and then improved implementations in the 386, 486 and Pentium processors. The key feature of the X86 protected mode hardware (see Figure 14.2) that provides error detection capabilities is memory segmentation. The X86 memory segmentation allows the software application to be constructed in such a way that logical application elements, known as "tasks" in a multitasking environment, can operate in completely separate memory spaces. The on-chip MMU contains the logic to assure that each task is only allowed to access its private code and data. Any attempt to access code or data outside its own memory space is done through a structured process which validates the authority to access that code or data through a hardware-based privilege-checking mechanism. Failure to stay within its pri-

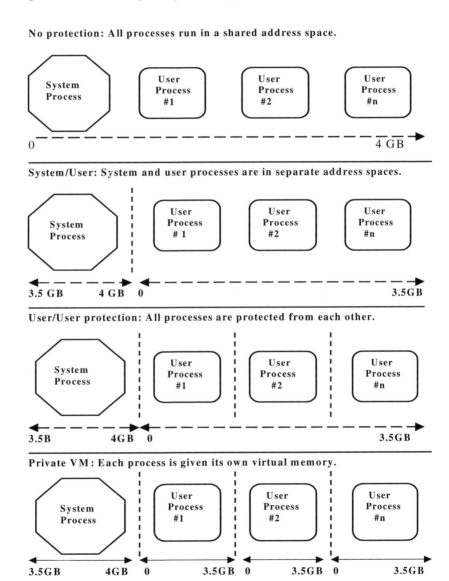

Figure 14.1 **Partitioning Tasks**. To prevent software crashes, OS should partition memory any of three ways: (1) all programs sharing the same memory addresses; (2) OS and programs in separate address space; or (3) provide each program with its own physical memory.

vate memory space or the use of an invalid privilege level to access global code or data are assumed to be software errors in the task's code and result in a processor exception.

This single segmented mode capability provides a wealth of software protection, isolation and containment features, all of which will be vitally important in the net-centric computing environment, including: (1) protecting error detection and correction code, since such code can be isolated against corruption by placing it in a separate segment; (2) automatic error detection, by placing each task in a separate private memory space, where the CPU can automatically detect a certain class of errors; and, (3) ensure error handler runs, since the errors detected by the hardware generate a processor exception, which in turn generates an exception handler, very similar to an interrupt service routine. Important for net-centric applications, where finding and repairing problems will be vitally important to network reliability, the memory protection hardware also provides information about where the error occurred.

As semiconductor processes have improved and CPU vendors have more silicon area with which to work, virtually every RISC vendor has a version in their family of embedded CPU offerings with on-chip memory management and/or protection. OS vendors are also building mechanisms into their software. It should come as no surprise that one of the first RISC processors to employ this technique was Intel's i960, for which one of the main applications was in data communications and networking.

In the i960, the protection portions of the architecture contains four features: (1) address translation for memory protection and virtual memory; (2) process or task management; (3) muliple processor support; and, (4) additional instructions to aid in the protection task. Although address translation is normally associated with systems that need virtually memory and paging such as workstations and servers, it has been used widely in embedded systems to provide a mechanism for isolating the memory spaces of processes and to protect the operating system from application programs.

Over and above what was done in the X86, the i960's major contribution to the protected architecture has been the concept of processes, adding an another level of protection (see Figure 14.2). Each process has an address space that is divided into four regions. Each process, by means of mapping tables specifies three regions 0, 1, and 2, and what functions and data they will contain. Region 4, however, is shared among all processes, because, independent of the process that is running, the system needs a uniformly addressable space where things not associated with a process, such as an interrupt stack and interrupt handlers, can be located. It might also be used for the operating system and data or code shared among processes.

Many of the OS companies whose software was originally designed to support the memory protection features of the X86 or i960, such as QNX and Microtec, are finding

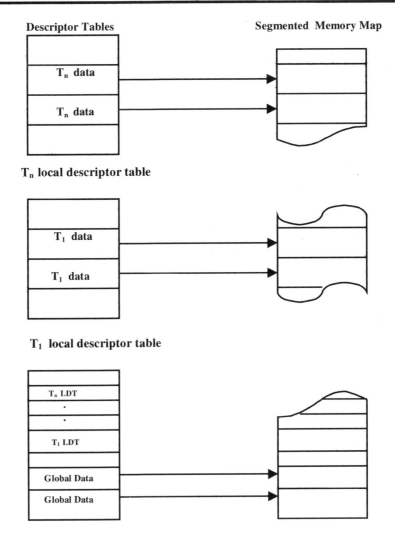

Figure 14.2 **x86 Memory Protection**. In this architecture memory is segmented so that logical application elements or tasks can operate in separate memory spaces so that faults and errors are confined and cannot spread to other elements.

that protected memory usage is expanding much beyond the original mission-critical and safety critical applications, such as avionics and medical instrumentation. What is now occurring to everyone is that with net-centric computing devices becoming an essential part of the information infrastructure, every application must be viewed as if it were mission critical.

In industry-wide standards such as POSIX, at least four levels protected mode operation and memory protection have been defined (see Figure 14.3): In addition to no protection, there are three other memory protection models: system/user; user/user; and user/user

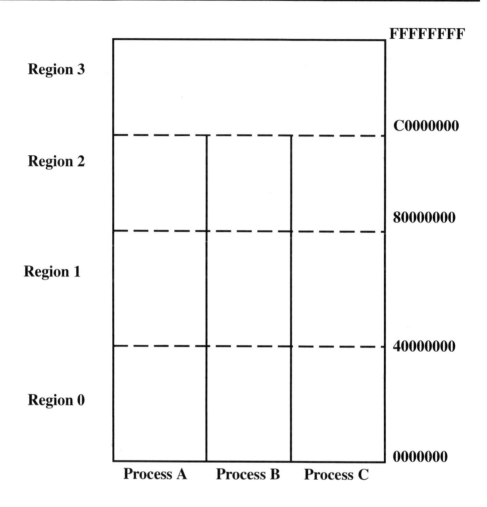

Figure 14.3 **960 Memory Protection**. To protect tasks, each process has an address space that is divided in four regions, three of which are isolated from one another and a fourth that is shared.

(private VM). The last three models relocate all code in the image into a new virtual space, enabling the MMU hardware and setting up the initial page-table mappings. This allows a program or application to start in a correct MMU-enabled environment.

In the no protection model, every process in the system is relocated during system build time to an absolute physical location. Without an MMU, there can be no distinction or protection between kernel and user address spaces, so all processes run in a common, shared address space. In the system/user protection model the application software and the operating system microkernel are relocated to the very top of the virtual address space. The page tables in this region are all marked as system, which allows access only by code that executes at ring 0 or 1. All other processes are assigned memory at low virtual addresses without the system bit set in the page tables. Since the microkernel runs at ring 0 and other less important functions of the OS run at ring 1, they are able to access all memory in the system. All other application processes run at ring 3, so the system address space is protected from errant user processes. However, in this model, user application processes are not protected from each other. The system/user model provides an "entry level" of system protection and requires a minimum allocation of page tables within the OS, making it portable for memory-constrained systems.

The next highest level, the user/user protection model, builds on the system/user protection model by also protecting user processes from each other. All page-table entries are marked as "system" except for those of the currently running process. Each time a process context-switch occurs, the kernel sets the system bit on the page tables for the process being switched from and clears them on the page tables of the process being switched to. In effect, the entire address range of the system is considered "system" except for the running process. To avoid having to set and clear a large number of 4096-byte page-table entries on every process switch, an OS that supports this model assigns virtual memory to processes in 4M blocks (the size of a page directory). Therefore, to protect or unprotect 4M of memory, we need to hit only a single page directory entry. Since the vast majority of processes are less than 4M, the performance cost of this protection is minimal. There's no cost when switching between threads in the same process or to any thread within the microkernel itself. The memory cost of this protection is one 4K page-table per process per 4M of memory used.

Finally, there is the user/user private VM (virtual memory) model. Here, each process is given its own private virtual memory, which starts at 0 and spans to 3.5 Gigabytes. This is accomplished by giving each process its own page directory. The performance cost for a process switch is similar to the previous user/user model, but the cost of a message pass will increase due to the increased complexity of obtaining addressability between two completely private address spaces.

Where on this scale do the most common operating systems fall. Some versions of UNIX, which had its start in the mainframe, minicomputer and superserver worlds, support all four, as do some real-time operating systems that either have traditionally supported the X86, such as QNX Systems Software Ltd., or operated in the high reliability telecommunications market, such as Anea OSE Systems, Inc. Most other RTOS support one or two, depending on which markets they are targeting.

In the desktop world, the implementation of the protected memory schemes has been uneven to say the least. One of the enduring mysteries of the personal computer industry is why Microsoft Corp., whose success has been inextricably tied to the X86 architecture, has until recently virtually ignored the protected mode mechanisms built into that architecture. Unlike competing OSes for the desktop in the mid-1980s, such as Digital Research's DR DOS and IBM's OS/2 both of which supported the system/user model, Windows 3.0/ 3.1 had no support for memory protection. In Windows 95, there was a move toward memory protection of a sort. Although it uses a segmented memory model, the operating systems processes are not really protected from the application processes. When an application goes, the system in most cases still crashes. What the event handler in Windows 95 did, however, was simply provides a message on the screen notifying the user of an error, and where it might have occurred. Windows NT, because it competes head to head against UNIX in workstations and servers, initially supported the system/user model. More recent versions are moving more toward the fully protected end of the spectrum. In the desktop environment, it is only with Windows 98 that Microsoft has chosen to finally support system/user memory protection.

However, in the net-centric computing world that is emerging this may not be enough. With the barrier between the network and the computer dropping away and, in some instances, the network becoming the computer, problems that affect one will affect all. Since it is not possible to write or test software to the point where one can guarantee no errors or bugs, protected mode features in OSes and CPUs will become the norm. And the first company to come up with a cost-effective and memory efficient way to provide the highest level of protection - user/user with private virtual memory - could become the Intel or Microsoft of net-centric computing.

Finding and Fixing

Once an error occurs and has been isolated and contained and the system or network has shut itself off gracefully, there is also the issue of finding the problem and fixing it. In many of the recent Internet blackouts involving problems with routers and switches, much of the downtime has been taken up by engineers first finding out what the problem was, the precise reason it occurred, and from whence it came. In the communications and networking environment, downtime is money, or rather, no money.

At the system level, telecommunications and networking providers have developed some very sophisticated methods for doing this. But what is the average net-centric system developer to do? Just as with memory protection, the answer may have been there all the time. In this case, it is the use of features contained in almost every RTOS offering on the market: resources such as trace event handlers and kernel profilers, both normally used in the debug stage, but removed once the product has gone into production and shipped. Increasingly a number of RTOSes such as QNX Software Systems Ltd.'s Neutrino are shipped with a simple trace even handler implemented as a resource manager. The trace program attaches to the trace event interrupt to catch trace events and make them available for reading on the OS's /dev/trace device. Events are maintained in a circular buffer that may be configured as small as 1 kilobyte or as large as 2 GB. When the buffer fills, older events are replaced by newer events, and the former saved to a disk or transferred over a communications line, reading and saving as they occur. Complementing this capability in the Neutrino OS is the ability to perform kernel profiling. Kernel profile events can be placed in all the major execution paths within the kernel. Kernel profiles can be examined for all kernel calls and interrupts of interest, such as every instance when a time period is created or arises, each time a timer fires and even the effect on a thread of a timer's firing.

The problem with such features, and a major reason they have not in the past been used in the field, is the impact on code size and performance. Depending on the number of kernel profile event calls installed, these calls will cause a noticeable increase in code size and a measurable decrease in performance. In the past, a number of companies have used the capabilities incorporated to help developers run instrumented kernels with profiling to collect debug information and then used them in run-time versions of their products, despite the differences in performance and code size. However, with recent improvements in OS design much of this difference is disappearing, making this alternative much more attractive to a wider range of designs. Whereas the versions of the earlier QNX 4's OS, instrumented with trace event handling and profiling, resulted in an 15% increase in code size and a 25% decrease in performance, in the newer Neutrino RTOS, this penalty is reduced by two-thirds. With an increase in code size of no more than 5% and losses in performance of no more than 8%, many more companies are going for the instrumented option that gives them that extra bit of security.

Tools for Monitoring: HTTP-Based Servelets

One of the simplest and most straightforward ways that a systems designer can add some degree of monitoring and remote debugging capability to a net-centric computer environment is to make use of a the new breed of embedded servelet programs being made available by a variety of software vendors (see Chapter 3).

In essence, what companies such as Integrated Systems (ISI), Microtec, PharLap Software, and Wind River Systems have done is to adapt the traditional TCP/IP-HTML Web page paradigm to the needs of remote diagnostics. This is done by turning on its head the traditional Web model in which one individual with his or her browser is used to access many servers on the network. In the alternative "many browsers, one servelet" model, the idea is to incorporate servelets (also called daemons, microservers and picoservers) directly into a remote location on the network such as a router and then to access the CPU intelligence inside when necessary for data, diagnostics, systems updating, and repair via any standard browser. Using Web technology as a diagnostic tool has the potential to change fundamentally the way various locations on the Internet are monitored and managed. Such communications have been possible before, but not without considerable communications infrastructure implemented and much more code than necessary in the embedded application to perform its real time tasks. Even using an X-windows GUI or writing your own text-oriented user interface in C, the amount of memory required in the system being instrumented ranged anywhere from 200 to 500 Kbytes at a minimum. Using the upside-down approach to implementing the TCP/IP protocol requires that the embedded servelet contain adequate code and routines to collect information, format it into HTML form, and present it to a remote browser when requested. Usually no more than 50 kbytes of HTML script is required, and often even less, depending on the sophistication of the application.

The basic function of an HTTP servelet is to respond to requests from a Web browser by sending an HTML page. However, there are important differences between a common desktop HTTP server, which acts mainly as a file server, and one that is embedded in a remote network device. One of the most obvious is that an HTTP servlet needs to be scalable to the specific lower-memory-space requirements of the remote device and be able to run out of ROM. Second, where desktop HTTP servers normally use the widely accepted Common Gateway Interface (CGI) to dynamically create pages to show status or display data at any specific time, CGI is often ruled out when the servlet is located in a remote device such as a network router or bridge. This is because using HTTP in this way requires a much tighter linkage between the browser and the servlet accessed. It also requires a much faster response time than those typical of a normal HTTP Web page. For this reason, CGI is ruled out, because it is big, slow, inflexible and only loosely coupled to the servlet.

As a result, many of the servlets developed by a number of software tool vendors use a variety of proprietary techniques for providing dynamic information. Because there is no standard as yet for this kind of application, a system designer who wants to build this kind of monitoring and diagnostics capability into a net-centric application must carefully assess the alternatives available and pick the one most appropriate in terms of performance and size. The system developer who wants to monitor events on a remote device simply specifies which variable functions are to be exported from the application running, say, on a router. The developer of the HTML content to be delivered back to the browser then uses this interface to include dynamic information, allowing pages with critical information about the status and working of the remote device to be generated and updated at any time.

Java-Based Monitoring

Building on the relatively closed environment that the embedded servlets provide, a number of companies are using the same Java language that they are employing for the network computers and Internet appliances to provide additional levels of diagnostic and monitoring capabilities. Most active are data communications and networking companies such as Cabletron Systems Inc., 3Com, Cisco and Newbridge Networks, Inc., among others. A good example of the kind of schemes being employed is the Java-based device management system (see Figure 14.4) developed by Cabletron to monitor its i960-based bridges, routers, and switches. As with many implementations, it makes use of an in-house developed HTTP server to serve up the Java classes on demand and provide a host environment for the execution of servelets on remote sites.

In the architecture it has constructed, any of a number of Java-enabled browsers can be used to monitor, debug, and reconfigure the underlying hardware and software. Within the browser, a device management console has been created. It is an HTML Frames-based application console built on top of Java's Abstract Windowing Toolkit, allowing it to be completely portable and window system independent. It manages two other components: scope objects and application applets. Presented to the system operator or manager as a hierarchical view, each Scope object contains a display component, an HTML reference to the provider applet and applet specific information.

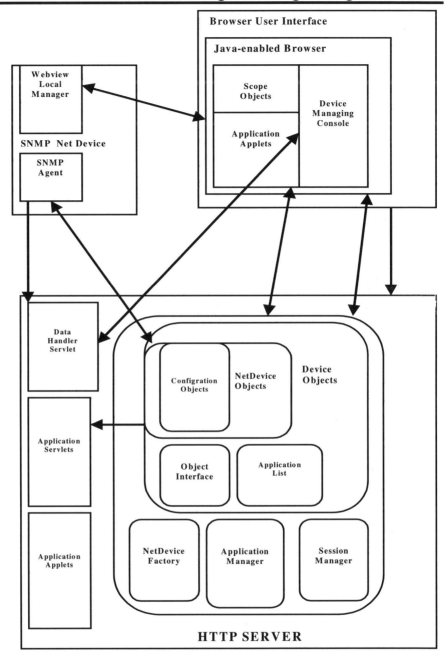

Figure 14.4 **Java-enabled Management.** Network system provider Cabletron has implemented a network management and repair system based on the use of Java and its downloading features to generate up-to-date monitoring of a network and its components.

The application applets contain information on specific tasks to be performed on the remote embedded server site on the network. At the specific site being monitored there resides an HTTP server that contains a number of components: the Device Object Factory, DeviceObjects, ConfigurationObjects, and a NetDevice Factory. The common currency, the means by which these various components interact, are NetDevice Objects, which provide a memory representation of a network device. When invoked by the browser with a request for an IP address, the DOF orchestrates the process of device identification and access to information. It then returns an HTML page to the browser containing the device management console, which contains specific information about the status and operation of the remote site and allows the user to execute, in the form of downloaded Java applets, the appropriate corrective action based on the information displayed. Rather than the TCP/IP protocol, the Cabletron system communicates between the browser components and the servelet components using SNMP (simple network management protocol) and uses Java's Remote Method Invocation mechanism to deliver information back and forth.

Building on this foundation of Java-based monitoring tools at the device level, Cabletron has begun to convert its existing C++ -based Spectrum network management system to Java and integrate it with CORBA to provide an enterprise- and network-wide fault and traffic monitoring system. Currently, Spectrum uses the Web and the Internet to monitor networks based on its i960-based routers, bridges and digital switches. The two core components of Spectrum are SpectroGRAPH and SpectroServer. Complementing their operation are three C++ -based applications: SpectroRx, Web Alarm View, and Web-Based Reporter. SpectroServer is a network manager, and SpectroGraph is a graphical user interface that allows clients to access the SpectroServer applications. Among the most important is SpectroRx, which provides users of Cabletron's hardware with automated fault resolution. It gives network technicians the ability to quickly isolate and fix faults by using knowledge based on a prior set of problems. Web Alarm View centralizes fault isolated alarm data from anywhere in a network using a Web browser, providing a quick reference point for how many alarms of various conditions exist in a network. The Web-based Reporter collects and brings information about network operation to the desktop. Via the Web, reports can be automatically scheduled to cover everything from router or server failures to unexpected Internet useage.

In the future Spectrum will revolve around the use of CORBA to provide the mechanisms by which objects located on the network can transparently make requests and receive responses. It is using Java to provide a mechanism by which applets can be downloaded to a trouble spot for monitoring or correction of the problem. Through the Internet Inter-Orb Protocol (IIOP), a platform independent messaging protocol, applications developed using Java can interact with other Internet-based applications. For example, an Alarm Manager Java applet accessed via a Web browser would use IIOP to communicate with a Corba-based network management application elsewhere on the network.

An on-going engineering initiative at Cabletron that will eventually be integrated into Spectrum is a Java-based SpectroRx tool, in which a Web browser would be used to access the remote SpectroGraph application where SpectraRx resides. When invoked, the Java-enable SpectroRx provides the remote user with a list of the faulty device hardware or software attributes, as well as suggestions to correct the problem based on previous experience. Another initiative that will eventually be integrated into Spectrum is an auto-discovery feature which uses both Java and CORBA to provide a fully distributed and extensible fault and traffic-load discovery system. Java-based agents will be used to perform the network discovery operations, and CORBA interfaces are used to communicate that information and allow remote users to assess and correct problems.

Building Fault-tolerant Networks

As effective as such procedures will be, additional efforts must be made to develop fault-tolerant software mechanisms that are effective in distributed computing and network environments. It is on the network where the most common mechanism of failure - outside of deliberate acts of sabotage - occurs: an isolated problem in a processor at one switch, bridge, or router causes the network to shut down, due to some software glitch that has been propagated. Even if it were possible to interconnect reliable processors with 100% bug free software, this would not result in a robust and fault-tolerant network of processors. The most that can be hoped for is a network in which individual nodes, be they switches, routers or bridges, are individually trustworthy and work well under most conditions. Until there is a better understanding of the interactions of large numbers of processors in a network of thousands or tens of thousands of nodes, everyone working on net-centric computer development will have to accept the fact that these systems may freeze up when anything happens to an individual processor.

Even with the limited understanding of such interactions, efforts are being made to develop fault-tolerant software that allows processors and networks of processors to restore normal operation, even when problems occur. The most widely used technique involves primary and backup systems, in which the backup operates in the background and is called into operation when or if the primary fails. This approach is difficult to implement and has its own problems. At the software level, it involves maintaining duplicate sets of programs and files on the two systems. One drawback of this approach is that switching between the two systems in an extensive network involves constantly updating files as they change. If this is not done just right, in the right sequence, and in a timely manner, the end result is a primary and backup that are no longer identical. The same problem occurs even if the problem is not with the primary but with the communications links between the two. If the backup takes over and continues operations until the link is reestablished, again the primary and the backup are not identical. An even more serious problem lies in the fact that the primary and the backup must be mirror images. If they are, then a software error or bug in one is most likely duplicated in the second.

To deal with such issues, program developers who work in the network and distributed computer environments are looking to a technique called "active replication" that is just now emerging from universities such as Cornell, and the laboratories of large telecommunications vendors. In this approach, a system's software is organized into process groups, which are sets of programs that cooperate together to achieve a particular goal. These process groupings are then replicated and distributed throughout the network. Additional redundancy is achieved, since programs typically can belong to several of these groups. Active replication is effective in making a system's software more fault-tolerant, because any group member can handle any request so that if one processor crashes, work can be redirected to another location. Even active replication has limits, however. One is that, if all members of a group handle an incoming message or process data in the same erroneous manner, all the members could in theory crash at the same time. A more common and realistic limitation is its effect on response time, something to which the average user is very sensitive. A rule of thumb in computer design is that if the average response time of a system to a user input goes much beyond 1 or 2 seconds, it is perceived as system that is not working properly. Because the distribution of replicas of members of program groups is, in effect, a form of load sharing, some processors could be slowed down by the additional load and others not at all. But the overall effect is a slow down in the entire network of distributed elements.

There are still a lot of unknowns to resolve that simply cannot be done in the laboratory, and the only way to get more experience with the operation of networks of distributed processors is to implement them and gain some practical knowledge that we can apply to problems.

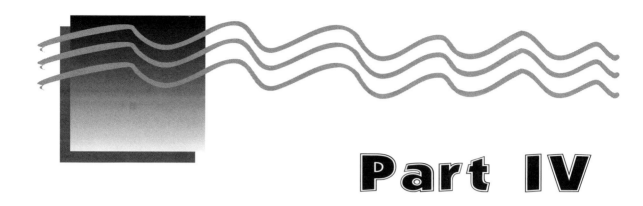

Part IV

The Future of the Internet

Chapter 15

Multimedia and the Future of the Internet

Even as engineers and designers build Internet appliances, Web-enabled settop boxes and network computers for use on the Internet and World Wide Web, the technological assumptions and conditions by which they made decisions are even now undergoing drastic changes. One of the major changes is the increasing capabilities of desktop computers and servers vis-à-vis multimedia and the second is the insistence on the part of the average computer user to make full use of these capabilities on the Internet and World Wide Web: adding audio, video and graphics to Web pages, sending multimedia emails, and using the Internet for video-on-demand services.

Currently, making the Internet and World Wide Web more multimedia friendly is being done on a piecemeal basis. Audio, video, and 2-D and 3-D graphics are all the focus of separate efforts. Only recently has the industry begun to make efforts to coordinate these efforts and to come up with a standardized scheme for synchronizing all of these elements. However, this may be changing. As of late 1998, a new standard may be in effect that will do what all the well-meaning bickering amongst PC hardware and software vendors and amongst organizations on the Internet and World Wide Web have not.

Called MPEG 4 (see Figure 15.1), it is an outgrowth of the standardization efforts such as MPEG 1 and MPEG 2 to come up with a common video-audio specification for desktop computer, consumer electronics and digital TV. Scheduled for final approval in Nov. 1998, MPEG 4 is nothing less than an effort to create a worldwide multimedia standard for networking multimedia in all its forms. In one sense, MPEG 4 represents a reversal, or at least a sharp turn in the direction that the earlier MPEG 1 and MPEG 2 were moving. Whereas those specifications aimed at improving quality of the image without regard to bandwidth issues, MPEG 4 goes off in a different and more fundamental direction. While it retains the same basic discrete cosine transformation core algorithms, MPEG 4 is more similar to the H.32X communications specifications in that it is an effort to further improve quality, but in the smallest bandwidth possible.

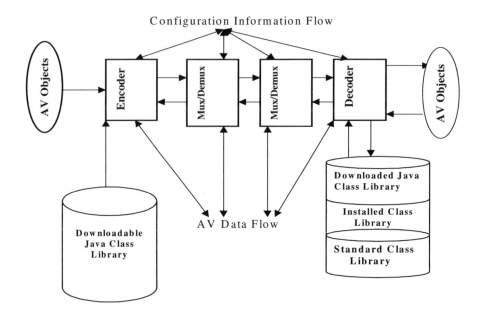

Figure 15.1 **MPEG 4.** An emerging standard for distributing interactive multimedia over the Internet and World Wide Web, MPEG 4 specifies the way in which audio, video, and 3D objects are encoded, decoded and synchronized.

In another sense, MPEG 4 represents a much more ambitious agenda. It is an attempt to place under one umbrella such things as (1) object-based representation of audio, video, and graphical sequences; (2) 2D and 3D graphics capabilities using extensions to the Internet's virtual reality modeling language (VRML); (3) a dynamically scalable compression scheme that makes use of Java code to download the appropriate class libraries from the encoder to the decoder; and (4) the use of a layered "sprite" based representation scheme for video similar to that used in such 3D schemes as Microsoft's Talisman to allow much better prediction and integration of synthetic 2D and 3D content with "natural" video, audio, and voice.

Before getting to the specifics of MPEG 4 and the impact it will have on the way net-centric computing is done, it is first necessary to look at the scope of the problem and current approaches to solving it.

Audio And Video On The Internet

Successfully transmitting video and audio over the Internet and even over intranets and virtual private networks poses major technology challenges. Video images, which consume massive amounts of bandwidth, must contend with limited network throughput, and with transport protocols that were not designed for continuous transmission. In a complex and disparate system such as a computer network, no single technological advance emerges as a panacea for transmitting video. But advances in several key areas are now appearing that together will combine to significantly improve the quality and consistency of real-time media over IP (Internet protocol) networks.

While most Web sites still deliver video and audio in the traditional downloadable variety, saving the data to the hard disk before displaying, a new streaming video technique is now gaining hold. It allows servers to send video to the desktop continuously in real time, without downloading it first to the hard drive.

The bandwidth pipe to the end user continues to expand. ISDN terminal adapters and pre-standard 56K modems are currently shipping. These technologies represent a 60% speed increase over the existing V.34 standard. Cable and ADSL technologies, available on a limited basis today, are capable of delivering multimegabit/sec. rates. However, the full benefit of ADSL and cable advances will be tempered by the need for infrastructure development on the part of telecommunications carriers and cable operators. Jupiter Communications, a market research firm specializing in interactive communications, projects that even by the year 2000 over 80% of home-based Internet users will still use analog access methods. In corporate environments, users have access to higher bandwidth, but this is typically shared by many users and many applications.

The fundamental way to address bandwidth constraint is through compression. An uncompressed full-motion, film-quality 30 frames per second quarter-screen color video image consumes over 70 Mbits/sec. So a great deal of compression is required to accommodate it to even a cable modems 10 to 30 Mbit/sec data rate. There are a variety of compression technologies in use for streaming video. Most of these fall into one of two categories: Discrete Cosine Transform (DCT)-based and non-DCT-based. DCT methods, including MPEG and H.263, take advantage of similarities between consecutive video frames, and only transmit differences as opposed to all the data in every frame. MPEG1 and 2 were designed for CD-ROM and commercial broadcast delivery, respectively. They offer excellent clarity and motion representation, but they do not fare well at transmission speeds significantly less than a megabit per second. By comparison, H.263 was designed for modem-based video conferencing and as a result performs much better at low to midrange speeds ranging from 28.8 through 500 kbit/sec.

The primary non-DCT method in use today is wavelet compression, which has its roots in still images. Wavelets work by coding successive layers of resolution for each video frame, and then playing them back at the highest level of resolution that the available bandwidth will allow. The advantage of wavelets is that a single video file can be played out over multiple bandwidths. The disadvantage is that wavelets are forced to transmit each individual frame in its entirety. This individual frame transmission causes difficulty at low bandwidth in attaining the frame rates necessary to convey motion.

As severe as the bandwidth constraints are, they are not as big a problem for video as are the Internet's standard transport mechanisms and protocols. The Internet is a vast router network that exhibits tremendous variability in terms of throughput, congestion, and end-to-end delay. Most network applications rely on the Transmission Control Protocol (TCP) to ensure reliable transmission of packets across the network. TCP addresses packet loss through the use of retransmission, causing the familiar pauses followed by bursts of data. This is acceptable for text, which needs to be delivered error free, but not continuously.

However TCP is inappropriate for video and audio media in which continuous playout is integral to comprehension, but in which the error-free transmission of every pixel is not. A better alternative is to use the User Datagram Protocol, or UDP, and combine it the new streaming audio and video techniques. UDP does not retransmit lost or error-filled packets and therefore provides a continuous play-out. Most importantly, UDP is much better suited for the new streaming media technologies that are emerging to allow better quality video and audio in real time, within the constraints of present Internet bandwidths. While most Web sites still deliver video and audio in the traditional off-line downloadable variety, saving the data to the hard disk before displaying, the new streaming video techniques allow servers to send video to the desktop in real- time, and continuously, without the need for the client to download to the hard drive, if one is available.

Depending on the choices made between video quality, real-time performance, and bandwidth, most of the streaming protocols now available on the Web deliver bit streams ranging from 8 to 600 kbits/sec., with compression ratios ranging from none at all to 200:1. To compensate for variations in throughput, even during a single connection, streaming products usually incorporate stream rate adaptation capabilities. This technique varies the rate of the video-audio stream in response to network capacity, while utilizing a data buffer at the client end to deliver a constant media stream to the viewer.

Because UDP makes no attempt to recover lost packets, there is a need for specialized loss compensation and coding technologies; a variety of techniques designed to reconstruct the original video-audio signal even in the face of network loss. For example, Motorola's TrueStream software uses Dynamic Video Image Correction (DVIC) and Audio Loss Interpolation (ALI) to compensate for network packet loss. DVIC is a form of forward error correction in which a small amount of redundant data are sent, allowing the client decoder software to intelligently replace lost packets and significantly mask their absence. ALI is an audio interleaving method that spreads continuous sound bits out across a number of packets so that the loss of any single packet will not be significantly noticeable.

Other important coding technologies include pre- and post-filtering and content-based coding. Pre-filtering simplifies an image by reducing the range of values for color and brightness, therefore making the image representable by a smaller set of data. Post-filtering conceals artifacts introduced by network impairments by visibly smoothing the border of a block of pixels. An example of content-based coding is Motorola's Content-Sensitive Bandwidth Allocation, in which bandwidth is borrowed from low-motion portions of a clip, such as talking heads, and allocated to needier areas such as those involving high motion or rapid scene changes.

In addition to the many improvements being made in the underlying technology of streaming video, other changes are necessary in order to simplify the process for the user.

Today, most streaming products require a specialized browser plug-in or viewer applications software. While such software is generally available for free on the Internet, users are understandably reluctant to download and install software each time they encounter a new content format.

Many Solutions, No Standard

The main problem with streaming video is that there is no clear standard and at least a dozen different and proprietary ways to achieve it. The main players in the mainstream of the market are Microsoft with its Netshow and Active Streaming Format (ASF), Motorola's Truestream, Progressive Networks' Real Video and Real Audio, VDOnet with its VDOLive, Vivo's VivoActive, Vosaic's MediaServer, and the Vxtreme Web Theatre, all of which require dedicated servers. High-end streaming video servers are also available from companies such as IBM, Oracle, SGI, and Tektronix.

Today, while there is no standard by which streaming servers and players from different vendors can interoperate, such efforts are beginning to emerge. One is the Real-Time Streaming Protocol, RTSP, an attempt to ensure interoperability in the same way that HTTP defines interoperability between Web browsers and Web servers. RTSP is currently an Internet Draft before the Internet Engineering Task Force (IETF). It is likely to become a full IETF standard by the middle of 1998.

Based on proposals from many of the streaming video suppliers, RTSP is most likely to be based on either Microsoft's ASF-based Netshow, Progressive Network's RealVideo and RealAudio, or Vosaic's Media Server. While Microsoft's Active Streaming Format stands a good chance of becoming the core of the RTSP specification simply because of the company's presence in the market, some of the alternatives have technical advantages that mean they should not be ruled out.

Progressive Network's advantage is that as the pioneer in streaming video, it dominates the market, with an estimated 60,000 servers deployed using the technology. It also has the most third-party tools available and is the most versatile of all the techniques now available. First, the RealVideo and RealAudio streaming protocols come bundled together. Second, at least two different compression-decompression schemes are available: the standard codec for typical 28.8 to 56 kbps modems and the ClearVideo fractal codec licensed from Iterated Systems, Inc., for higher-bandwidth environments. Progressive's technology is also the most adaptable. It can stream via IP Multicast, UDP, TCP, and HTTP and supports anywhere from 60 to 100 streams simultaneously.

Vosaic's MediaServer technology has three key aspects that also make it a likely candidate to serve as the model for the RTSP standard. First, MediaServer owes its origins to the same University of Illinois-NCSA developers who created Mosaic which became the foundation of both Netscape's Navigator and Microsoft's Internet Explorer. Second, Vosaic has developed an adaptive transport protocol, VTP, that ensures delivery of packets, whatever their content, to the end user. Where TCP/IP sends each packet and makes sure each plays, and UDP sends the packets but leaves it to the network to determine when they play, VDP monitors server, network and client to ensure that each packet that can play can be delivered. Third, Vosaic's Media Server is written in Java and makes use of the dynamic downloading features of the language to speed up the process of streaming the data for display in real-time.

Fourth, whereas many of the other streaming protocols are based on more efficient proprietary schemes, MediaServer is totally standards based, using MPEG 1 and 2 and the H.273 video compression specifications for all components. Fifth, of all the protocols, it does the best job of synchronizing. It separates the audio and video elements, indexes each frame for synchronization and sets a priority on smooth audio and video flow unimpeded by fluctuations in the data stream due to network traffic.

3D On The Internet

Another multimedia data type that can be expected to change the way we use the Internet and World Wide Web, and the way designers and engineers build their computer systems is 3D. By far the most widespread protocol for handling 3D graphics and animation on the Internet has been the Virtual Reality Mark-Up Language (VRML). Formalized in 1995, VRML 1.0 focused on static scenes and hyperlinks. Using a VRML browser, a user can move about a static 3D scene, examine its objects and jump to links.

Central to VRML is the concept of the "world," which is nothing more than a scene or a graphics frame. Each VRML scene has a fixed number of items of specific types and qualities. In addition to describing the content and layout of a world, a VRML document can also include linkages or anchors to other Web documents. Each VRML scene has a point of view, which is called a camera, through which each scene is viewed. VRML allows a scene to include several different predefined viewpoints.

The mechanics of retrieving a VRML scene from a remote location are much like that with an ordinary Web browser. The Web server that receives a request for a VRML scene attempts to fill the request with a VRML document that contains coordinates rather than actual images. The scene is re-created on the browser using the coordinates supplied by the VRML document. Once a document has been received by the VRML browser, it is parsed. From the parsed description, a rendering program creates the visual representation of the objects described in the VRML document. A VRML scene, or world, can be distributed. In other words, it can be spread across the Web in many locations, much in the same way that an HTML document or page can contain text from one location and graphics from another.

VRML has very quickly moved from static to much more realistic dynamic 3D scenes, first with a number of proprietary extensions and most recently with the approval of VRML 2.0 which supports dynamic object behavior, multiuser interactions and multimedia components such as animation, sound, and video.

Particularly compelling to the engineers and systems designers sitting on the International Standards Organization's MPEG 4 committees is the incorporation of a Java API into VRML 2.0. It allows the downloading of 3D images based on data from throughout a network, using its Script nodes, the building blocks of VRML. In the creation of a multiuser environment, using Java's dynamic downloading capability it is possible to assign avatars, representatives of individual users in the game or environment. These avatars incorporate a range of properties, such as movements by the user (eventIn), movements of current avatars (eventOuts), transitions to new avatars (eventOuts), and garbage collection of avatars who have left the virtual world. The Java class referenced by a particular script can then do the detail work involved in the exchanges between the user and the world and between avatars, transmitting changes to and from other users or a multiuser server over the network.

With this enhancement to VRML 2.0, there are a number of other ways for VRML and Java to interact. First, it is possible to use the VRML 2.0 Java API to link into a VRML world from outside the VRML file. Second, it would be possible, with VRML browser, or an add on, to link into a VRML world from a Java applet.

MPEG 4 Brings It All Together

Whereas MPEG1 and 2 were aimed at CD-ROM and broadcast applications with bandwidths in the 1 to 4 Mbit/sec range and the H.32X specifications were targeted at about 28.8 to 640 kbits/sec., MPEG 4 is designed to operate over a range from 2kbits/sec. mobile applications to 4 Mbit/sec. broadband Internet applications. And where the earlier specifications were delivery oriented, MPEG 4 is designed to be interactive, allowing active viewer participation.

To allow a multimedia video data stream to be easily scalable to the requirements of the receiving systems and the variety of interconnect protocols, the MPEG 4 specification will probably incorporate a layered representation scheme similar to that used for 3D by Microsoft in its Talisman 3D architecture in which not all segments in an image are treated equally.

In Talisman, 3D effects are achieved through the use of four techniques: layering, image compression, chunking and multipass rendering. To reduce or eliminate the use of the traditional frame buffer, each object in a frame is stored in a separate image layer that the system can then combine to form a final image. This allows the graphics subsystem to access each object independently and then only if the object changes. An advantage of this technique is that many compute-intensive 3D operations can now be simulated using 2D operations on 3D transformations. A tradeoff of using this approach is that it requires trading geometric detail for higher animation rates, making it more appropriate for 3D games rather than professional or engineering 3D.

For compression and chunking, Talisman makes use of concepts developed originally for video compression algorithms such as JPEG and MPEG. In the Talisman approach, each image is broken up into 32 by 32 pixel chunks and sends them through the 3D pipeline chunk by chunk. And especially useful for reducing the bandwidth requirements for pixel intensive operations such as texturing and rendering, Talisman uses a JEPG like technique, Texture and Rendering Engine Compression, to squeeze file sizes down to reasonable levels. Then chunk by chunk, the rendered images are fed back through the texture processor for use in generating a new layer.

A similar methodology is being applied to video under MPEG 4. Unlike previous DCT algorithms, which broke an image up into blocks, each of which was encoded, decimated and decoded, under MPEG 4 different components of an image are treated separately: the background; static elements; moving images; hybrid images such as the human face, with some static and some dynamic aspects, an so on. Each is separated out and encoded and compressed according to different rules and different levels of efficiency. Borrowing from 2D-3D graphics, sprite coding will probably be used to encode the background images and some synthetic or static objects at reduced bit rates.

Moving objects within a scene, however, will use traditional core DCT algorithms for compression and motion prediction. As far as audio is concerned, MPEG 4, unlike its predecessors, will deal with speech and audio together, with very low bit rate speech processed using speech synthesis. Right now, at least three methods are being considered, with the final version containing elements of all three.

In its details, the video layering scheme that will be implemented works within a layered video object coding system that uses sprite and affine motion models. In this scheme, a video frame is sorted into a linear combination of layers, which come in two types: sprite layers and ordinary layers. Ordinary layers are those aspects of a scene that can be encoded with conventional motion-compensated transform coding, such as H.263. Whereas ordinary layering is applied to such things as stationary backgrounds, sprites are large objects built up from pixels in potentially many frames which are transmitted first and then warped and cropped to form a part of an object in a later frame. In such a scheme, the sprite for a stationary background is the combination of its visible parts in all the frames in a sequence. During decoding, the sprite is warped into the current frame with some additional information about the trajectories that the sprite is taking. The use of the sprite provides two important advantages in the MPEG 4 layering scheme. First, for objects that undergo rigid 2D motion, sprite compositing is adequate to generate estimated objects so that motion estimation encoding need not be performed. Second, sprites can be used for parts of an object in a particular frame that can not be predicted from a previous frame. Once identified, these parts can be encoded and compressed using techniques that are more efficient that the traditional motion compensated coding methods.

Integrating Synthetic and Natural Coding

Perhaps the most important aspect of MPEG 4 is its Synthetic/Natural Hybrid Coding (SNHC) scheme. SNHC defines how interactive 2D and 3D information - including synthetically and naturally generated audio-visual information - is compressed, multiplexed and transmitted. SNHC eliminates current video clips in on-screen windows in favor of fully integrated visual objects, each with its own audio or video channel. At the same time, SNHC enables interoperability of media streams and software downloads for hardware- and software-based decoders, thereby supporting true on-the-fly programming and upgrades.

SNHC is concerned with the compression of specific media streams (such as geometry, text, control parameters, or text to speech) beyond traditional audio and video. SNHC manages the representation and coding of synthetic objects and their natural audio-video counterparts and the spatial-temporal composition of these natural and synthetic objects to create semantically meaningful scenes.

In earlier implementations of MPEG, all the details of a scene, such as a newsroom with the anchorperson, the background, and whatever inset might be displayed beside the anchor, would be dealt with as a single frame. Compression would take place on a pixel-by-pixel basis or by comparing major elements of one frame to those of the frame preceding it. SNHC provides for the coding of each object in the frame - the anchor, the background and the inset - as separate objects. Furthermore, synthetic objects, such as inset photos or captions, are compressed in a manner apart from the natural elements of the scene, including the image of the anchorperson and the sound of his or her voice. These synthetic objects are interoperable with the natural objects in the scene and can be manipulated independently of all other objects. In addition, SNHC includes algorithms specifically designed to compress facial and body animation and synchronize speech at rates well below earlier compression schemes. MPEG 4's low-bit-rate audio capability enables speech encoding at an 8-kHz sampling rate with an associated data-transmission rate of 2 to 6 kbits/sec., but can also support data rates up to 64 kbits/sec.for higher quality audio.

Besides compressing each object separately, SNHC enables scaling of mixed media types based on storage media, communication bandwidths, and the rendering power of terminals with variable animation capabilities, graphics displays, and audio synthesis capabilities. This approach allows further reduction in the bandwidth of mixed media, offering trade-offs in quality versus update for specific terminals, and allowing the use of different distribution methods for content that exploit spatial and temporal coherence over buses and networks. This combination of features allows playback of a scene at various quality levels in environments ranging from high-performance networked desktop systems to video telephones. One thing that has been determined is that MPEG 4 will incorporate much of what has been done vis-à-vis standardization with regards to the Virtual Reality Modeling Language. Currently, MPEG 4 will support all the VRML 2.0 functions in its architecture.

MPEG 4's Object Model

To allow developers and content providers to work with a multimedia environment integrating all the different multimedia types, the MPEG 4 coding scheme uses an object-based model, in which any number of audiovisual objects (AVOs) can be encoded and sent as separate streams. Audio objects are specific audio channels, such as a single trumpet in a group of instruments or one voice on a video teleconferencing link. In video, however, objects are much different, incorporating both space- and time-related elements: shape, texture, and motion and information on how each changes over time.

To provide a mechanism for manipulating AVOs, the MPEG 4 specification includes an object-oriented system description language (SDL) that allows systems to exchange data and interact and modify the transmitted information on the fly. C++-like in its structure, SDL provides the linkage between AVOs such as audio frames, sprites, 3D objects, natural or synthetic objects and the representation models used for transmission (splines, waveforms, etc.). To do this, SDL incorporates a class hierarchy definition, as well as three language specifications: the readable language specification, the binary language specification and the syntactical description language. The first defines a readable format for transmission to the decoder, while the second provides a binary executable format for scripts or descriptions that can be understood by the decoder. The third is used to describe the bit-stream syntax and is basically an extension of the MPEG 2 syntax.

Beyond these system-level functions, SDL was originally to have a role in the composing and decompression process. It was originally going to be used to send downloadable code to the decoder during the preliminary configuration phase of the multimedia transmission process. What this requires is the ability to download code that is executable on any platform. But rather than reinvent the wheel and develop a new platform-independent mechanism, it was decided to remove this requirement from SDL and go with an already available language, such as Java.

Using downloadable Java-based code (see Figure 15.2), default composition information is included in the AVO streams sent out by the encoder, specifying the actions to be performed by the objects. During this negotiation phase, the multimedia decoder can interact with the encoder, specifying what its processing capabilities are, how much memory it has, and the pixel quality required. Based on this information, the encoder and decoder choose which methods will be used for the transmission.

During this part of the negotiation process, Java applets helpful in the decompression process can be downloaded to the decoder to be executed afterward. Based on this downloaded information, the decoder can then interact with the multimedia stream to change the appearance of the decoded sequence such as the 3D localization of sound, the pitch of speech, or an object's motion characteristics. MPEG 4 will also have a new multiplexing scheme that will allow bit streams to be composed from different sources, say, locally from a CD-ROM and remotely from the Internet, allowing the end user to retrieve MPEG 4 images, combine them with local content and display the result.

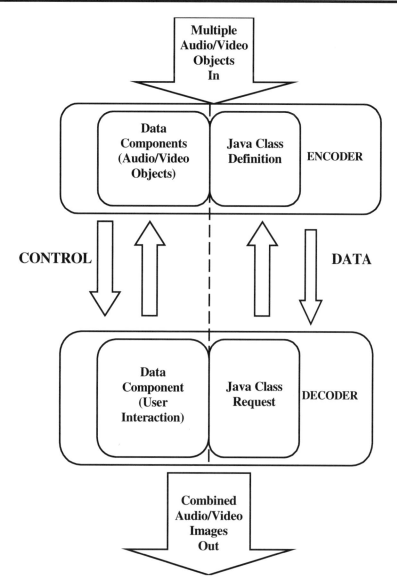

Figure 15.2 **Java-based.** To enable MPEG 4 to operate in synchronization between multiple nodes on the Internet, Java class definitions and applets are downloaded from the multimedia encoder to the receiving decoder.

In addition to the flexibility this gives as to how multimedia information is transmitted and displayed, the use of downloaded Java applets during the negotiation phase makes MPEG 4 dynamically scalable and capable of adapting to the capabilities of the decoder. With appropriate Java-based composition code, the multimedia encoder can build its bit streams so that a low-performance decoder would be able to decode only a part of the data and get lower-quality services, while more advanced CPUs and decoders could make use of the full capabilities of the bit stream and offer better service.

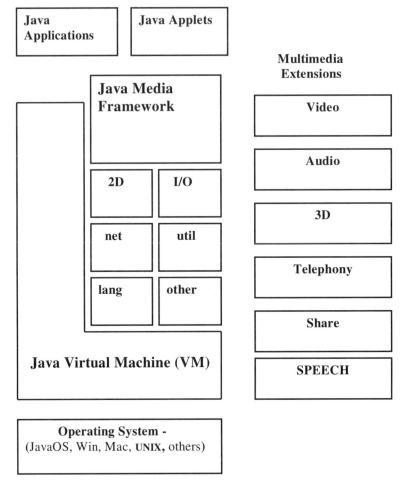

Figure 15.3 **Synchronization.** The JavaMedia API and extensions like Framework provide the mechanism by which applets downloaded from the MPEG 4 encoder to the decoders synchronize the accompanying multimedia.

Synchronizing Audio, Video and Synthetic Objects

Increasing the chances that Java will be the defacto language used in MPEG 4 for downloading configuration applets is the fact that recent extensions to the Java API libraries (see Figure 15.3) solve one of the last remaining problems with the new specification: synchronization and multiplexing of decoded natural audio-video. A number of companies have been working on this problem and have submitted proposals to the MPEG 4

committee. The Davic consortium, for example, has developed a solution based on MPEG 2 and VRML, while Microsoft, Sun and Silicon Graphics each have built proprietary solutions.

With a recent extension to the JavaMedia API called Framework, the Sun proposal appears to have moved to front and center. If there is anything that would limit MPEG 4's future as the over-riding standard for multimedia in a networked environment it is the ability to synchronize the various elements in a typical multimedia transmission of the future. Most of the efforts to date have attempted to pull all these elements together, but have been biased in one direction or another. One approach builds on VRML and adds video, but is weak on audio; another is strong on the traditional 3D and video, but not so good on describing 3D objects in a virtual environment; and none, to date, really addresses the issue of scalability, adequately.

This is why with the addition of Framework, a Java-based downloadable coding-decoding mechanism will probably be the direction in which the MPEG 4 standard will eventually go. Essentially, the Java Media Framework (JMF) API is a collection of classes that enables the synchronization, display, and capture of time-based data within Java applications and applets. Being jointly developed by Sun Microsystems, Inc., Silicon Graphics Inc., and Intel Corporation, JMF has been released as three APIs in three stages: Player, Capture, and Conference. Java Media Player was the first release of the JMF technologies and supports the synchronization, control, processing, and presentation of compressed streaming and stored time-based media, including video, audio, and MIDI across all Java-enabled platforms.

The Java Media Player receives and plays media from any source across all Java-enabled platforms and separates the issue of data type from the transport protocol, simplifying playback of media stored locally or on the network.

The JMF, along with Java 2D, Java 3D, Java Animation, and Java Speech, constitute what are called by Javasoft and Sun the Java Standard Extension API, at least until it is decided to include them in the Base API. An Extension API, once defined, can be added to and modified, but must maintain backward compatibility by accepting all legacy calls.

Java 2D extends the set of classes in the Abstract Windowing Toolkit (the Java .awt package, and is based on the Adobe Imaging Model for device independence. The latter is a single comprehensive model for line art, text, and images that uniformly addresses color, spatial transforms, and compositing. Extending AIM, Java 2D is able to handle arbitrary shapes, text, and images, providing a mechanism for doing transformations such as rotation and scaling. It also allows control of how graphics primitives are rendered with features such as stroke width, join types, and color and texture fills. As far as text is concerned, Java 2D allows a range of capabilities, from simple font definitions and use to character layout and font encoding.

Java 2D is designed to allow Java-enabled layout, compositing and Web design tools to be competitive with traditional desktop publishing tools. An additional benefit, however, is that it is programmable. To deal with problems of color conversion, Java 2D allows color spaces to be defined that map color values from one space to another, making it possible to guarantee that an image will appear the same on an RGB monitor as it does on a gray scale monitor or in the cyan, magenta, yellow and black scheme used by the printing industry.

Java 3D, from the point of view of the MPEG 4 specification, has a number of features that will make it useful in any interactive multimedia application or applet. The most significant thing is that Java 3D is an amalgam of the best features of many of the existing 3D APIs, including Microsoft's Direct3D, Silicon Graphics OpenGL, and Apple's Quickdraw 3D. Moreover, it operates at a much higher level, more like a high-end image generator. Graphics elements within Java 3D are treated as separate elements and connected in a tree-like structure called a scene graph, which contains a complete description of the entire scene, including geometry, attributes, and viewing information needed to render the image.

There are three rendering modes. Two of them are immediate mode, which takes the basic points, lines, and triangle information from an application and renders at the most efficient level, and retained mode, which requires that the application define in general terms how the scene graph is constructed and how objects within it can change.

The third rendering mode is the most interesting, especially when looked at in the context of MPEG 4 and 3D schemes such as Talisman. It is a compiled-retained mode, which works on subsections of the scene graph. The Java 3D API controls the actual organization of the scene graph, as well as the way each node connects to others and the way the scene graph is traversed. Contained in the API are the semantics of the scene graph, but the API can be extended by applications that add their own rendering methods or defaulting to the intermediate mode.

The Animation API handles sprite-based animations in fixed and sequenced formats and includes a number of sophisticated functions that would be useful in the context of MPEG 4, including sprite collision detection and image transformation effects. With this API, instead of sending an animated sequence frame by frame from a server to the client, the sequence can be contained within a single image map, portions of which can be clipped and displayed sequentially to create the desired animation. While this requires some additional coding, it eliminates the need to create a new animation application.

Tying all these APIs together with audio, video, and MIDI, JMF provides all the synchronization and timing required to make them all coexist in a complex multimedia sequence, such as MPEG 4. It does this by encapsulating video, audio, and MIDI data types in a uniform media form, or envelope, that can be played back in the Media Player portion of the API. The relationships between the various media types can be defined in a number of different ways, independently of their format. This envelope enables the interoperability of dynamically composed media modules over a network or the Internet. Because the JMF treats audio, video and MIDI streams as processes, it is possible to use it to construct applications that manage latency across a network. One of the most important aspects of this capability is that it allows composition of the media to be delivered over the network and controlled locally at the client, although the actual audio and video data are transported from a distant server.

Chapter 16

New Processors for the New Internet

It is clear that as the Internet and World Wide Web evolve toward what might be called the "MultiediaNet," the microprocessors that power the servers, the connected PCs, NetPCs, NCs. and Internet Appliances will undergo radical changes.

Many of the new CPUs now being built have already added features and instructions to accelerate multimedia operations. Except for users in particular applications who have an overriding need for multimedia, the average consumer, in the past, has had no compelling reason to use multimedia. So multimedia capabilities in traditional stand-alone PCs were added only if they did not significantly increase the cost of the system and then mostly for their entertainment value.

This situation, where CPU and systems vendors have to push the technology into the market, may be changing. As bandwidths increase and standards such as MPEG 4 become widespread, multimedia capabilities will be absolutely necessary on the Internet and World Wide Web. And this will turn the market from one in which multimedia technology is pushed into the hands of users, into one in which the technology is pulled into the market by this need.

Consider, for example, a net-centric device accessing the MPEG 4-enabled multimedia-rich network of the near future. A good example might be a Web-enabled set-top box. In its all-digital form it allows a user to not only access any of numerous cable or satellite channels, but also to access the Internet and view real-time video, audio, or 3D; hold a video conference; or play a game on a 3D site on the web.   These kinds of real-time interactive operations will require entirely different capabilities than are available even on current multimedia-enabled desktops. Not the least of these is the ability to manipulate multimedia data types in real time, and handle not just one data stream, but multiple ones of different types.

Previous generation CPUs were able to process noncontinuous media, such as the input from a word processor or data sheet or the packetized information from the Internet, in 32 bit chunks. What they did not do very efficiently was handle so-called continuous media, such as digitized video, audio and 3D. These new media types can typically be represented by 8- and 16-bit data types, extracted from analog signals in the time domain. In the network environment, the lack of efficiency is even more pronounced. Even though many of the schemes to up the bandwidth of existing telephone and network wiring involve the use of compression and decompression schemes similar in many respects to those for multimedia data types, they depend on bit-data-type processing done in real time.

What multimedia and these compressed network data streams share in common are input data streams that are often large collections of small data elements such as pixels, vertices, or frequency-amplitude values. Usually present in very large quantities, these data types tend to require identical processing and need to be processed very quickly to preserve either communications or video quality.

The Need For Parallel Processing

To get some idea of the computational chore involved, it is instructive to look at the MPEG algorithm, the worldwide standard for compressing, decompressing, and displaying video. It is based on the discrete cosine transform, a method of representing visual values and converting them into digital equivalents. Of the six steps involved in the conversion and decompression of a video stream, the inverse DCT (IDCT) step, in particular, even after optimization, consumes the largest chunk of execution time, almost 40%. The next two largest time-consumers are the display step, followed by motion estimation and compensation, both of which also contain some opportunities for parallelism. During compression, it is motion estimation and compensation that eats up processing power, followed by the inverse DCT.

The two inherently serial steps, decoding the MPEG headers and Huffman encoding and decoding, are relatively insignificant in terms of execution time. The IDCT step is a prime candidate for speedup because it contains a great deal of parallelism that can be sliced in many different ways in its processing of 8 by 8 blocks of pixels.

At 30 frames per second, the rate at which humans see "good-" quality video, each 8 by 8 IDCT block can be decomposed into eight independent, one-dimensional IDCTs on the rows, followed by eight independent 1D IDCTs on the columns. This translates into sixteen 1D IDCTs per block or almost a million (950,400) 1D IDCTs per second. Each 1D eight point IDCT itself has room for a number of different parallel operations — within a 1D IDCT, across 1D 8 by 8 IDCTs, across 2D 8 by 8 IDCTs of either one color component or multiple color component blocks, or even across frames.

With so many possible pathways within the IDCT, it is obvious that parallel processing of some sort would be extremely beneficial in improving MPEG and other multimedia applications. The nature of the operations within the DCT-based MPEG algorithm indicates that the best solution would be the use of single-instruction, multiple-data (SIMD) parallelism in which a control processor dispatches common instructions to multiple data processors, each of which performs the instruction on its own pair of data items.

But the dilemma is how to introduce SIMD structures into a CPU architecture without violating traditional CISC or RISC design principles The trick is to implement it so that, from the point of view of a traditional 32 -or 64-bit-wide instruction, what is seen is a standard RISC architecture, but when viewed by multimedia-optimized subwords, it looks like a SIMD machine.

Fortunately, the SIMD architecture can be added on top of existing architectures with modifications, which, while not simple, do not require substantial changes in the underlying design or unnecessarily increase the die size of the ICs. Because of this, most of the major 32- and 64-bit CPU vendors have been able to come to market with multimedia enhancements to their original designs: Intel Corp., Sun Microsystems, Silicon Graphics, Digital Equipment Corp., and Hewlett Packard.

The advantage of a SIMD architecture is that from a control point of view, as well as a software development point of view, programming issues, while more complex, are not beyond the ken of present techniques. Rather than build a separate SIMD execution engine, most of the CPU vendors have, in essence, piggybacked the SIMD pipeline onto the existing architecture. The most logical place for this in most cases is in the floating-point portion of the design, where the internal data paths already range from 64 to 80 bits. This has allowed the design of SIMD-like structures using subword parallelism that makes it possible for a processor to cram anywhere from eight to ten 8-bit-wide subwords into a single instruction cycle. The main differences between the various CPU vendors is in the way that they reconfigure the internal register file structure to accommodate SIMD operations and the multimedia instructions that they choose to add.

Adding Multimedia Instructions

The most well-known of the enhancements is the set of multimedia extensions collectively called MMX by Intel Corp. Initially deployed in the Pentium and Pentium II processors, it is also being used in the server-oriented 32-bit Pentium Pro as well as the next generation 64-bit Merced. Every x86-based processor Intel produces in the future will include multimedia extensions. This technology has been covered briefly in Chapter 1.

Here the focus will be on some specific instructions included in this architecture and what instruction mixes are being added to the other competing 32- and 64-bit processors. The big questions that need to be answered are (1) what is the most appropriate mix of instructions for the desktop and (2) what are the most appropriate instructions to include for networked multimedia operations on network computers, servers, routers, and switches?

As far as the desktop is concerned, Intel has put a great deal of effort in picking instructions for the desktop presentation of multimedia off of CD-ROMs and some low-bandwidth Internet connections. But are there additions and alterations that need to be made in this new environment of the connected PC? Before looking at that issue, it is important to review what Intel has done right, at least as far as the standalone desktop environment is concerned.

In the mix of instructions that Intel has selected as appropriate for the desktop, the arithmetic and logical instructions operate on MMX operands the same way as arithmetic and logical instructions do with integer operands. The difference is that MMX instructions operate on the new SIMD data types and apply saturation if desired. For example, a Packed ADD of two Packed Byte quantities will add eight sets of bytes to another eight sets of bytes to produce a third set of 8 bytes. Without MMX instructions, the addition would have to progress using the integer method eight times. All arithmetic and conversion instructions may choose to saturate their results if desired. Saturation causes the result to be clamped to its maximum or minimum value, so the result will not wrap around. The maximum value is determined by the size of the data type and whether signed or unsigned operands are used.

This capability is important in the processing of multimedia operations. For example, when a programmer might want to brighten each pixel because of a light source, saturation would be used to ensure that if a pixel were already very bright making it brighter would not wrap it around to become dark. If a byte pixel were value 08h, and 10h were subtracted from it to make it brighter, saturation would clip it to 00h (most white) instead of letting it wrap-around to be FEh (very dark). Without saturation, the programmer would have to ensure that the value did not wrap around by using standard compare and branch instructions. Saturation therefore saves program code, instructions, and most importantly branches, as branches are a performance limiter in highly pipelined CPUs.

Conversion instructions are important in packing and unpacking, a process that is virtually synonymous with compression and decompression algorithms. Packing data moves data from the source to the destination, where the destination is a smaller data type. Unpacking data does the reverse, moving data from a smaller to larger data type. These instructions may be used to save memory and transfers during multimedia operations while maintaining precision. A common sequence during the IDCT operation where these instructions are highly effective is as follow: 1) pixel data are stored in byte quantities; 2) data are loaded as packed bytes; 3) packed bytes are unpacked into packed words; 4) the application performs higher-precision operations on the packed words without regard to over/underflow; and finally, 5) the significant portions of the data are packed back into packed bytes and stored back to memory. Packing and unpacking data actually take data from the source and the destination, apply saturation if desired, and store it to the destination. In packing, the upper half of the destination comes from the source, while the lower half comes from the destination. In unpacking, the source and destination are interleaved into the destination. This allows zeros or other numbers to be filled in where necessary.

Also playing an important role is the comparison instruction. Like the arithmetic and logical instructions, it performs SIMD-like parallel compares on all elements of the data type at once. However instead of setting CPU flags to indicate true/false, the comparison instruction sets the bits in the destination operand to all 'ones for true and zeros for false.

The destination operand then becomes a mask operand of ones and zeros which can then be used to alter the results of other operands. This type of comparison is used to speed up program execution by eliminating branches, the one aspect of modern high performance RISC design that is often a crippling bottleneck for multimedia operations..

Multiply-Add, sometimes known as MAC(Multiply Accumulate), is an instruction included to speed signal processing algorithms often used in 3D operations. The Multiply-Add instruction multiplies two sets of 16-bit precision inputs to produce four 32-bit results, then adds them to produce two 32-bit end results. A subsequent add will produce a single, scalar result. Since Multiply-Add can be pipelined every clock and has a low latency, it is a very powerful instruction.

There were a wide range of useful extensions added by Intel in its MMX extensions, but as the previous examples illustrate their focus is on acceleration of multimedia operations mainly involving decompression.

The Compatibility Issue

In its efforts to accelerate multimedia Intel had to make decisions on the instructions to add in the context of compatibility, a problem none of the other CPU vendors has. Specifically, all existing Intel architecture software had to run without modification, including preemptive operating systems, of which Windows is the most glaring and maddening example, which can without warning terminate an application, and swap the application's state in and out of memory. To solve this problem the MMX technology is implemented using the floating point register set, but with the MMX registers mapped into the 64-bit mantissa of the 80-bit floating point registers. This allows the FPU tags to be used indicate valid MMX register data. And since all major operating systems are aware of the FPU unit, the MMX State is automatically comprehended and addressed.

In the initial implementations of the MMX extensions, all instructions were designed to execute in one clock cycle except multiply and multiply-add. Added to the chip are a multiplier, two ALUs, and one shifter. The multiply unit is a new three-clock latency, single-clock sustained multiplier. It is capable of 16 by16 bit multiplies, or two 16 by 16 bit multiplies followed by an add (multiply-add). The ALUs can each perform adds, subtracts, and compares of SIMD data types in one clock cycle. The shifter performs SIMD shifts, packing, and unpacking also in a single clock cycle. Even though a specific implementation of the MMX architecture on an Intel CPU could vary slightly, the data types, instructions, and operations were designed to execute, for the most part, in a single cycle, in parallel, and pipelined.

Some Intel Omissions

Intel left some important instructions out of the first MMX specification, limiting its usefulness in future networked interactive multimedia operations. A rather large one concerns instructions that are most valuable to applications, such as video conferencing, soft modems, and video processing as well as net-centric computing based on MPEG 4. One example is a distance instruction, which is primarily used in motion estimation - the principal component of video-compression algorithms. The distance instruction compares a 16x16-pixel target block against reference blocks by calculating the sum of the absolute differences. The algorithm determines the closest match by finding the block with the smallest absolute difference. The baseline MMX capability of subtracting subword data values is the primary component necessary for the distance calculation. The amount of extra logic circuitry to support a full distance operation is trivial and it should come as no surprise that Intel's competitors such as Cyrix spotted this error and made sure to support distance instructions in their implementations.

Another useful operation not included in Intel's MMX original suite is an average calculation, which adds the packed data values of source and destination operands and shifts the result right by 1 bit. Again, this is a simple instruction to implement, given a base line capability of executing subword-parallel instructions. The average instruction is most useful in the execution of the motion compensation algorithm found in MPEG decoders. This one instruction replaces nine MMX instructions when averaging pixel values from bidirectional frames. An average instruction can be used for simple interpolation as well as for implementing low-pass FIR filters, which remove high frequency components by smoothing, which is especially important in many communications and networking applications.

Finally, the original MMX lacks a magnitude instruction, which is useful for finding the largest or smallest number among a group of numbers. The instruction compares the absolute value of the packed words in the source A operand to the absolute value of the packed words in the source B operand and sets each subword in the destination to the subword of the largest magnitude. On an x86 processor that does not have this instruction, some signal-processing algorithms would have to use conventional integer instructions for signal measurement and scaling.

Applications that must find a minimal Euclidean distance among a group of distances greatly benefit from the magnitude instruction. An example is the Viterbi decoder that serves as part of a modem receiver algorithm. Other applications, mainly in telecommunications, utilize the instruction as well. Examples include pitch extraction in a vocoder algorithm and frequencydomain examination in a DTMF detection algorithm.

The National/Cyrix Alternatives

Given the accelerating momentum in the direction of a multimedia rich Internet and World Wide Web, it would seem that Intel and its MMX technology could drive vendors of alternative CPUs out of the market for net-centric desktop systems and Internet appliances. This is not the case.

Intel may be able to maintain its lock on the desktop with a new generation of multimedia-enabled X86 based systems. The company may even be able to maintain a dominant role in the much lower cost NetPCs. But it will have to share the market for networked multimedia with a number of other companies. One group of CPU makers, Advanced Micro Devices, Cyrix Corp. and National Semiconductor, Texas Instruments and SGS Thomson among them, have specifically targeted the low- to mid-range of the market for connected PCs and NetPCs with alternative architectures compatible with both the Pentium and Pentium Pro. Moreover, some have not only adopted the MMX enhancements, but have incorporated some extensions of their own to further improve performance.

Even with non-MMX-based x86 and Pentium processors, these companies have already given Intel a run for its money targeted the low end of the PC and NetPC market. National Semiconductor, for example, has published a platform specification, called Odin, built around a highly integrated version of the 486 microprocessor. And after things have sorted out following its acquisition of Cyrix, National is expected to continue to go after this market. It also expects to makes some inroads into the NC market as well.

With the acquisition of Cyrix, National now has a number of building blocks with which to continue this strategy into the Pentium, Pentium II, and Pentium Pro world, with and without multimedia data types.

The MediaGX

In the near term a good alternative for adding multimedia capabilities to the $500 NetPC or network computer is Cyrix's MediaGX (see Figure 16.1), which National Semiconductor is using as the core of a new line of highly integrated PC-on-a-chp family.

It is an x86-enhanced CPU that is targeted at the under $1000 system that Intel is aiming at with its NetPC specification. Not only does it integrate a graphics accelerator with an advanced DRAM controller on board the CPU chip, but it also boasts some other unique features: it is cacheless, and rather than separating graphics and main memory, it integrates the two into a unified memory architecture. It also incorporates a proprietary Virtual System Architecture (VSA). Similar in concept to virtual memory, VSA provides a mechanism for trapping accesses to nonpresent hardware and providing the requested function with software. With the further integration that National has in mind, under $500 multimedia-capable NetPCs and NCs are within sight. The development of VSA provides a means to eliminate legacy hardware and gives the system designer the freedom to innovate without sacrificing compatibility.

Attempts in the past have attempted to retain many of the performance enhancing features of the Pentium architecture, but without substantially increasing cost. The most direct way has been to build processors and systems without second level (L2) cache. Another alternative is the use of a Unified Memory Architecture (UMA) to decrease the amount of DRAM in a system and moving the functions of peripheral devices into software. These strategies have merit, but all have significant problems.

As regards to cacheless systems, certainly many PCs, even fifth-generation systems, have been shipped without L2 caches. In fact, the L2 cache is optional on almost every motherboard. Unfortunately, systems without an L2 cache suffer a serious performance penalty. A well-designed motherboard with high-speed EDO DRAM gets the performance penalty down to about 15%, but a typical system without an L2 suffers around a 25% performance hit.

The currently proposed UMA approaches also have performance problems. In UMA systems, the CPU and graphics contend for DRAM bandwidth, thus increasing the average time it takes for the CPU to access DRAM and cutting performance by 15 to 25%. Finally, moving peripheral hardware functions into new application programming interface (API) software potentially decreases system performance by taking cycles away from application software. As troubling is that the new multimedia APIs are tied to a specific operating system, ignoring backward compatibility with legacy software. Thus, each of these three cost reduction strategies has problems, but all show promise if their problems can be overcome.

The performance problems of systems without an L2 cache arise from the relatively long amount of time it takes to access DRAM. A CPU in a PC with a 66-MHz system clock accesses DRAM in six clocks, so long as the access hits an open DRAM page. This translates into an access time of 90 nanoseconds (ns). A DRAM databook, on the other hand, specifies the page-hit access time of 35 to 40 ns, less than half the time actually observed in real systems. The reason it takes so long to access DRAM in a traditional PC is that the standard system architecture inherently wastes time.

This wastage is most obvious each time a read cycle is performed. The CPU, running in a clock-multiplied mode with an internal frequency of 100+ MHz, has to synchronize the memory access request to the system clock. Synchronization consumes a core clock or two unless the memory access happens to be perfectly synchronized to external clock, something that occurs rarely if at all. Next, the access is driven on the bus pins at the beginning of a system clock cycle and is sampled by the chip set at the end of the cycle. The request then flows through the chip set where the DRAM address control lines are driven. Finally, the DRAM returns the read data through the chip set and back onto the system bus, where the data are sampled by the CPU at the end of the next system clock cycle. Much of the time in this sequence is spent in useless delays: synchronizing to the external clock, driving the request to the chip set, and waiting for the data to come back through the chip set.

In the MediaGX, the delays associated with such cacheless systems is the integration of the memory controller into the CPU. When the CPU needs data, it drives the DRAM signals at the next available core clock edge, rather than waiting to synchronize to an external clock domain. Likewise, data returned by the DRAM are sampled at the end of a core clock. Using a very high frequency clock to drive and sample DRAM lines enables memory access timing that closely matches the theoretical performance of the DRAM. Accesses that hit open DRAM pages can achieve DRAM access times of 35 to 40 ns, getting data to the CPU in less time than the 45 ns it takes to get data out of a pipelined burst SRAM in a standard system with a 66MHz system bus.

A second modification that is often made to improve performance without affecting cost is to reduce the amount DRAM in a PC by using a unified memory architecture where system memory and the graphics frame buffer share one DRAM array. The principal problem with this scheme is the performance loss from sharing DRAM bandwidth between CPU accesses and graphics refresh. Refreshing a 1024 by 768 by 8b pixel screen consumes at least 57 Mbytes/sec of bandwidth, thus reducing the bandwidth available to the CPU by a significant fraction of the realizable DRAM bandwidth. From the CPU's point of view, the contention makes the DRAM appear slower than it should be, reducing performance by about 20%.

The way that the MediaGX deals with the UMA performance problem reduces the bandwidth consumed by the display refresh by a factor of 10 or more. This is done by incorporation of advanced lossless compression circuitry into the CPU (see Figure 16.1) to create a compressed version of the frame buffer and then service the display refresh from the compressed data. This method requires the refresh controller portion of a UMA system's graphics unit contain hardware that losslessly compresses graphics data as they are read from the frame buffer during a screen refresh. If a given line of the screen

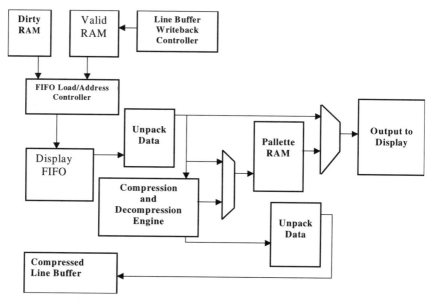

Figure 16.1 **Integrated Graphics**. Targeting low-end graphics and multimedia-intensive, under $1000 NetPCs and network computers, National/Cyrix Media GX incorporates a graphical compression engine to reduce the CPU to memory bandwidth in cacheless unified memory architectures.

can be compressed to some threshold, the compressed version of the line is written back to a separate compressed frame buffer also stored in DRAM. Subsequent screen refreshes are sourced from the compressed data and run through a decompressor before heading to the display. In this scheme, two frame buffers are maintained in a system using display compression. Applications and drivers write and read data from the uncompressed frame buffer, maintaining software compatibility. The compressed frame buffer is used solely for refreshing the display. Whenever an application writes to a portion of the frame buffer, tags associated with the modified lines are set to indicate that the lines need to be recompressed. The newly modified lines are read from the uncompressed frame buffer during the next screen refresh and compressed as they are displayed. This compression mechanism reduces the bandwidth consumed by screen refresh during a typical GUI session by a factor of 12 or more, virtually eliminating the bandwidth contention problem and thus the UMA performance problems.

As effective as the use of display compression is in improving UMA performance, the MediaGX also incorporates an advanced arbitration schemes to allocate bandwidth. This allows the CPU to provide extensive read and write buffering throughout the system, but it required the integration of the memory and graphics controllers to reduce overhead in switching control of the DRAM bus from one chip to another.

The VSA Alternative

As regards the migrations of functions out of the peripheral hardware and into software running on the CPU to reduce costs, a proposal from Intel promotes the concept of creating new APIs as a mechanism by which hardware functions can be implemented in software. Moving functions from hardware into software has a long and distinguished history in the computer industry, but there are a few problems with it. The principal ones have to do with support for legacy devices that don't comprehend the new APIs, the performance problems inherent in the implied multi-layered driver structure, and the OS-specific nature of this solution. The bottom line is having a new API for audio doesn't remove the need for audio hardware, unless the API can work in every operating system the consumer may use.

The VSA scheme employed in the MediaGX achieves the same goals, without the disadvantages. Similar to the way virtual memory is managed, VSA provides a mechanism for trapping accesses to non-present hardware and providing the requested function with software. The "virtualization" of software in this way operates in a privileged context, completely invisible to applications and the operating system. VSA makes it possible to virtualize almost any device regardless of the operating system being run. VSA is implemented with a greatly improved System Management Mode (SMM). Upon receipt of a system management interrupt, the CPU saves the machine state and switches to a new context, SMM, where software can run without interfering with the operating system or applications.

With VSA, SMM mode is triggered if software attempts to access any non-present, virtualized device, and the access is handed off to software that provides the function implied by the access. Using software to replace hardware can drive costs down, but if not done carefully it can drive performance down as well. CPU core enhancements, such as new instructions and L1 cache modifications can be made to reduce the performance impact to an acceptable level. With a VSA-capable CPU and suitable core enhancements, an industry-standard PC audio card can be cut down to a codec and a few kilobytes of code.

To support VSA, the MediaGX incorporates a much improved System Management Mode, with far lower entry/exit overhead and the ability to nest SMIs. Other features reduce the execution time of the virtualization code itself. Any instruction whose operands hit in the L1 cache executes without pipeline stalls, as though its operands were in registers. Moreover, in the MediaGX, a small (zero to four kilobytes) portion of the 16K L1 cache can be statically or dynamically locked down such that the contents of the locked region will never be invalidated or evicted from the cache except under software control. Software that is aware of this feature can effectively extend the register set of the machine by storing variables in the locked cache region. The locked cache can also be used to store inner loops of performance-sensitive code, CPU state information pushed during SMM interrupts, and occasionally portions of the stack. One of the cost advantages of a MediaGX is the elimination of a separate graphics controller and frame buffer DRAM. The MediaGX graphics pipeline is tightly coupled to a CPU core that has been enhanced to further accelerate other graphics operations. Instructions were added to the CPU to perform block transfers of data within system memory and between virtual memory and the graphics pipeline or frame buffer. Such instructions are particularly useful for rapidly displaying text or bit maps stored somewhere in virtual memory, and for manipulating blocks of compressed data. Very flexible, this mechanism allows the graphics pipeline to operate in concert with a block transfer instruction to perform accelerated rendering of bitmaps in virtual memory.

Going after The High End

Holding a lot of promise for high end connected PCs and servers, even more than Intel's current mix of MMX instructions, is Cyrix's multimedia-enhanced sixth generation Pentium Pro alternative, the 6X86 family. Originally, codenamed the M2, this processor provides an MMX implementation that is fully compatible with Intel's approach. Yet it improves on many of the standard instructions, with enhancements such as lower-latency multiply, fused multiply-add for better multiply accumulate (MAC), one-clock Empty MMX state (EMMS), and other improvements that help deliver the performance demanded by multimedia applications.

Similar to the Intel approach, the Cyrix-National implementation also reuses existing circuitry in the CPU's floating-point unit allowing implementation of the complete MMX specification with less than a 1% increase in silicon area required by Intel. But where Intel chose to use the logical floating-point register state for both floating-point and MMX operations, the Cyrix design goes in a different direction. Because the Intel approach requires that the programmer avoid using floating point instructions at the same time as MMX commands, Cyrix has chosen a much less complicated approach. What Cyrix has done is reconfigure existing floating-point resources in addition to the registers. Since the

X86 architecture does not support simultaneous floating-point and MMX operations, why have idle hardware lying around when existing floating-point data paths can accommodate MMX operations? In this architecture, the carry chain of the adder that sums floating-point mantissas is broken into partitions that can propagate the carry or not, depending on hardware controls. This feature allows the same hardware to execute either floating-point or MMX operations.

The floating-point mantissa multiplier provides another level of reconfigurability. In a single cycle, the current multiplier computes a 24 by 72-bit product. This rectangular aspect ratio lends itself to efficient partitioning to compute parallel integer subword operations. In effect, the multiplier array is divided into eight 16 by 12-bit subarrays. A pair of subarrays generates partial products, which are summed in a final carry-save-add (CSA) stage to produce two partial products, a carry product, and a sum product. These final products are summed by the same reconfigurable adder used for floating-point mantissa addition and for MMX ADD and SUBTRACT operations.

The unique aspect of this reconfigurable adder is the addition of a crossbar switch (see Figure 16.2). Each switch latches data operands to pipeline operations, although the latch can be bypassed in certain circumstances. When an ADD or SUBTRACT operation occurs, for example, Switch 1 is bypassed and data operands feed directly from the register file to Switch 2 in a single clock. This feature allows the architecture to support pipelining for all MMX operations since all arithmetic operations execute in a single cycle except for Multiply and Multiply-Add. The architecture supports single-cycle throughput by forwarding the results of an operation to either of the crossbar switches so they can be used in a subsequent operation.

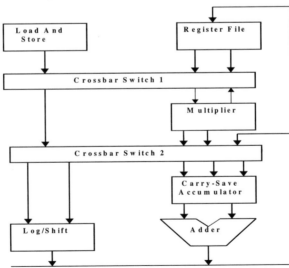

Figure 16.2 **Crossbar Enhancements**. A key element in the National/Cyrix approach to integrating multimedia instructions into the X86 architecture is to use a pair of crossbar switches to allow the CPU to handle floating-point and multimedia instructions at the same time.

The ability to store critical code loops or data values in local high-speed memory would give a big performance advantage for MMX operations, but only if the execution pipeline could access this RAM quickly. In the Cyrix design this is done by implementing lockable on-chip cache and pipeline support. To understand how this works to the advantage of the Cyrix approach, let's look at the Intel implementation of the MMX pipeline where instructions execute similarly to other x86 operations. An MMX instruction consists of an opcode byte, followed by a MODRM byte followed by displacement or immediate bytes. By comparison, in the Cyrix design, these instructions flow through the pipeline, and MMX operations share the first five stages of this pipeline with all other x86 operations. The sixth stage (the IQ stage) loads MMX opcodes and any memory operands into an MMX instruction shelf located in the floating-point unit. Instructions are dispatched from this instruction shelf in order, depending on the resources availabile. Once dispatched, a cycle is taken to access the register file, followed by a single execute cycle-two if the instruction involves a multiply operation. Finally, each instruction is completed with a writeback cycle. Although the pipeline is rather long for an MMX instruction, all exceptions are known and acknowledged in the pipeline's IQ stage. If no exceptions are pending, the instruction can be sent from the IQ stage to the instruction shelf, freeing the pipeline. This capability can improve performance when conventional x86 integer operations are tightly mixed with MMX operations.

Again, unlike the Intel implementation, the M2 processor's superscalar hardware can simultaneously execute an MMX operation with any other x86 operation. While only one MMX instruction can execute within a single clock cycle, the single-cycle throughput of all MMX arithmetic operations provides a high level of performance. Furthermore, the ability to pipeline access to the on-chip cache allows all load-execute, read-modify-write MMX operations to execute in a single clock, a significant advantage, given the x86's limited register set..

Understanding this situation led Cyrix to configure the on-chip cache so that lines can be locked, a feature not included in the Intel implementation. This creates, in effect, a scratchpad RAM capability and provides a deterministic access that boosts performance, given that the Cyrix implementation of the pipeline provides access to the stored values as fast as if the data were in a register. And important for networked multimedia, it is also possible to execute code deterministically for real-time data processing.

Multiplying Its Advantage

Even though the multiply-accumulate operation is a cornerstone of signal- and media-processing algorithms, the MMX specification as originally defined by Intel does not provide a true multiply-accumulate operation. The specification does call for a multiply-add, in which pairs of 16 by 16 multiplies are executed in parallel and the pair of 32-bit results accumulated in a destination register. To do a series of these operations, however, each multiply-add instruction must be paired with another add instruction to accumulate the intermediate results from the multiply-add in a true accumulator fashion. Moreover, the 32-bit results of the MMX multiply are summed in a 32-bit packed data format that does not accommodate overflow. Typical signal processors provide accumulators of larger precision than that of the multiplier to handle intermediate data overflow.

By comparison, the Cyrix architecture executes the MMX multiply and multiply-add instructions quite efficiently, allowing single-cycle throughput with only two cycles of latency. The architecture also allows a multiply followed by an add to be fused such that the two instructions execute in just two cycles, thus completing within the two-clock latency of the multiply alone. This is done by combining the add with the last cycle of the two-cycle multiply. Programmers can put together a sequence of multiplies whose results are subsequently added to an accumulator to perform four subword parallel multiply-accumulates in just two cycles.

An application that requires both MMX's subword parallelism and floating point's dynamic range and/or precision can take best advantage of the available architecture by breaking out the two types of code into separate subroutines. Assuming each subroutine returns the relevant results or places them in memory, each MMX routine can complete by executing an EMMS operation, which empties all registers so that a subsequent floating-point routine can execute from a clean slate. While Intel's Pentium implementation requires 53 clock cycles to end each MMX routine with an EMMS instruction, the Cyrix processor takes only 1 cycle. The Cyrix technique works quite efficiently, provided the application does not have to save the entire MMX or floating-point state when switching between floating-point and MMX code, as the switch then takes numerous clock cycles to save all 108 bytes of the floating-point logical state.

In the future, Cyrix-National engineers are looking further extend the concept of SIMD (single instruction, multiple data) processing to include floating-point data types as well as integer types. A 64-bit operand, which can hold packed single precision floating-point data types, supports all floating-point requirements in today's applications and those of the distant future. Adding support for floating-point data types to the MMX specification could double or even quadruple the x86's floating-point processing power (measured in FLOPS). Provided a 200-MHz processor could execute two MMX instructions in a given clock and the operands were of packed single-precision floating-point type, the change would yield a peak throughput of 800 MFLOPS - a dramatic improvement in the capabilities of the conventional x86 floating-point architecture.

AMD's MMX Candidate

Unlike Cyrix-National, Advanced Micro Devices, Inc. (Austin, Texas) is sticking relatively close to the Intel approach in its implementation of the MMX extensions. But it also has added some features that will aid in multimedia operations, especially in relation to 3D, an important part of the MPEG 4 specification.

In both the current K6 and the new K8 it has added 24 new multimedia instructions, targeted at eliminating the one weakness in the MMX suite: its inability to accelerate 3D operations efficiently. AMD has added a multiply-add instruction similar to that in many multimedia coprocessors, as well as the ability to multiply a pair of 32-bit floating-point values, then adding the result to another floating point value in a single operation. This is very similar to what Silicon Graphics has done in its set of multimedia extensions. Rather than piggyback operations on the floating point registers as is done with in the

MMX extionsions, AMD has added a separate execution unit specifically for handling 3D operations.

Multimedia-Enhanced Servers

In the media-rich Internet and World Wide Web of the future, servers as well as clients will need multimedia capabilities. While Intel Corp. has established a leadership position in the low to medium performance end of the desktop market, it is not clear how well its MMX technology will fit the requirements of the server market, even though the extensions are also being added to the company's low-end server engine, the Pentium Pro.

The main players in the Internet and network server markets (Digital Equipment, Hewlett Packard, Silicon Graphics, and Sun Microsystems) are not only active in the server market with 64-bit processors but with multimedia-enhanced versions as well. And the instructions they have chosen to implement seem more appropriate to the requirements of networked multimedia than Intel's current MMX mix. Hewlett Packard is already in its second generation of multimedia-enhanced processors for workstations and servers: MAX1(motion acceleration extensions) for 32-bit PA-RISC CPUs and MAX2 for its 64-bit PA-RISC processors. There is also Silicon Graphics' MMDX enhancements for its MIPS architecture; Digital Equipment's Motion Video Instructions (MVI) for the Alpha; and Sun's Visual Instruction Set (VIS) for the 32 bit Sparc and 64 bit UltraSparc processors. Intel, by comparison, is not expected into the market with a 64-bit multimedia-enabled version of its architecture much before late 1999.

Because they are targeting mainly workstations and servers, as well as bridges and routers, the multimedia-enhancements to the DEC, HP, SGI and Sun architectures are focussed not only on decompression of video and audio, but compression as well.

DEC's MVI Enhancements

Typical of the approach being taken to provide the capabilities necessary for networked multimedia in the age of MPEG 4 are DEC's MVI enhancements, first implemented on DEC's 64 bit 21164PC, which is part of the technology transferred to Compaq Computer Corp. after its acquisition of the company..

The MVI extensions makes it possible to achieve high quality compression in software on Alpha CPUs. MVI accelerates compression using the prevalent ISO and ITU-T video compression standards including MPEG-1 (VHS-quality video); MPEG-2 (broadcast-quality television); H.261 (ISDN video conferencing); and H.263 (Internet video conferencing. MVI incorporates three instruction classes: the Pixel Error instruction (PERR), MAX/MIN instructions and PACK/UNPACK instructions. Unlike video decompression, where the largest consumer of CPU cycles is the Inverse DCT algorithm, the situation is the reverse in MPEG video compression. Here, motion estimation is the largest consumer of CPU cycles It represents more than 70% of compute cycles consumed by the encode function. The PERR instruction (see Figure 16.3), greatly accelerates motion estimation by speeding up the macroblock search that is integral to this process. PERR computes the sum of the absolute value of the differences of eight pairs of bytes. The PERR instruction replaces nine lines of code in a motion estimation loop. The MAX and MIN instructions allow efficient clamping of pixel values to the maximum values allowed in different standards and stages of a codec. They enhance performance through eight-way parallelism. The PACK and UNPACK instructions expand and contract the data width on bytes and words. These functions promote pixels to 16 or 32 bits, allowing the pixel data to be operated on with greater precision.

Motion Estimation Loop (cycles-instructions per function)

Alpha CPU Instructions Without Media Extensions	Operations Performed	Alpha CPU Instructions With Media Extensions
	Move larger bytes into r1 and smaller bytes into r0 to get positive result for subraction.	
1. cmpbge r0,r1,r3 xdr r0,r1,r24 2. zapnot r24,r23,r23 3. xor r0,r23,r0 xor r1,r23,r1	------------------ subtract pixels ------------------	
4. subq r1,r0,r0	Expand byte results into words, align them and then sum.	------------------ PERR ------------------
5. zap r0,#85,r1 6. zapnot r1,#8,r0 7. srl r1,#8,r1 8. addq r0,r10,r10 9. addq r1,r11,r11		------------------ Total: 8 pixels processed in one instruction cycle
Total: 8 pixels processed In 9 instruction cycles		

Figure 16.3 **Motion Estimation**. Key to the Alpha's ability to compress multimedia data efficiently is the use of instructions such as PERR which reduce the number of cycles to execute by almost 10:1.

MVI takes advantage of the Alpha architecture's 64-bit design, which easily accommodates all precision requirements for multimedia standards. In contrast, the design of the MMX extension for x86-type processors has a precision shortage. For example, MMX's 16-bit multiply-add cannot provide the minimum 19-bit precision required to comply with the IEEE IDCT requirements and falls short in AC-3 audio decode where 20-bit matrix transforms must be used. Wheras Intel with its MMX makes double use of the Pentium's existing floating-point registers, DEC's MVI is implemented in the 21164PC's integer unit. This approach is more efficient than a floating point unit implementation because MVI instructions operate on fixed-point pixel data , making use of existing logic functions, including AND, OR, NOT, and vector compare.

Implementation in the integer unit was possible because the Alpha architecture contains thirty-one 64-bit general-purpose integer registers. These registers provide sufficient storage to support the chip's issue bandwidth of 500 million MVI instructions per second concurrently with 500 million additional integer instructions per second. Because MVI is located in the integer unit, there is no need to save and restore register contents to use MVI operations, thereby enhancing performance in motion video compression and improving context switching time, which is important in networked multimedia applications. This approach also maximizes the performance of concurrent 3D graphics and video by avoiding contention with floating point registers. Such concurrency will be important in the MPEG 4 environment, where it will not be uncommon to be dealing with multiple VRML-based 3D images from multiple locations on the network at the same time.

Intel's MMX, which shares registers with the floating point unit, does not enjoy these advantages. The MMX architecture can cause 3D data and motion video to conflict with each other in emerging applications such as MPEG 4, concurrent 3D/motion video games and video conferencing with 3D white boards, reducing image and audio quality. This is a serious consideration that the designer of MPEG 4 net-centric multimedia hardware and software should keep in mind.

Also important to keep in mind is that the limited floating point performance of X86 microprocessors compared with Alpha microprocessors can significantly limit 3D graphics performance (25 Linpack 100 by 100 MFLOPs for the Pentium 200 compared with the 500 MHz Alpha 21164's 250 Linpack 100 by 100 MFLOPs)..

The HP Entry

Rather than add specific functions or instructions, the approach taken by HP in its PA-RISC-based servers and workstations is to add capabilities that enhance the operation of the processor in general, as well as multimedia operations in particular. Specifically, this involves finding the most frequent operations, breaking them down into simple primitives, and accelerating their execution. First employed in the company's 32 bit PA-7100LC and more recently in the 64-bit PA-RISC 2.0 architecture, this involved implementing in silicon a set of special parallel subword instructions to perform general-purpose operations such as add, subtract, average, and shift_and_add in parallel. Then the existing microprocessor data paths were modified to enable parallel operations on subwords; that is, data narrower in width than the width of a word in the microprocessor.

The means by which HP implemented SIMD is also significantly different from the way Intel did it. In the PA-RISC architecture, the necessary SIMD substructure is implemented by means of parallel data items which are no more than subwords packed into standard-sized words. The data processing elements are just partitions of existing or new functional units, and the control processor is the normal instruction fetch and dispatch unit. The parallel memory subsystems feeding the data processors are the usual word fetch mechanisms in a standard microprocessor: from the single memory to general registers and functional units. The SIMD instruction here is the standard 32- or 64-bit- wide RISC processor instruction. With this approach, the basic microprocessor requires no other pipeline, register or memory changes.

The actual operations that these parallel subwords perform is determined by the frequency with which they occur in easily parallelizable computations. The multimedia extensions incorporated into the PA-RISC 2.0 fall into several broad categories: parallel addition and subtraction, parallel multiplication, parallel averaging, and saturation arithmetic.

Compared to Intel's decompression-oriented mix and DEC's compression oriented mix of multimedia instructions, HP has taking a much more balanced approach, trying to accelerate both equally well. In the IDCT stage of the MPEG decompression, for example, fully 35 million of the total 206 million instructions required are additions and subtractions. By blocking the carry from the low half-word to the high half-word, when parallel adds and subtracts are executed the 32 bit ALU in the 32 bit PA7100LC can perform two parallel half-word adds or subtracts in the same time - a single CPU clock cycle - that it takes to perform a single 32 bit add or subtract. In the 64-bit PA-RISC 2.0 processors, each ALU can perform four adds or subtracts in a single cycle. Since most PA-RISC processors have at least two ALUs, this means that four 16-bit ALU operations can be executed in the 32- bit architecture and eight in the 64-bit environment.

Essentially, at an insignificant increase in hardware cost, this approach increases by four to eight times the peak execution bandwidth for multimedia video applications that use parallel 16 bit arithmetic at an insignificant increase in hardware cost. While many multimedia applications, such as audio and modem code, use multiplications as frequently as additions, this is not the case for many video algorithms. So, rather than incur the expense cost of including a 16 by 16 bit multiplier on the CPU chip to complement the parallel 16-bit add and subtract options, HP designers found that it was more cost effective to do this in microcode. In the PA-RISC architecure, it is possible to multiply by constants without requiring a full multiplier by the simple expedient of using compilers to do so as a series of shift and add instructions. For multimedia acceleration, the already existing shift_left_and_add instruction in the PA-RISC has been complemented by two additional instructions: parallel 16 bit shift_left_and_add as well as parallel 16 bit shift_right_and_add. Since the ALU is already partitioned to allow parallel adds and subtracts, it was only necessary to partition the preshifter input to the adder, which did not cause any significant cycle-time impact.

Multimedia For Free?

The impact of this approach to adding multimedia capabilities to the central processor is that it makes possible high-quality, full 30-frames per second full motion video at essentially zero incremental cost. The implementation of these additional instructions did not affect the processor's cycle time in either the 32- or 64-bit case and added less than 0.2 % to the silicon area of the CPU. In the case of the PA-7100LC, the area used was mostly previously empty space around the ALU, so these multimedia enhancements can be said to have contributed to more efficient space utilization, rather than adding incremental chip area.

In the earlier 32- bit design, the multimedia-enhanced PA-7100LC outperformed the older PA-7100 based systems significantly. For example, in the older systems, running at 99 MHz, a full-screen 24 bit color video clip achieves only an 18.7 frame per second rate. By comparison a multimedia-enhanced version produces 26 frames per second (fps) at 60 MHz and 33.1 fps at 80 MHz. Even when high quality audio is added, performance on the 7100LC based systems did not suffer significantly. Even when the highest fidelity stereo MPEG audio is used, a 60 MHz multimedia-enhanced system achieved a rate between 15 and 20 fps, an 80 MHz CPU achieves 24 fps and a 100 MHz version reaches 27.4 fps.

In the 64 bit PA-RISC 2.0 architecture, the parallelism has been doubled, with a resulting improvement in performance. Multimedia performance on such architectures can be expected to improve even more as the performance of the general purpose processors increases. With PA-RISC processors, there has been a doubling in performance every 18 to 24 months. This would imply that larger windows, multiple smaller windows or MPEG 4 video streams may be decoded in the future by such multimedia-enhanced CPUs.

Sun's Visual Instruction Set

Whereas HP focused on a multimedia instruction set appropriate for both high end workstations and servers, Sun Microsystems in its next-generation 64-bit UltraSPARC has focused specifically on adding multimedia capabilities useful in servers in the low- to mid-price and performance range. Not only has it added many of the same kinds of multimedia enhancements that HP and DEC did, it has also added a number specific to particular kinds of networked multimedia applications.

It also includes an instruction for motion estimation instruction known as the Pixel Distance Calculation or PDIST. The most time-consuming aspect of the MPEG compression algorithm is trying to analyze the motion of the image by comparing each part of the current frame against the previous frame. Specifically, motion estimation searches a reference frame for the closest match to a 16 by 16 block of pixels in the target frame. The closest match is determined by finding the block with the smallest absolute difference between the target and reference frames. The single PDIST instruction compares two sets of eight 8-bit values in parallel and performs a partitioned subtraction of two 64-bit registers. The processor then takes the absolute value of each of the differences and adds it to the accumulated difference value.

Tallying thoee up, PDIST performs eight subtracts, eight absolute values, and seven additions for a total of 23 fundamental operations. PDIST also replaces numerous shifts required to align these pixel values. On the other hand, a traditional microprocessor takes about 47 instructions to perform this same computation. However, the improvement is even greater when the accumulating error for a 16 x 16 block is taken into account. Whereas this takes 32 PDIST instructions in addition to the necessary load instructions, it would normally require 1,500 conventional instructions on a standard microprocessor. The more than 10 times performance improvement in this case makes possible real-time H.261 video conferencing on the desktop, as well as MPEG 4 processing, also in real time.

Other interesting additions to the VIS are the block load and block store (BLD, BST) instructions which permit the system designer to easily comply with the extraordinary data throughputs needed for video, graphics, and networking operations.. These instructions are used to transfer blocks of 64 bytes between memory and a group of eight floating point registers in the processor. By using BLD and BST, the system designer can move up to 600 MBytes of data per second across the processor-memory bus — 300 MB in and out of the processor. That large amount of data is transferred by a single processor execution unit on the bus. Moreover, BLD moves that volume of bits while circumventing the normal cache structure. Rather than overwriting to the cache to swap memory pages in and out, BLD leaves the cache memories untouched and executes 64-byte loads and stores directly into main memory. This operation allows the processor to act as the video processor by blitzing data on and off the screen. These two instructions also enhance crucial network operations. Within the network layers, data can be moved many times before they reach their final destination. Block movement of data can thus be much faster and cleaner. Block load-store also makes it easier to share applications and data between multiple users. Copies of applications or data structures can be quickly generated even though someone else is also using them. This feature is especially valuable to collaborative work groups in which many users work with the same data and applications.

In addition to these three major instructions, there are another 25 plus on-chip instructions tailored for graphics manipulation. They are provided via the implementation dependent opcodes made possible by the SPARC V9 architecture. All instructions are pipelined and operate on data formats of 8-, 16-, and 32-bits.

Critical in the efficient execution of these visual instructions is the unique design of the UltraSparc's combined floating-point and graphics unit (see Figure 16.4). The FGU is composed of five different functional blocks for : (1) floating-point divide-square root, (2) floating-point addition and subtraction, (3) floating-point multiplication, (4) graphics addition, align, merge, expand, and logical operations, and (5) graphics multiply, pack, compare, and PDIST. Graphics operations, in this case, rely on the integer registers for addressing image data such as address calculations for loads and stores and the floating-point registers for manipulating image information. Division of duty between the integer and floating-point registers increases graphical throughput by providing the maximum number of registers and instruction parallelism.

Graphics data are stored in the FGU registers for three reasons. First, graphics applications can use the integer registers for addresses, loop counts, and similar data, leaving all 32 FP registers for graphics data. Since many of these programs use a larger number of registers, this is an important advantage. Second, this makes better use of the chip's instruction-issue rules, which permit only three integer instructions per cycle, but allow four if at least one is an FP or branch instruction. Graphics code can ideally issue two graphics instructions per cycle to the FGU along with one load or store and one integer arithmetic instruction, possibly for loop control or address generation. Third, a few of the graphics instructions take multiple cycles to execute, although they are pipelined and can be issued one per cycle.

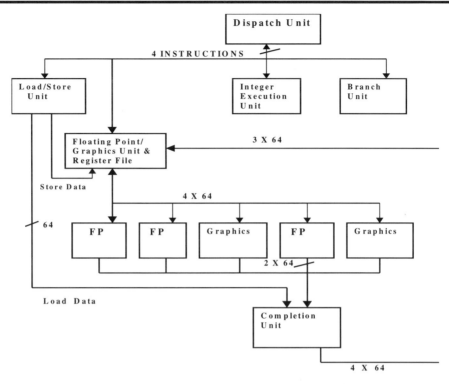

Figure 16.4 **Sparcling Multimedia**. Sun accelerates multimedia in UltraSPARC with use of specialized floating-point/graphics unit (FGU) that combines integer and floating point calculations necessary for specialized audio and video instructions.

Other major VIS instructions include FPMERGE, FEXPAND, ARRAY, and Partial Store (PST). FPMERGE takes two sets of 4 bytes and interleaves them into a 64-bit value. This instruction can convert pixels from standard to packed format and can also be used in discrete cosine transform (DCT) functions to transpose matrix values. FEXPAND coverts 8-bit data to the Fixed16 format, eliminating the need for all intermediate results to be rounded and clipped. The ARRAY instructions for 8-, 16-, and 32-bit operations perform data conversion to resolve another design issue relating to visualization of a 3D data set using a 2D slice of arbitrary orientation.

Intel's Once And Future 64-bit Solution

To handle the demands of an increasingly multimedia-oriented Internet and World Wide Web will require much more than even this new generation of servers and desktops. It will require a fundamental change in architecture, especially as bandwidths grow larger as alternatives. But with performance boosts on the order of 1,000 time current capabilities in the offing on the network side, comparable performance boosts on the processor side would have to be on roughly the same magnitude, from today's hundreds of megahertz to throughputs in the tens of gigahertz.

Many of the alternatives outlined in this chapter are impressive in their ability to boost performance, but there is an upper limit to the kind of improvements possible using existing RISC architectural concepts, even when complemented by SIMD. By early 1998, clock rates of 300MHz were possible on mid range desktop systems and in the 500 to 750 MHz on some server-oriented processors. And as we have learned in earlier chapters, a shift in processing technology from aluminum to copper interconnects may give the industry enough performance to push clock rates over the 1GHz mark, more than enough to match the performance parameters of the network.

But what is also necessary is a clean, and revolutionary break from existing RISC and CISC architectural approaches, away from superscalar, pipeline, and branch prediction as well as incremental mprovements in cache memory design. Impressive as all these are, they are still basically evolutionary in nature.

What is needed is already well-known, and computer architects have been conducting research on and improving on some of these techniques for years. Some of these techniques have already found their way to the market in the form of multimedia coprocessors, as we shall see in the next chapter.

But the risk that a manufacturer of a processor for a mainstream desktop, workstation, or a server takes is leaving behind a large number of users of the existing architecture, who cannot or do not need to make the transition. Despite the diversity of applications, one goal is shared in common: to bring as much parallelism to the chore of processing these data-intensive applications as quickly as possible and at a cost in silicon die area and in dollars that is as low as possible. The trick for makers of mainstream processors is how to make as much of an architectural break as possible, but without leaving their existing user base behind. Surprisingly, the first company to achieve this is the one that has the most to lose: Intel Corp.

First solid details of its next generation 64-bit processor started to emerge in late 1997. Expected to be in volume production in late 1999 or early 2000, it was code-named Merced, but it is also known as IA-64 for Intel Architecture-64. It uses a new technology called Explicitly Parallel Instruction Computing or EPIC. As with its RISC brothers and its 32-bit Pentium-PentiumPro predecessors, the IA-64 architecture pulls out all the stops and employs a variety of techniques such as instruction predication, branch elimination, and speculative loading to extract more parallelism from program code.

In terms of instructions, the IA-64 is likely to include many more multimedia- oriented instructions that even current MMX-enabled processors. But to execute them with as much parallelism as possible, Intel has turned to EPIC which it says is not a very long instruction word (VLIW) architecture. However, Intel "doth protest too much" for it very much looks like the VLIW architectures incorporated into some multimedia coprocessors such as the Philips TriMedia..

Paradoxically, Intel also denies that it is RISC- like, although it uses fixed 40-bit-wide instructions that are packaged into 128 bit VLIW bundles. This is very much different from the more CISC-like architecture of earlier X86 instructions in which each instruction is a single unit that can vary from 8 to 108 bits, requiring that the CPU to tediously decode each instruction while scanning for instruction boundaries. To process these VLIW instructions, the IA-64 will incorporate multiple instruction execution units, each of which will have 128 integer-type general-purpose registers and 128 floating-point registers. The size and number of registers will go a long way toward executing multimedia type instructions more efficiently.

Interestingly, the IA-64 architecture may have some features that will make it much more efficient in executing Java-type byte code more efficiently and safely especially in its use of speculative loading, which allows data to be loaded from memory before the program needs it. In many respects, speculative loading is similar to the TRY-CATCH structures in Java, except that in the IA-64 it works at the machine level. In Java, if the TRY statement attempts a risky operation, where there is some uncertainty about completion, use is made of a CATCH statement. If TRY succeeds the program continues normally. If unsuccessful, CATCH grabs it and stops the program from crashing.

As of the middle part of 1998, it was not clear what mix of multimedia instructions Intel was going to add to Merced. It is probably a given that the mix will be much more oriented toward MPEG decompression and 3D operations, given the requirements of MPEG 4. Its close relationship with HP in the definition of Merced would indicate that much of HP's PA-RISC experience with multimedia would be taken into account.

Whatever the final mix of architectural enhancements and instruction modifications, the move from 32 to 64 bits in the Intel architecture is likely to accelerate the trend toward the use of 64-bit processors in the servers, routers and switches on the Internet. This alone would considerably improve the Web's ability to handle networked multimedia.

AltaVec: The 128-bit PowerPC Solution

Except for some manufacturers of the MIPS architecture, most of the companies building 64-bit multimedia-enhanced CPUs have focussed most of their attention on the desktop, workstations and servers. Left out in the cold, so far, are the nuts and bolts of the Internet and World Wide Web: the routers, bridges and switches.

Not for long.

Capitalizing on its strength in low-end enterprise networking and data communications, Motorola Inc. has developed a multimedia extension to its PowerPC family of processors targeted specifically at the requirements of routers and switches. It is called AltaVec. Unlike extensions developed for servers focused on compression of multimedia data and those for the desktop designed for decompression, the AltaVec architecture (see Figure 16.5) contains features and instructions aimed at improving the I/O capabilities of routers and switches that will have to move multimedia throughout the Internet. In addition to routers and switches, initial targets for the architecture include IP telephony gateways, high data rate cable and multichannel modems. To this end, in addition to specially selected multimedia enhancements, this new architecture incorporates wider data path and field operations as well as logic to accelerate such operations as memory copies, string compares and page clears.

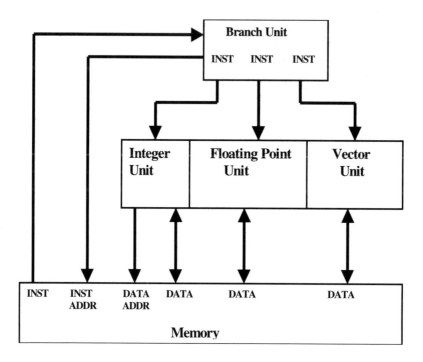

Figure 16.5 **Motorola's AltaVec.** Targeting network routers and switches, rather than servers and network computers, AltaVec expands the PowerPC architecture with a 128 bit vector execution unit that operates concurrently with the existing integer and floating point units.

Unlike the VLIW route that some other CPU vendors have taken, Motorola has chosen to use a short vector parallel architecture. To maintain compatibility with the rest of the PowerPC, all of the multimedia operations are handled by a 128-bit wide vector execution unit that operates independently, but in parallel with the standard 32/64-bit PowerPC architecture. Operating concurrently with the existing integer and floating point units, the VEU is able to execute up to 16 operations in a single clock cycle.

Where traditional long vector architectures used in supercomputers are designed to handle hundreds of elements at a time, the AltaVec works with vectors that are only 4, 8, or 16 elements long. This is because the longer vector words, while useful in scientific calculations, are of little use in applications involving communications land multimedia.

Like the other standard processors that have incorporated multimedia extensions, the AltaVec architecture uses the single instruction, multiple data (SIMD) format on multiple data elements. But where these other architectures overlay SIMD on the existing RISC architecture, using in many cases the same registers and other resources, the AltaVec approach sets aside a separate execution unit totally dedicated to SIMD operations. Moreover, because the vector execution unit (VEU) is a 128-bit wide computing engine, it is much more flexible and efficient in the way it handles many multimedia data types. This means it is not only able to operate on many 8, 16, and 32-bit data elements in parallel, but on multiple 64- and 128-bit data elements concurrently as well.

A good indicator of the power of this approach in networked multimedia applications, in particular, is how it is used to form the permute operation, one of the more powerful of the 162 new instructions defined for AltaVec. Particularly useful in communications operations, the permute operation is used to arbitrarily select data with granularities as small as a byte from two 16-byte source registers and move it into a single 16-byte destination register. This is enormously useful in applications where 8- and 16-bit data must be reorganized in memory before and after processing. Compared to the four or five operations per byte necessary to perform permutations on traditional 32- or 64-bit RISC CPUs, this single permute operation, executed on the 128-bit wide VEU can be done in a single clock cycle.

This is because instead of sharing a set of 32- or 64-bit registers with other non-multimedia operations, as in existing implementations, this is done in a single separate register file with 32 entries, each 128-bits wide. These 128-bit wide registers are loaded and unloaded through the use of vector store and vector load instructions that transfer the contents of a single 128-bit register to and from memory.

Because of the care Motorola has taken in the definition of the AltaVec, the PowerPC could find itself in virtual ownership of a signficiant portion of the router and switch market, especially as the Internet and World Wide Web transistions to the more multimedia-friendly megabit/ and gigabit/second communications technologies.

Chapter 17

Achieving Real-time Networked Multimedia

There is a real concern that, while alternatives such as Intel's MMX on the desktop and others for servers are sufficient for current network-connected configurations, they may not be up to the capabilities of high-bandwidth networked multimedia. This is because the latter has a number of real-time and deterministic constraints that must be met. In the Internet of the future with higher bandwidth connections, such processors will have to be able to juggle the requirements of the desktop as well as the network I/O interface.

Many networked multimedia applications, such as video conferencing inherently require real-time responses. Even in the present environment of 33- to 56-kbit/sec. bandwidths to the home, it is in the nature of many multimedia applications that a non-real-time response is simply not viable. This is why many software programs for video decode and display, when faced with making a choice between maintaining a certain throughput in frames per second with no loss in quality and dropping a frame here and there, usually opt for the latter.

In the relatively near future, the network interface is apt to undergo improvements in performance that are on the order of a thousand-fold, from modems that operate in the kilobits per second to ones that operate in the millions of megabits per second. For the approach being promoted by Intel and other CPU vendors (doing multimedia acceleration in software on the main processor) will require a comparable speedup in processor clock rates. While CPU performance on the desktop has moved from 100 to 300 MHz in main-stream processors and may achieve 500MHz before the end of the decade of the 1990s, it will not be enough to handle the requirements of the Multimedia Net.

One possible approach, at least on the desktop, is an evolution similar to what has occurred in servers. Just as servers are now adding specialized I/O processors to handle the high-bandwidth transmissions in and out of a system, it is likely that the desktop, too, will require some sort of I/O processor. Its job would be to offload the main processor of much of the networked multimedia processing, which will involve juggling with and manipulating multiple, high-speed data channels. In the interactive multimedia environment assumed by MPEG 4, the processor could be dealing with multiple data inputs from the Web: audio objects from one location, video from another and 3D from still a third, with the user interactively pulling them together and organizing them on the screen.

Even without the need to handle networked multimedia in real time, the processing load that will be placed on the main CPU in both servers and desktops can be expected to increase as well. For example, with the complex multimedia applications that are under development, a multimedia PC must simultaneously be able to provide full-screen, full-motion video with MPEG 2 video compression- decompression and better sound with Dolby AC-3, along with telephony and video conferencing capabilities, which will require performance in the billions of operations per second: . MPEG 2 decoding requires 1 to 2 billion operations per second (BOPS), 3D rendering requires 1 BOP for 1 million triangles per second, and real-time MPEG 1 encoding requires tens of BOPS.

There are a number of multimedia coprocessors available that would fill the bill: Chromatic's mPACT, Samsung's Multimedia Signal Processor and the Philips Trimedia processor, among others. On the face of it, with clock rates for Pentium II and Pentium Pro architectures with MMX moving toward 500 MHz and even 1GHz, these multimedia coprocessors would seem to be destined for a marginal role on the desktop. However, in addition to focusing impressive processing power on the problem of multimedia acceleration, many of these coprocessor designs have also been optimized for real-time operation.

In particular, these media processors are well suited to performing context switching in real time, which are difficult for general-purpose architectures such as the x86, and made even worse with Intel's proposed multimedia extensions. Context switching is one of the most publicized of issues relating to MMX. The new MMX-enhanced host processors must not break backward compatibility with existing application code. For this reason, Intel has architected the technology to switch between the CPU's floating point unit and MMX processing at alternate times. Thus, the CPU is either in MMX mode or floating point mode - but never both at the same time.

The time it takes for the CPU to switch from one of these modes to the other is on the order of 50 clock cycles for the P55C(the original name for the Pentium II), during which time the host is unable to process either multimedia data or floating-point data. Even though this switch time is expected to be reduced in future MMX implementations, it can never be eliminated without breaking backward compatibility with existing applications.

This inability to process floating point and multimedia operations simultaneously can significantly affect the performance of some multimedia tasks, such as 3D graphics processing, for example. The 3D rendering pipeline has processing requirements which differ greatly depending upon the specific task being accomplished. For example, the high-level stages of the pipeline - including geometry processing, lighting, and transformations - require intensive use of floating-point processing. Once triangle vertices are determined, setup calculations can be accomplished in either fixed-point or floating-point arithmetic. However, to provide the high precision required for the complex scenes of 3D games or OpenGL-based scientific environments, floating-point processing is necessary. Because of the simultaneous requirements of rendering and floating-point operations, it is difficult to process the entire 3D pipeline on a host CPU without greatly reducing performance. This is one reason why Intel has suggested in its Graphics Controller '97 Specification that setup and rendering calculations be done on external graphics controllers. Thus, MMX serves as a complement to external 3D hardware, rather than a replacement for it.

Multimedia coprocessors effectively have zero context switching time, since they are capable of carrying out multiple operations per clock. It is not uncommon for a multimedia coprocessor to sustain 32 operations per clock, which are asymmetric; that is, different logical operations, including multiply accumulate, decimal shifting, adds, divides, etc., may be processed simultaneously on the same clock cycle. Thus, multimedia coprocessors have an intrinsic architectural advantage which allows them to perform at several times the compute power of standard or MMX-enhanced scalar processors built with comparable CMOS process technology. Additionally, all of this zero-context-switch processing can happen in standard PCs running Windows without requiring application vendors to optimize their code.

Chromatics' mPACT on Multimedia

The mPACT coprocessor, designed by Chromatics Research and built by Toshiba, for example, can easily support all seven functions included in the broadest definition of an interactive networked multimedia system: a powerful GUI accelerator, 2D and 3D graphics support for games, MPEG video, audio, a range of telephony options, FAX/modem/ISDN/ADSL capabilities, and desktop video conferencing.

The mPACT media engine(see Figure 17.1) is designed to do one thing very well - crunch multimedia data - and do so in various combinations simultaneously. To do this, the mPACT media processor combines elements of supercomputing and digital signal processing (DSP) technologies, including a very high bandwidth, parallel architecture, very long instruction word (VLIW) and single-instruction/multiple data (SIMD) execution as well as vector processing to achieve the billions of operations per second throughput necessary to perform several multimedia functions concurrently in real time.

Data moves into and out of the engine simultaneously over five high-speed I/O buses at up to 500 Mbytes/sec., which means the processor is never waiting to send or receive data between peripherals, host bus or RDRAM. At the same time, a 792-bit-wide data highway is moving up to 8 billion integers per second between hundreds of arithmetic logic units working in parallel. As a result, it is able to achieve 2 billion integer operations per second for most functions, and up to 20 BOPS for the critical motion estimation function used in video encoding and video conferencing.

Inside the mPACT multimedia coprocessor, a programmable core performs all the multimedia data processing required to execute the various audio, video, communications, and graphics functions. This core is surrounded by interface controllers for each of the five

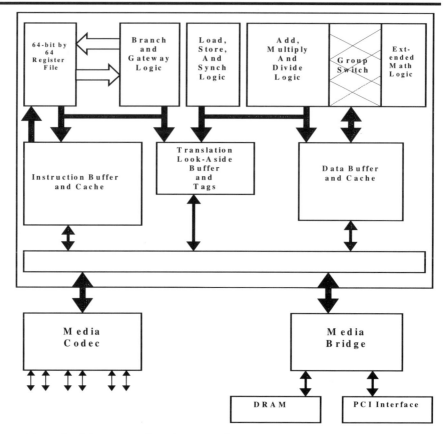

Figure 17.1 **mPACT core.** To allow the core to focus on multimedia, core is surrounded by coprocessors that use five independent I/O buses for RDRAM, display, video, PCI, and peripheral.

high-speed external I/O buses — RDRAM (500 Mbytes/sec.); PCI (133 Mbytes/sec.); display (270 Mbytes/sec.); video (27 Mbytes/sec.); and peripheral buses (10 Mbytes/ sec.).. A single multimedia memory buffer, typically comprised of 2 to 4 Mbytes of Rambus DRAM (RDRAM), acts as the central store for all multimedia data and mediaware modules. RDRAM provides the necessary bandwidth with a single-chip, low-pin-count solution. Using direct memory access (DMA), the interface controllers move data into and out of the RDRAM buffer on a microsecond-by-microsecond basis where they are made available to the processor core.

The supercomputer-like architecture of the processor core combines five groups of parallel processing units as well as an eight-port SRAM, a vector register file, an instruction decode, and the RDRAM interface. Four of the processing units contain a total of 50 arithmetic logic units (ALUs), while the motion estimator used for video encoding contains 400 ALUs. The eleven outputs from the processing unit groups and the SRAM read ports are each 72-bits wide, and are arranged in parallel to form the 792-bit results bus.

The results bus ties these outputs back to the 72-bit wide processing group input bus, SRAM write ports and instruction decoder input. This is done through the use of a data crossbar arrangement that effectively gives any input access to any output. If a result is to be used in the next cycle for another ALU operation, it passes through the crossbar into the appropriate ALU. If it is not to be used in the next cycle, it passes into one of the write ports of the SRAM, where it waits until it is needed and then passes out through a read port. Reading and writing to the results bus can occur concurrently. By making the SRAM multiported and high speed (4 billion integers per second) execution can occur directly from SRAM with zero wait states. Data moves between the on-chip SRAM and the external RDRAM buffer over the 500 Mbytes/sec. Rambus interface.

As mentioned earlier, to make the most efficient use of this parallel architecture, the mPACT multimedia engine employs a combination of VLIW, SIMD, and vector processing. Use of VLIW helps reduce code size and eliminates problems with instruction caching, since more information can be included in the instruction word. Use of SIMD allows multiple pieces of data to be processed as a result of a single instruction.

Each 72-bit instruction word is fetched on every clock cycle, and each instruction word contains two instructions or opcodes. Each opcode, in turn, can operate on 2 to 16 integers simultaneously, enabling as many as 32 integer operations to occur in parallel for each clock cycle. In addition, vector instructions can automatically perform multicycle operations on an array of operands, providing much greater efficiency than if these operations were performed using traditional program loops. The VLIW engine also increases processing efficiency by using all fixed point operations, relying on the system processor to perform floating-point calculations.

RISC+ARM=Real-time Multimedia

An architecture developed by Samsung Semiconductor takes a different approach to the problem: the multimedia signal processor (MSP). The MSP combines a traditional RISC controller with a SIMD processor, relegating some functions to dedicated logic that cannot be handled effectively in a programmable environment (see Figure 17.2).

The hardware architecture of MSP consists of three components: (1) a conventional microcontroller, in the case of the first implementations, an ARM CPU; (2) a SIMD Vector Processor (VP); and (3) special-purpose hardware engines. The ARM controller performs three tasks: (1) running the RTOS; (2) overall system management and control; and (3) certain high-level, low-volume scalar media processing. The VP performs one task: high performance signal processing. The special purpose hardware units each perform just one type of task: fixed-function processing that cannot be performed efficiently in the other two units.

The use of a traditional RISC CPU as a controller and the SIMD architecture of VP provides a familiar, but powerful software programming environment. The software interface is clean, simple, open, and consistent and because it is based an ARM, familiar to a

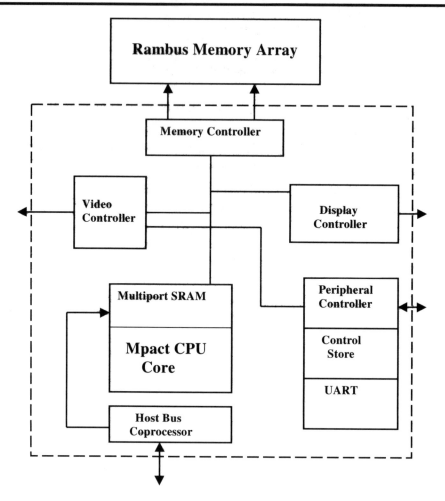

Figure 17.2 **Samsung RISC+DSP**. Rather than use an esoteric VLIW scheme, Samsung combines a traditional DSP processor for multimedia with an ARM CPU core for control.

lot of net-centric systems developers. And since MSP does not have the hardware architecture, OS or binary compatibility baggage of legacy X86 architecture weighing it down, it provides a clean SIMD design.

While all but one of the MMX instructions are similar to MSP instructions, MSP provides nearly twice a many multimedia instructions in a vector engine that has a bus that is four times wider. Moreover, similar to the Alpha CPU implementation, it supports simultaneous integer and floating point processing.

As for the constraints of real-time operation that networked multimedia place on the net-centric computer, the MSP approach provides the right architecture for deadline- driven preemptive scheduling by combining a very simple ARM RISC core with a 256-bit SIMD engine. The use of two processors allows it to run two independent instruction streams concurrently, one for RTOS and system management on ARM and one for signal processing on VP. Since the load on ARM is interrupt and system intensive, latencies are minimized by doing preemptive scheduling. This works out very well since the entire register state on ARM is only 16 words. The VP, on the other hand, is very fast at signal processing, but its state is also large.

So, for its most efficient utilization, cooperative real-time context switching is achieved through collaboration with the application programmer. In other words, VP context switches are made at programmer-defined clean points where the VP state is relatively clean. This model is further tuned for real time by providing programmers with the ability to inform the scheduler of the real-time requirements of their applications using time constraints. The scheduling decisions are made using this information, and VP context switches can be extremely fine grained since a clean point can be performed in code every 100 microseconds. This makes for a networked multimedia system tuned for hard teal-time.

The MSP architecture also makes real-time scheduling and debugging much easier in a multimedia environment. With a simple SIMD architecture such as the one used in the MSP, many of these problems are solved by its very design. It is only necessary to schedule one SIMD instruction with an I/O or branch instruction every clock. The simple conventional 32-bit instructions make programming in assembly very easy and produces compact code. It is also possible to do most of the coding in C or C++.

In its implementation of the MSP, Samsung has made a few simple extensions to C and C++ to define new data types for vectors and matrices, as well as special MSP SIMD operators to the C language. Using these extensions, the application developer can write C code to process vectors in parallel in an otherwise fully sequential C/C++ program. Furthermore, since MSP executes one SIMD instruction per clock, reporting precise exceptions is an easy task.

Philips Tries All Media

A third alternative that attempts to satisfy the dual requirements for both real-time control and real-time multimedia processing is the TriMedia architecture from Philips Semiconductor. It combines a digital signal processor with a general-purpose combine engine based on VLIW techniques. Its role is to subsume all the specialized video and audio compression chores, perform many graphics and video processing functions as well as many of the complex mathematical computations that must be done in real time.

The first TriMedia chip, the TM-1, is a single-chip multimedia coprocessor with an integrated, glueless PCI-bus interface (see Figure 17.3). TM-1 integrates a number of independent DMA units, a glueless SDRAM interface, and a high-performance processor core, which Philips calls the DSP/CPU. All the internal elements communicate via the on-chip, 32-bit data highway. It can acquire multimedia input data streams, process them, and output multimedia data streams, all in real time.

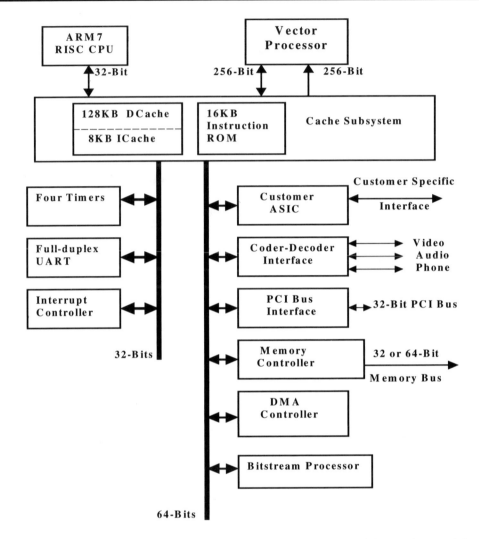

Figure 17.3 **TriMedia Architecture**. In addition to a variable length instruction word data processor (VLD), the TM-1 integrated direct memory access units, a DRAM and PCI interface, as well as an image coprocessor and serial interfaces to cameras and data communications links.

The two keys to the real-time responsiveness of TM-1 are the independent DMA input and output units and the internal data highway bus. In general, a responsive, real-timemultimedia system needs to perform several functions simultaneously: receiving uncompressed video and audio data, compressing and decompressing data, sending uncompressed video and audio, and communicating compressed data. TM-1 uses dedicated DMA engines to perform the simple but urgent tasks of video and audio capture, video and audio output, and serial communication with modem or ISDN analog interface hardware. In addition, a VLD (variable length decoder) DMA engine processes variable-length MPEG Huffman codes, and an ICP (image coprocessor) handles image scaling and color conversion independently.

An internal data highway bus controls access to local synchronous DRAM (SDRAM). The highway arbitration unit implements, in hardware, some of the load balancing essential for handling audio and video in real time. Finally, an arbitration unit allows each bus master to use the bus according to bandwidth allocations programmed by software. Each master is guaranteed a (programmable) minimum bus bandwidth and a maximum latency. All unused bandwidth is available to the DSP/CPU core within one cycle of a request or to any other master within a few cycles of a request.

The processor at the core of the TM-1 is based on a 32 bit VLIW (very long instruction word) architecture that combines DSP capabilities with the programming ease of a general-purpose CPU. It provides traditional microprocessor operations plus dozens of specialized single-instruction, multiple-data (SIMD) operations that directly accelerate multimedia algorithms. For example, the SIMD operation for motion-estimation uses 8-bit operands and performs eleven operations in parallel: four 8-bit subtracts, four 8-bit absolute values, and three additions. Using this combination of the VLIW architecture and specialized multimedia instructions and operations, up to 3.8 billion operations per second can be achieved. This means that an MPEG-1 video and audio decode can be performed with only of 23% of a TriMedia processor's resources with no performance impact on the host processor. Other multimedia algorithms are likewise accelerated.

A third generation TriMedia design makes this approach even more attractive. In its TM-2000 Philips has taken the architecture to 64-bits and developed a unique five slot instruction architecture. This new multislot scheme grew out of research into improving the instruction set, particularly in areas relevant to graphics processing, like data matrix manipulation. Attempts to do that resulted in some operations having three or four argument registers and/or two result registers, which is beyond the scope of a single slot. By allowing operations to occupy two or more slots, the functional units dedicated to each slot can connect to the multiple register-file read and write buses.

Beyond the Desktop

Beyond the desktop, the Multimedia Net will also affect the way the many other net-centric computing systems are configured: non-x86 network computers, Internet appliances and Web-enabled set-top boxes. The new generation of multimedia coprocessors could certainly fill many of the requirements of these systems. Not dependent on the architecture of the desktop, these multimedia coprocessors could be used with any non-x86 architecture, either as simply another component, or in a customized implementation that integrates the control CPU with the multimedia coprocessor. One alternative that come the closest to this ideal is the Samsung MSP that combines an ARM RISC CPU with a media processor on the same chip. This combination could be very effective in a MPEG 4 networked multimedia environment where Java is likely to be used as the basic language for negotiation between sender and receiver. The fact that the ARM architecture is exceedingly Java-friendly would be a very important benefit.

The MIPS Alternative.

Another possibility is the MIPS architecture from Silicon Graphics MIPS subsidiary, which has now gone independent as an embedded CPU vendor, using the ARM licensing model. In addition to its use in the Silicon Graphics' high-end graphics and video workstations and servers as well as in consumer-oriented video game systems, this architecture has been adopted by a number of vendors as the vehicle for implementing a variety network computers, Internet appliances and Web-enabled settop boxes..

Recognizing the direction that multimedia was going on the Internet, the company has also come up with a set of architectural and instruction set extensions that will allow the MIPS design to transition easily to the world of networked multimedia. In particular, in its 64-bit MIPS V architecture, the now independent company has developed two different sets of enhancements, one general purpose, and the other specific to multimedia processing.

First, the MIPS V instruction set starts with the MIPS IV instruction set and adds support for SIMD single-precision floating point to support the front-end of image and video synthesis using the preferred 3D data types. Second, it has added a modular application specific set of extensions called the MIPS Digital Media Extension (MDMX) to the MIPS instruction set architecture.

MIPS V borrows an innovation from the DSP world, and provides an accumulator with extra precision to support the width required by intermediate calculation. But unlike DSPs, MIPS V implements a vector accumulator to take advantage of the SIMD aspect of the instruction set. The MIPS MDMX also provides a few other features to better support multimedia processing.

MIPS V is the new baseline instruction set for the company and its many semiconductor partners. Completely compatible with MIPS IV, it is aimed directly at the front-end algorithms of 3D graphics processing, including 3D geometry transform and clip, and lighting and shading calculations. These algorithms are usually done in single-precision (32-bit) floating-point. In a 64-bit processor, single precision operates at the same speed as double precision, while using about half of the capabilities of the hardware.

By packing two 32-bit floating point values into a single 64-bit register, and providing new instructions to operate on this paired-single format, MIPS V provides twice the floating point operations per second with only a small amount of new hardware. The paired-single format was added to basic MIPS IV repertoire of computational operations, such as multiply, add, subtract, and compare instructions. For example, a paired-single add adds the high halves and low halves of its two source registers in parallel and puts the two sums into the destination register as a pair. In effect MIPS V implements two-element vectors.

The MDMX instructions features two formats: Oct Byte (OB) for eight unsigned 8-bit integers and Quad Half (QH) format for four signed 16-bit integers, each packed into a 64-bit word. MDMX shares thirtytwo 64-bit wide registers and 8 condition code (CC) bits with the existing floating-point unit (FPU). Data are moved between the shared register file and memory with existing floating-point load and store double-word operations, and between floating point registers and integer registers using other instructions.

There are two parts of the instruction set. The first is a conventional register-to-register set of instructions. For exampl,e operations such as add, subtract, logical, shift, min, and max operate on two source registers and the result is clamped and stored into destination register. Other instructions exist for handling unaligned data, permuting, comparing, or doing conditional selection. The rest of the MMDX architecture is centered around the vector accumulator. This feature allows MMDX to realize the full potential of the SIMD paradigm by computing directly from the packed representation. There is a single 192-bit register which serves as eight 24-bit vector accumulators for OB data and four 48-bit vector accumulators for QH data.

An unusual feature of the architecture is the use of a vector element as a scalar in a SIMD operation, which is much different than most of the other SIMD implementations. It draws on the experience programmers had with such high-end computer designs as the Cray which includes both vector/vector and vector/scalar forms of all operations. Because this vector/scalar combination arises naturally in many algorithms, the company found it critical in implementing extremely efficient motion estimation searches. In addition, the element select feature allows several constants to be packed into a single register, thereby saving registers, extremely helpful when processing DCT coefficients.

Meeting the Requirements of MPEG 4

While the architectural modifications discussed so far address many of the requirements needed for real-time net-centric computing in an age of networked multimedia, left unaddressed is the requirement in standards for MPEG4, not only for real-time multimedia processing, but real-time processing of Java commands.

While compiling Java to the native format of a specific processor is certainly one way to make Java more real-time, this would negate the advantage of Java and its "write once, run everywhere" capability. This latter characteristic is particularly important under MPEG 4, where Java will probably be used to communicate between multimedia encoders and decoders during the negotiation stage.

On the desktop, it is possible that an MMX-enhanced CPU could handle both the multimedia compression and decompression as well as the Java negotiations, but only within tightly constrained conditions. One of the promises of MPEG 4 is the ability to incorporate into a single MPEG 4 session not only several different kinds of multimedia types, but also several different data streams, each from different resources throughout the network. This means that an architecture is required that both handles not multimedia in a real-time and deterministic manner, but also performs the stack-oriented processing required by the Java virtual machine in real time as well. With the continuing escalation of processing requirements for multimedia and the multi-threaded nature of MPEG 4 multimedia data streams, there is some doubt that even a 500 MHz Pentium II-style CPU could handle both adequately.

On the desktop and in the network computer there are some alternatives. One would be to use a multimedia coprocessor to handle the majority of the multimedia operations and use the main processor to handle the Java-based negotiations. Another is to expand the capabilities of the graphics processing subsystem, adding in the ability to do the stack-oriented Java negotiations before the media stream actually reaches the processor. This approach is being pursued by graphics chip companies such as Cirrus Logic Inc. which in late 1997 became a licensee of the ARM architecture. As we have learned in earlier chapters, ARM has some stack-oriented features and is reasonably good at executing Java.

Net-centric devices such as thin client-based network computers, Internet appliances, and Web-enabled set-top boxes will require another solution, one that combines real-time multimedia processing and stack-oriented processing into the same architecture. There are currently a number of alternatives that come to mind, including multimedia-enhanced versions of the ARM architecture and Sun's microJava and picoJava architectures. In addition to the approach taken by Samsung which combines an ARM CPU on the same chip as a multimedia engine, Advanced RISC Machines and its partners are also working to incorporate multimedia capabilities into the basic architecture. Another possibility is the StrongARM CPU acquired from DEC when it as dismembered and parceled out between Intel and Compaq. Intel is planning to use StrongARM rather than any of its existing architectures as a way to break into the potentially high volume market for net-centric IAs and set-top boxes. Because StrongARM is a highly effective Java execution engine, the addition of multimedia instructions to this design is not an unlikely possibility.

Chapter 18

Moving from GUI to NUL..to XUI?

Within the space of 20 or so years, we have all moved into a computing environment in which we have access to several gigabytes of local storage and a network-connected environment with literally hundreds and thousands of terabytes of information at our fingertips.

But what is becoming clear is that unless something is done we all run the danger of "getting lost in hyperspace." As attractive as the idea of instant access to the volumes of information over the World Wide Web is, the still somewhat esoteric and complex methods of searching and visualizing it could eventually turn off many sophisticated, and not so sophisticated, users. Something has got to be done to make it easier for everyone to find their way to, interact with, navigate among, and make use of the multiplicity of text, graphics, audio and video data now available on literally hundreds of thousands of Web pages around the world.

How does one avoid getting lost in the hypertext and hypermedia world of the Internet and the Web? Well, how does one avoid getting lost in an unfamiliar city or territory? First, one has to have a good guidebook to find the places to go. Then, one needs a map to find the way to the destination. It is also useful to have some means of recognizing when you get to a destination and some means of putting it into context. Knowing your location at the corner of 5th Avenue and 31st Street in New York City is one thing, but knowing that there is a deli up the street or that you are one block over from a restaurant or a hotel is a lot more comforting.

On the streets of a city, where we must navigate in two dimensions, doing all the above can be trying at times. Imagine doing it in three-dimensions or in the world of information "hyperspace" where there are dozens of dimensions or ways of measuring where you are.

In this hypermedia world , where our view is via the computer monitor, and more recently, the TV screen of the Web-enabled settop box, it is clear that we need to make a shift from a graphical user interface to a new paradigm, what some have termed the "network user interface."

First Attempts

Numerous attempts are underway to come up with a different user interface appropriate to net-centric computing. But most are simply extensions of existing desktop paradigms, rather than something completely new.

On the network-connected PC side of things, Microsoft is adding browser-like navigation for both internal files and external Websites to its Windows 97 and NT operating systems. IBM has added net-centric features to its OS/2 Warp 4. On the network computer side, Oracle has developed NC Desktop, Sun has HotJava Views and WebTop (see Figure 18.1), Lotus has its Kona Desktop, Netscape has developed Constellation, and Santa Cruz Operation has Tarentella.

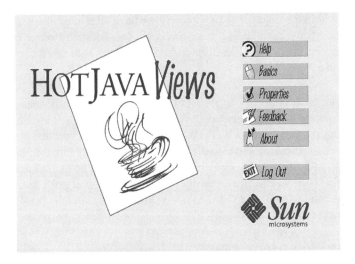

Figure 18.1 **Simplification By Elimination**. Current approaches to creating a network user interface, such as Sun's HotJava Views, try to simplify the interface by eliminating functions as well as the user's freedom of choice. © 1998 Sun Microsystems. Used by permission.

They all share a number of common characteristics. First, they essentially erase or at least blur the differences between local resources on the desktop and those on the network. They do this by integrating the network functions into the common desktop operations, rather than launching them as separate functions. These would-be NUIs present a universal view of all of a systems resources, both internal and external. They also make use of new Web push and pull" technologies to automatically update desktop content. Third, they provide mechanisms - useful in the corporate intranet environment - for storing or mirroring a client's files and local state on a server.

What they all also share in common is the misplaced belief on the part of the designers that ease of use means building interfaces that achieve simplicity by limiting the user's ability to make choices. Some discard features such as double clicking, overlapping windows, hierarchical menus and button bars. Others, adopting the maddeningly frustrating "Microsoft knows best" attitude, give the user only a minimalist GUI, with only a few tasks graphically represented.

While some of these proposed candidates for a network-user interface do represent some significant and needed modifications, they are hardly the revolutionary changes that are needed. They also all seem to miss the point. It is as if automobile designers, in trying to make it easier for the driver of a car to navigate around a city, rearranged the knobs on the dashboard and eliminated the speedometer and gas gauge.

Granted, automotive engineers can not do much about the outside world that the driver sees through the windshield. But in the virtual world of the World Wide Web, there is a lot that developers of computer hardware and software can do about the way the average user views the terabytes of data on the Internet, and navigates his or her way through the thousands of Web pages at each site.

Making the interface to the Internet and World Wide Web easier to use and to navigate does not mean elimination of complexity by the elimination of features, it means substituting simplicity for complexity.

Is Z-GUI The Answer?

Among the many organizations and research groups studying the problem of making the Web easier to use are the people at Xerox Palo Alto Research Center in Palo Alto, California. In the late 1970s and early 1980s, it was they who pioneered the now common WIMP graphical user interface (windows, icons, menus and pointers) but failed to commercialize it. Beginning in the late 1980s they began studying the problems of dealing with huge amounts of information, mostly in a closed network environment of connected workstations, with hundreds of gigabytes of data accessible online. More recently they are looking to use the same technology to simplify viewing and navigating around the World Wide Web.

Pushing the concept of the traditional x-y two-dimensional GUI to the extreme, they have developed what they call Z-GUI, where the Z stands for the third dimension. It, in essence, uses 3D analogs of the physical world to make information stored in a computer system more intelligible, navigable, manageable, and understandable.

The new desktop visualization paradigm has two major components. The first is wide widgets and the second is 3D workspaces. The first paradigm turns the "going in order to see" methodology that we have gotten used to with WIMP into a more natural "see and go to" approach. Unlike the real world where you usually want to see where you are going before going, the traditional GUI requires the user go to a task bar and unfold a menu in order to see the options concerning where to go.

Taking advantage of some of the advanced graphics in today's desktops, the Xerox user interface, what I call the XUI (or zooey), uses a "focus plus context" technique similar to the way we view the real world. Just as we view a scene in front of us, focus in on one particular feature and then take some sort of action, the XUI presents the entire panorama of choices, but with only those at the point of attention in focus. To view in detail what the other choices are is simply a matter of changing the center of attention, just as we do visually, moving our head or our eyes to focus on something that was in the blurry periphery of our vision. Used with 3D workspaces, using wide widgets, objects become the framework for structures and spaces - typically based on an underlying information spine like a hierarchy, a table, or some other set of associations.

Using the two concepts, Xerox scientists have created a number of interesting ways to view and categorize large amounts of data, including Perspective Wall, Cone Tree and its derivative Hyperbolic Tree, BiFocal Display and 3D Rooms. These have been combined into a system called the Information Visualizer, a system that uses 3D and interactive animation to work with large information spaces.

Perspective Wall

Using the 3D and animation features of today's desktop hardware and software, the Perspective Wall (see Figure 18.2) uses the physical metaphor of folding to distort an arbitrary 2D layout into a 3D visualization, the wall. The wall has a panel in the center for viewing details and two perspective panels on either side for viewing context. The perspective panels are shaded to enhance the illusion of 3D. The vertical dimension of the wall is used to visualize layering in an information space. Extending the concept, the wall holds "cards" that represent files in a computer system that are structured by modification date (horizontally) and file type (vertically). To enhance the focus of the foreground panel of the wall, it makes the neighborhood of the detailed view larger than the more distant parts of the contextual view.

Smooth transitions among views are achieved by allowing the user to adjust the wall much like a sheet in a player piano moving selected notes to the center of the view. When a user selects an item, the wall moves that item to the center panel with a smooth animation, helping with the perception of object constancy. Another feature of the Perspective Wall is the ability to adjust the ratio of detail and context, much like a rubber sheet. The Perspective Wall works with any 2D layout that can be described as a list of 2D vectors and 2D positional text. The wall makes it easy to visualize the results of an information search because it shows all similar items simultaneously and in context.

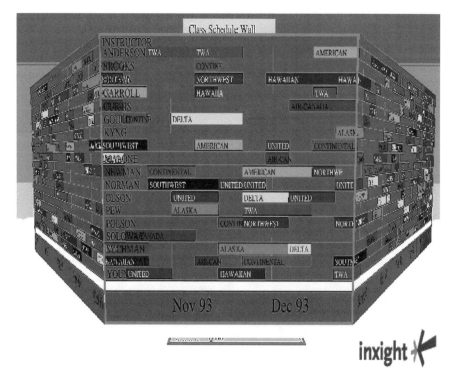

Figure 18.2 **Perspective Wall**. Using the metaphor of folding to create a distortion of a 2D layout of files or Web locations, the 3D-like wall allows a user to view particular locations in detail with less important locations on either side. *Source*: Xerox Corp., InXight.

Cone Trees

Another useful way of visualizing large arrays of complex information is the traditional tree structure, which can be as complex or as simple as needed. In a simple Cone Tree (see Figure 18.3), data are represented as a collection of index cards, hanging off the branches like greeting cards hanging from a Christmas tree. The top of the hierarchy is placed near the "roof" of the 3D space in which the tree is visualized. Below it, like layers of branches and leaves on a tree, are the "children" of each preceding layer. The body of each cone tree is shaded transparently, so that the cone is easily visualized, but does not block the view of the cones behind it, or cards on the tree on the other side. When a node is selected with the mouse, the Cone Tree rotates so that the selected location or card and the related cards and branches connected to it are brought to the front and highlighted.

The Xerox researchers have also incorporated techniques to allow the information cone trees to be manipulated - "pruned" if you will. These include the ability to view parts of trees, to restructure trees dynamically, and to search through underlying information on the trees. Also incorporated are ways to prune the trees and to allow them to grow, either by mouse gesture or by menu.

Bifocal Display.

Another 3D construct that was developed originally for offices and organizations that contain information subdivided into a hierarchy of journals, volumes, issues and articles, the Bifocal Display is a data visualization technique that could be very useful in the Web environment. It integrates detail and context by displaying information items positioned in a horizontal strip. Bearing some resemblance in concept to the Perspective Wall, the strip is a combination of a detailed view and two distorted views. Items in the foreground are detailed and on each side are strips distorted horizontally into narrow vertical strips, much like books on a library shelf. The vertical strips contain information about the data such as a title, author or keywords, very much like the information displayed on the spine of a book. The detailed view might contain a page from a journal or book, and the distorted view might contain the years for various issues of a journal.

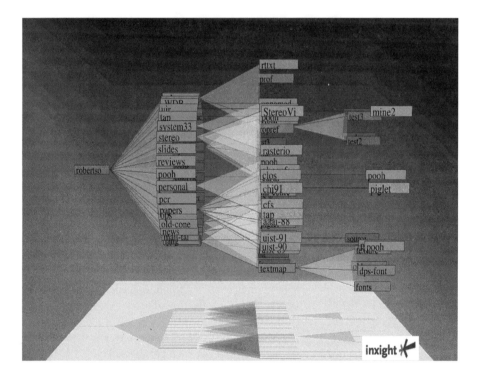

Figure 18.3 **Cone Tree**. In this 3D view, data or files or Web sites or Web pages could be represented as a collection of index cards hanging off the branches like greeting cards on a Christmas Tree. *Source:* InXight Software

3D/Rooms With a View.

When trying to visualize the information contained on the various sites on an intranet or on the World Wide Web, the big question is a familiar one: How do you avoid the problem of not seeing the tree because of the forest? If there is no way of organizing the information cone trees, the user of the Web would be no farther along than he is now.

Xerox researchers have also addressed this problem in a number of ways. One of the most interesting is to organize and compartmentalize information trees and walls into 3D rooms and then organize the rooms into structures that allow the viewer to select from among an array of rooms.

In 3D/Rooms, a user is given a position and orientation in the room. He or she can move around the room, and zoom in to examine objects such as Cone Trees and Perspective Walls and other physical analogies appropriate to the virtual world of information. The information seeker look closely at objects, look behind himself, walk around an object, and walk through doors to other rooms.

A radical departure from the now familiar desktop full of windows, Rooms required the development of a new set of algorithmic building blocks. There is a strolling metaphor for random walking or exploratory movement of the use, and a point-of-interest movement algorithm for rapid and precise moves to objects of interest. Still another algorithm allows 3D objects to be moved rapidly using the mouse.

To view all of the workspaces containing objects, 3D/Rooms also contains an overview mode that allows the user to view all the rooms simultaneously. Like a ghost or immaterial spirit, 3D/Rooms allows the viewer of the information to move from the view of the "building" and appear in a particular room instantaneously, moving about in it and manipulating informational objects.

Another in a wide array of visualizing mechanisms is what I call the "satellite view." It uses the color capabilities of modern graphics systems to create a distant view of a vast array of information as if from a satellite. Just as satellites are able to view Earth with light from different portions of the spectrum and spot resources of interest, this view allows one to look at the information as an undulating plain moving off into the distance. Resources of interest can be seen by the differences in color. The spectrum is determined by the search criteria used, with hot spots of great interest or which exactly meet the search criteria in red. Cooler spots which are less of a match are in blue, and areas with no relevant information represented in various shades of gray and black.

Representing a paradigm shift as significant as the move from text-based displays to the now familiar desktop/windows, Xerox is determined not to loose out on the benefits of the this new way of viewing information, the way it did with WIMP. Through InXight, one of a number of subsidiary companies that it has created to commercialize it's research, Xerox is rapidly moving this technology into the computing mainstream. It has taken many of user interfaces, rewritten them using component technologies such as ActiveX and JavaBeans, and now sells them as a line of user interface controls it calls VizControls. To make it even more adaptable to the World Wide Web, it has incorporated an extensible markup language (XML) interface, which, as we shall see later in this chapter will change fundamentally the way data are represented, stored and searched on the Web.

In addition to selling them in a package it calls Visual Recall, InXight has licensed these visualizing components to a wide range of companies. For example, a 2D version of the ConeTree developed by InXight, called Hyperbolic Tree (see Figure 18.4), has been licensed and used by companies as diverse as Comshare, Fujitsu Ltd., Microsoft Corp., Softquad, and S3. About 20 to 30 companies have licensed various other aspects of these technologies.

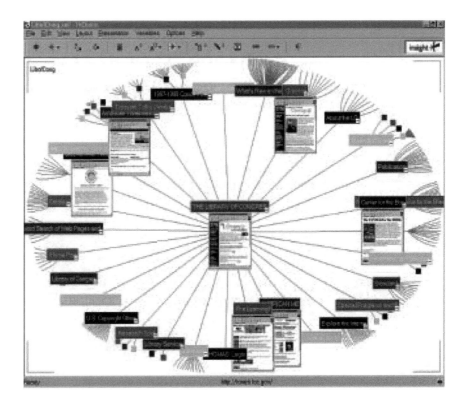

Figure 18.4 **Hyperbolic Tree**. One of the first commercial applications of the Xerox "conext+focus" scheme, the hyperbolic tree allows a user to move about a dense array of Web locations without getting lost, but retaining a sense of location. *Source*: InXight Software

Finding Out Where To Go

Other research that Xerox is doing, and that InXight is commercializing, deals with another network user interface problem that is just as frustrating as "getting lost in hyperspace." This is the problem of finding out where to go. The best analogy in the common place world of physical objects is the familiar map you use to find your way to a particular location. To find out what sector on the map to search, the map usually contains an index of street locations. Depending on the size of the city, there might be two to a dozen listings for the name of the throughfare you want, but, unfortunately, listed as a street, an avenue, a lane, a circle and so on. How do you determine which is the correct one?

Obviously, on the World Wide Web surfers face the same problem, but multiplied by several orders of magnitude. Admittedly, the familiar Web searchers that are easily accessible from almost any browser are powerful, but the results that the user gets can run into the thousands of listings.

But the problem does not end there. Similar to finding your way in a city with a map that is written in a foreign language, all of the major search engines - Altavista, Excite, HotBot, Infoseek, Lycos, OpenText, WebCrawler and the World Wide Web Worm, among others - to one degree or another require that the user learn and become proficient in AND/OR/NOT Boolean logic. This logic scheme, second nature to the engineer but esoteric to the

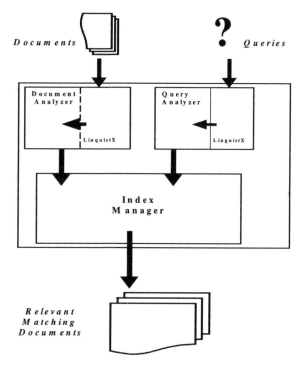

Figure 18.5 **Natural Language**. As an alternative to the Boolean logic used by most Web search engines, the Xerox LinguistX algorithms allow a user to ask for and search for data using natural language phrases and words.

average computer user, is used to sort out the good from the bad and the useful from the useless. And depending on the Web search engine, the accuracy of the results can vary widely.

What is required, and what researchers throughout the computer industry are looking for, is a means by which search requests can be entered in natural-language form, analyzed efficiently, and the results displayed in a clear and precise way. While researchers at Xerox and InXight are not alone in their efforts, the tools they have developed are indicative of the direction in which such work will take the World Wide Web.

For example, they have developed LingusitX (see Figure 18.5), a suite of natural language processing components that allows users to enter queries in ordinary language and search natural-language documents without the use of esoteric Boolean logic. The power of this suite derives from four high-speed text-analysis algorithms: the language identifier, the stemming and lexical analysis analyzer, the contextual tagging and phrase extractor, and the summarizer.

The first algorithm identifies and sorts documents according to the language and character coding in which they are written. This algorithm enables Web crawling and other document-collection software to either omit or perform special processing on documents in virtually any language. The second algorithm does a number of things. Stemming relates all possible forms of a word to their base form, providing for more efficient indices and more accurate text relevance analysis. Lexical analysis is a more advanced form of stemming that identifies grammatical nuances, such as the tense and mood of a word, in addition to its root forms.

The contextual tagger enhances text-based information analysis by identifying the grammatical category of words by their context. The phrase extractor recognizes multiword concepts, instead of keying on the individual words, a source of many of the low-quality searches and summaries obtained with current Web search engines. Finally, the summarizer examines the content of a document in real time to identify key phrases and extract sentences to form a summary. It enables users to get the sense of a document's content without reading the whole document.

In addition, InXight researchers have come up with a number of other algorithms to aid in the actual search for the right documents. One is called Scatter/Gather, which allows a user to rapidly assess the general contents of a very large collection of documents by scanning through a hierarchical representation. This acts as a dynamic table of contents. Using this algorithm, a system first "scatters," or clusters, the collection of documents into a small number of document groups and presents short summaries of the groups to the user. Based on these summaries, the user can select one or more groups. The selected groups are gathered or "unioned," together to form a sub-collection. With successive interactions of the same algorithm, the system reapplies clustering to scatter the new subcollection into even smaller types. By browsing the collection in this manner, the user gains an understanding of the contents and can choose whether to explore further or to try another collection of documents.

Combining the search algorithms with some of the 3D visualization tools Xerox-InXight researchers have also developed "Butterfly," a method of visualizing citations among bibliographic records. It does this by combining searching and browsing with a visual interface that looks like the wings of a butterfly. The upper part of the interface focuses on searches and the bottom on browsing, with results displayed on each side, on the wings of the "butterfly." With it, a user can start with a set of queries to find articles or Web pages of interest and then browse references or citation links to related information. The results of the queries, which can be executed in parallel, are displayed in a pyramid of objects. Each query result is visualized as a horizontal layer in a colored pyramid, with different hues to indicate the source.

XML To The Rescue

While work on these new concepts began in the late 1980s, they have found use only in closed environments containing a homogenous set of documents, such as the local area network or Intranet for a large corporation or educational and research institution. This is because in a homogenous environment complex hypertext structures can be built using "metadata" extracted from the documents. Metadata consists of information about content, sources of information, the form the information is in, and so on.

Besides making searches much more efficient and accurate, metadata can be used to drive rendering and visualization tools such as those developed by Xerox. Using this metadata, visualizations can be derived that map sources into spatial and graphical elements. The metadata can also be used to allow users to select sources, as well as build a graphical memory of sources, allowing much easier navigation amongst various locations. Also, the metadata can be used in the rendering of the items retrieved from the sources.

As useful as such tools would be on the Internet, they have not been implemented because the World Wide Web contains a serious flaw. Because of the heterogeneous nature of the documents and Web sites and because of the limitations of the current method of organizing that information - the HyperText Markup Language - such metadata have not been available.

That is changing however, with the development and approval by the World Wide Web consortium and the backing by major participants such as Microsoft, Netscape, and Sun of XML, or the Extensible Markup Language. To understand the significance and impact of the emergence of XML into wide use, it is necessary to review the history of the implementation of hypertext protocols such as the Standard Generalized Markup Language.

In the early days of personal computing, starting in the mid-1970s, hypertext programs were developed to aid users in organizing and accessing growing amounts of information in textual form on their computers. Using hypertext links, information could be linked in a variety of ways. These linkages allowed the association and connection of files, parts of files, paragraphs and words with other files and parts of files in a "knowledge base."

319

These linkages were associative in that users could link files and parts of files via structures and associations other than the hierarchical and tree structures normally used in
computers to organize data. As the power of the CPUs and the amount of mass storage
increased, the sophistication and flexibility of these linkages increased through the use of
the metadata information contained in each file.

The Meaning Of Metadata

Without metadata, search engines and other data-filtering techniques have to rely on
brute-force methods of selection, such as keyword or even content searches, to isolate
information. Even then, they can miss the boat entirely (returning information about
chocolate chips instead of computer chips).

 The main concept behind metadata is the idea that certain groups of people have similar
needs for describing and organizing the data they use. Some data types are relatively
universal (<First Name>, <Address>, <City>, and so forth). Others are industry- or even
company-specific (<price>, <manufacturer>, <componentID>). Microprocessor and semiconductor manufacturers, for example, have a whole set of data types and acronyms
understandable only to hardware engineers. The use of metadata tags such as those
above, allows each of these data types to be easily recognized and used to create sites
optimized around both the data and the people using it.

Tags have associations and structure. A product has a price, a tax rate, a delivery charge
and so on. All these are tags. By defining the structure with metadata tags, finding and
manipulating, acting on, and interacting with data is much easier.

Through the use of such metadata, sophisticated hypertext programs and techniques -
such as the Standard Generalized Markup language (SGML), Augument, Discover, Fress,
IDEX, KMS, Hytime, Hyperties, Hypercard, Hypertrans, Hyperplus, KMS, MaxThink and
Zog, among others - could be used to access and link data, both through the use of
keywords and hyperlinks as well as graphically.

By the early 1980s, the Internet had evolved to connect together various academic,
corporate and governmental institutions and their growing hypertext-based databases.
But what worked in each individual database — containing homogenous information
from which metadata could be easily derived for building sophisticated hypertext linkages - did not work in an inter-networked collection of diverse and heterogeneous databases throughout the world. Not only was the sheer size of the combined databases
overwhelming to the existing hypertext programs, but there was no way to coordinate and
"rationalize" all the specific metadata descriptions of each database into one coherent
whole.

Not wanting to give up the simple and easy way hypertext programs were able to organize
and present textual information in easy-to-read pages, researchers such as Tim Berniers-
Lee came up with a "dumbed down" version of these sophisticated hypertext languages.
Called the "HyperText Markup Language," or HTML, metadata has no relevance. HTML

concerns itself only with the presentation of information in "Web page" form over the Internet. The companion mechanism by which this works over a network is the Hypertext Transport Protocol, or HTTP.

Because HTML does not contain any information about the context or meaning of the data displayed, sophisticated graphical navigation tools such as those developed by Xerox PARC cannot be used, since there are no metadata hooks by which they can anchor themselves. More importantly, without such contextual information, searches of the Internet can be frustrating because there is no way to discriminate between the usage of a phrase or word in one context versus the usage in another context. For example, consider the word "bill," which can be a name, a charge, paper currency, a law, or the beak of a bird. To do a search for the word "bill," without the use of arcane techniques such as Boolean logic often results in thousands of "hits." Even with the use of some of the natural language searching techniques that Xerox PARC and other institutions are investigating and putting in, say, the whole name of the person, would still result in many hits, some totally unrelated to full name of the person.

The Emergence of XML

XML, which has been approved by the World Wide Web Consortium as a supplement to HTML, rather than a replacement, solves such problems. XML differs from HTML in three significant ways. First, information providers can define new tag and attribute names at will. Second, document structures can be nested to any level of complexity. Third, and most importantly, any XML document can contain an optional description of its grammar for use by applications that need to perform structural validation.

Also unlike HTML, XML data elements have well defined tags that describe their content, enabling applications and autonomous Web agents to extract useful information from them. In addition, the available tags are not fixed but can be defined in a Document Type Definition (DTD) for a given application.

The DTD describes which tags are allowed in the document, as well as defining a hierarchy that determines where tags occur. For example, "paragraph" may occur in a "chapter," but not vice versa. Although the XML standard is still relatively new, mainstream DTDs for describing software components, for example, have already been deployed for applications needing standards-based open access to information. Since XML provides a mechanism for tagging data with fields that describe the data, the applications can pinpoint the exact data they need and share it in a manner that is easily brokered between applications.

In the applications and tools that companies such as Xerox and InXight are developing to help users navigate through hyperspace, XML would have major benefits. In particular, the inclusion of XML tags would allow users to switch between different views of data on a Web site or database, without requiring that the data be downloaded again in a different form from the Web server. One simple illustration is how it could be used to generate a dynamic table of contents that changes as the data on the Web site changes. Using XML in combination with a Z-GUI or XUI, it would be possible to present the user with a table of contents of the virtual document or newspaper. Acting as a window into the data, a specific location in the collection can be expanded with a mouse click to open up a portion of the table of contents (TOC) or an index and reveal more detailed levels of the document structure.

Dynamic views of this kind can be generated at run time directly from the hierarchical structure of the document. While this can be done currently with object- oriented technology, it is possible only with great difficulty. This is because the Web latency built into every expansion or contraction of the table of contents would make this procedure sluggish in many user environments.

With an XML-enabled database it would be possible to use Java to download the entire structured TOC to the client rather than just individual server-generated views of the TOC. Then the user could expand, contract, and move about in the table of contents in a much more speedy way since it is being run directly on the client.

Some engineers at Sun actually implemented an application like this as part of a Java-based HTML help browser, but the limitations of HTML required that they use some clever, but awkward, workarounds. Because ordinary HTML lacks the necessary metadata structure, it was not possible to reliably generate a TOC directly from the document, so the table of contents was constructed by hand. Nonstandard tags were invented for the purpose, and then the TOC was wrapped in a comment structure within an HTML page to hide the non-standard markup from Web browsers. A Java applet downloaded with the HTML document interpreted the hidden markups and provided the client-based TOC behavior.

In an XML-enabled environment, the process of generating this table of contents would be much easier. Neither the manual creation of the TOC nor its concealment would have been necessary. Instead, standard XML editors could be used to create structured content from which a structured table of contents is generated at run-time and downloaded to browsers. There it would automatically create and display the table of contents by using either a downloaded Java applet or a standard set of JavaHelp class libraries.

The ability to capture and transmit semantic and structural data made possible by XML greatly expands the range of possibilities for client-side manipulation of the way data appear to the user. For example, a manual on line that covers a number of different processors, say the microJava, an ARM chip or an x86 could be made to appear like a manual for microJava only, the ARM only, or the X86 only, just by clicking a "preferences" switch.

Bringing Back The Links

A major benefit of the addition of XML to World Wide Web is that the rich set of linkages that was common to early hypertext programs can now be implemented across the Internet. Currently, the situation is like a city with a network of streets in which traffic is allowed to only run one way and no one is allowed to take shortcuts through any alleys, make right turns or authorized U-turns. When the city is small, navigating around using only one-way streets is awkward, but not impossible. However, as the city gets larger and the number of locations to be visited and the number of intersections grows, traffic could be brought to a standstill.

This is essentially the situation currently on the World Wide Web, where despite its name and all of the publicity that has surrounded HTML, this so-called "hypertext markup language" actually implements just a tiny amount of the functionality that has historically been associated with the concept of hypertext systems. Only the simplest form of linking is supported — unidirectional links to hard-coded locations. This is a far cry from the systems that were built and proven during the 1970s and 1980s. In a true hypertext system of the kind envisioned once XML is implemented, there will be standardized syntax for all of the classic hypertext linking mechanisms. Instead of being stuck in a "global village" of unidirectional links, quite suitable when it was a few thousand Web sites, the Web surfer, with hundreds of thousands of Web sites available to visit, will now have a number of different ways to get from "here" to "there."

For example, in addition to one-way "streets," there would now be bidirectional links. Presently, a surfer who clicks on an embedded URL link to another and wants to go back to the original document and location must go to the top of the browser and hit the BACK or HISTORY icons and then scroll down to the location from which he jumped. With a true bidirectional link, the surfer could go to the new location, check out the document or paragraph within a document and jump back to the precise location from which he came.

Certainly, it is possible to move back and forth between two links in a HTML-based Web page as long as it is the same document. If the link is between two separate documents on the same Web site, it is possible to move back and forth between two specific locations, but only if the Web page designers have added additional code, in C or some other programming language, outside the HTML specification. And it is also possible to move between two Web sites that share the same proprietary syntax in their Web pages.

Under normal circumstances, however, it is not usually possible for a Web surfer to move back and forth between Web page files and URLs bidirectionally with the same link. It is certainly possible to hide the fact that a button might contain code for two links, one in and one out. But that is like requiring that the driver in a city with one-way streets make five or six right turns around the block to get to the same place she could have with a single left turn.

Another common traditional hypertext link is the one-to-many link, which, when combined with bidirectionality would allow a user to bounce back and forth from the main document to associated pages, graphics, or active programs. In true hypertext systems, it is also possible to link one word or phrase with another word or phrase in another document, or with the entire document. It is also possible to create transclusion links, in which the link target document appears to be part of the link source document. With this combination, it would be possible to create sophisticated "virtual documents" that appear to the user to be one document in one location but are actually spread out throughout the Web. The same set of linked pages, in total or separately, could also be part of multiple virtual documents. Organized one way, a set of pages could be an issue of EETimes for a particular date, or some or all of them could be part of another document, say a special report on a particular topic.

Another type of link, the N-ary hyperlink, can be used to create a ring of connections, that could take the surfer from the site or node he entered through a set of predetermined links and back to the same location. Using the N-ary and/or one-to many hyperlinks, a reader could go to the opening page of the EETimes Web site, and select from a set of date buttons for each Monday of the week during a year. The reader could then browse through the entire issue online sequentially, or go to a Table of Contents inserted into the set of linkages, and then jump to specific stories.

What XML would also allow are "beaten path" links. These are similar in concept to bookmarks in a standard Web browser, but much more powerful and flexible. They would enable a surfer to create his own path through a complex set of Web pages, at either one site or many, using mechanisms such as backup stacks, path macros, and hypertext "bread crumbs." This could be useful in several ways. First, serving the same functions as the GO or HISTORY drop down menus on a browser, they would allow a user to backtrack to his path through the World Wide Web. Or they could be used to create a customized "virtual document," with breadcrumbs used to mark a user's path through a complex set of Web pages or sites.

These are just a few of the kinds of hyperlinks that builders of traditional hypertext programs have created over the years. Rather than introducing additional complexity into the network user interface, the many types of linkages possible with the adoption of XML actually makes "driving" on the Internet much simpler.

If a city planner or traffic engineer took the attitude of the designers of the so-called "revolutionary" NUIs discussed at the beginning of the chapter, simplicity would mean moving to a small town where there were fewer choices. Citizens of the "virtual global village" that is the World Wide Web are like those in any large city. Rather than become confused with two-way streets, alleys, cul-de-sacs, and bus lanes, a typical driver uses only what is necessary to accomplish the job at hand. The traffic engineer of any city of substantial size would be out of a job if he took the attitude that current designers of the so-called next generation of network user interfaces do.

The Challenge of Networked Multimedia

As we move into the much more complex world of networked multimedia the problems of searching, navigation, manipulation and display, also get much more complex. On the one hand, there are technologies that are impelling the Internet and World Wide Web toward ever increasing amounts of video and audio data. These include the emergence of MPEG 4 as the defacto standard for transmitting video, audio, 2D and 3D graphics; increasing bandwidth into the home and in the corporate intranet; and, of course, the increasingly multimedia-capable processors for the desktop, Web-enabled settop boxe and network computer. On the other hand, there is a growing wealth of visual and audio information.

But the technologies for searching, browsing, and presenting multimedia video, audio and 3D graphics on the desktop are inadequate. Even with the inadequacy of current techniques, more and more video material, in particular, is being digitized and archived worldwide. But the methods for indexing and viewing such archives are still very primitive. The Internet movie database called IMDb on the Web (http://www.uk.imdb.com/) is indexed by hand with textual information written by humans about the movies, sometimes with a short clip included that has been selected at random. A number of long term efforts are underway to create methods of searching for, previewing and acquiring video and audio that have proximate analogs to the way Web surfers use the various text search engines on the World Wide Web.

One system, called WebSeek is based at Columbia University. It uses a combination of keywords and key images to narrow a search to a specific image or video clip. Similar to text-based Web search engines, it searches the Web constantly for images and video files. In a search, WebSeek first scans its database for the file names that may contain an acronym such as GIF or MPEG that indicates video or still images. It also looks for words in the file name that will help identify the subject of the files. The builders of the system have developed a sophisticated set of algorithms so that, when a likely set of hits have occurred, it then analyzes the images. Based on the prevalence of different colors and where they are, WebSeek then makes selections from among the various images. Each of the likely images is then compressed so that it can be represented as an icon for display of the results. For a video, it extracts key frames from a number of different scenes. After the initial text-based search, the user can select a category from a menu, where WebSeek displays a sampling of icons for the particular category. To narrow the search, users can click on any of the icons that most closely represent the image they are looking for and using previously generated image analyses, look for matches of the image that have similar color or shape attributes.

Other similar efforts are underway at the University of Chicago, the University of California at San Diego, Carnegie-Mellon University, the Massachusetts Institute of Technology and the University of California at Berkeley. At least two companies - IBM and Virage - have developed software they are selling for use on corporate intranets. There are also at least two companies, Excalibur Technologies and Interpix Software, who are supplying software to a number of the Web-based search engine firms.

While such concerns are not important to the casual user of the Web, and the even more casual browser of video libraries, such capabilities are extremely important to major entertainment companies who have big plans for the Internet, via the Web-enabled settop box or digital TV. One possibility is a digital TV magazine, where instead of reading short, and often misleading, text descriptions of upcoming programs, the viewer could look at abstracts selected by an image-based retrieval system. For video-on-demand systems, the early prototypes of which we are seeing in hotel rooms all over, the TV viewer or Web surfer could view selected video abstracts.

Deploying this technology in a widespread way on the Internet and World Wide Web is a challenge. While visual information retrieval systems are becoming more common on the Web, several critical barriers remain. These include the heterogeneity of the visual information and the retrieval systems, especially as far as their underlying metadata definitions are concerned; and the lack of interoperable retrieval methodologies.

The main advantage of HTML - a common tag set that is understood by all browsers - has been the reason that the World Wide Web has proliferated so quickly. But as the Web moves toward a more multimedia-oriented future, that same commonality is also its biggest flaw. New methods of searching and displaying multimedia data, no matter how effective a particular one is, will not proliferate widely, because they use techniques that are beyond the capabilities of HTML. It is the classic "chicken and egg" dilemma: Do you stick with a method for Web page generation and manipulation that is common to all systems, but is ineffective, or do you throw it out? Or do you build a parallel system, requiring that Web users maintain two different sets of tools for generating content, searching for it, and displaying it?

XML overcomes this chicken-and-egg dilemma because it is self-descriptive. Using XML with multimedia data, a document description can be provided in a unique Document Type Definition (DTD) and attached to each file, where it serves as the lexicon and the translator for the multimedia document.

Adding DTDs to the images and video representations provides the intelligence needed to make a number of critical decisions. For example, rather than the browser reading the images as passive arrays of pixels, it would know to view them as active arrays that include procedures for generating, manipulating, and describing their contents. XML-based DTDs will allow the design of video retrieval systems, which, when combined with MPEG 4's object-based representation of images, would allow the creation of both text and video abstracts of the contents of a file.

Riding the seesaw: implications for the future

The relationship between hardware and software in the computer industry is somewhat like riding on a seesaw. As the power of the CPU and the speed and size of the memory increase, this pushes software designers to develop more sophisticated applications to take advantage of the additional performance. These, in turn, push microprocessor designers to develop better and more powerful hardware to run the applications faster.

It should then come as no surprise that this relationship continues in the age of net-centric computing. The difference now is that it is a three-way relationship between the bandwidth available on the Internet, the performance and capabilities of the underlying computer hardware, and the power and sophistication of the net-centric software applications that they run.

For example, before the World Wide Web attracted the notice of the mainstream of the PC and mass consumer market in 1993-1994, multimedia capabilities were a hard sell. Video and audio capabilities were not features for which the average consumer was looking. So, microprocessor makers and system vendors found that they were able to sell the technology only if it did not substantially increase the cost of the system. 3D capabilities were of interest only to the margins of the market — avid video game enthusiasts on the one hand and professional engineers and others who needed 3D drawing capabilities on the other.

The World Wide Web and its potential is changing all that. As bandwidth increases, and standards such as MPEG 4 make networked multimedia a reality, multimedia capabilities are features for which the average consumer will be willing to pay. Similarly, with the need for much easier to use network user interfaces emerging, capabilities such as those implemented by Xerox and InXight to help the average user avoid getting lost in hyperspace may be the "hot app" that hardware and software vendors have been looking for to move 3D into the mainstream.

As Internet bandwidths increase, multimedia capabilities will not be something that companies have to push into the market with expensive advertising and "cut-to-the-bone" prices. It is already having an impact not only on how manufacturers such as Intel design their processors, but also in the increased the number of vendors and manufacturers who are offering alternative architectures.

While Part IV in this book has focused mainly on the impact of networked multimedia capabilities on the architecture of the processors, it is also changing the way engineers think about memory. The are looking beyond the existing DRAM technologies to new ones such as extended data out DRAMs, synchronous DRAMs, and Rambus DRAMs that can provide the performance to match the 300-500 MHz CPUs that are now available. They are also rethinking the interface between the DRAM in main memory and the CPU and looking for ways to eliminate the serious bottlenecks. Not only are new bus architectures, such as Intel's Advanced Graphics Port scheme being proposed and implemented, but engineers are also looking at integrating the DRAM in main memory onto the same chip as the CPU. This significantly reduces system cost and also allows the use of 128-, 256- and 512-bit wide memory-to-CPU interfaces that would not be possible otherwise.

Even more fundamental changes are in the future, especially as the computer becomes less of a desktop word processing and spreadsheet machine and more of a communications, information-retrieval and Web-browsing engine. If the main concern of users is in finding things on the Web faster and with more accuracy, it may be that current DRAM architectures and CPUs will be replaced with ones based on associative techniques that are better at searching and comparing.

Interesting questions all. But the answers will have to be the subject of some other book.

Appendix A

References and Recommended Reading

For the technical meat and insight in this book I must give thanks to the hundreds of engineers and software designers I have interviewed and with whom I have "jawed" on the subject of net-centric computing. In addition, following is a list of references I have used extensively in the preparation of this book and which I can recommend for reading on all aspects of net-centric computing:

1. NCs, Connected PCs, NetPCs, IAs and Set-top Boxes

Connected PCs stretch bandwidth limits, by D. Craig Kennie, page 76, *Electronic Engineering Times,* Nov. 18, 1996.

Cost cutting key to network computers, by Bill Jackson and Paul Cobb, page 96, *EETimes*, Nov. 18, 1996.

Embedded Web Servers in Network Devices, by Ian Agranat, page 30, *Communications Systems Design*, March, 1998.

$500 Internet Browser, by Brad Reed, page 68, *Circuit Cellar*, August, 1996.

Hard Choices: What the PC, NC and NetPC have to offer, by Richard Comerford, page 20, *IEEE Spectrum*, May, 1997.

Inside the NC, by Peter Wayner, page 105, *Byte Magazine*, November, 1996.

Inside the Web PC, by Tom Halfhill, page 44, *Byte Magazine*, March, 1996.

Internet appliance designs require focus, by Soohoo and Steven Cole, page 82, *EETimes*, Oct. 27, 1997.

Internet Connectivity, by Jack Ganssle, page 34, *Embedded Systems Programming*, January, 1998.

Internet smart phone dials new markets, by Robert Simon and Thomas Brenner, *EETimes*, page 101, Nov. 18, 1996.

Net Appliances must be built for growth, by Jay Johnson, page 80, *Electronic Engineering Times*, Nov. 18, 1996.

Net-centric applications, by Bruce Cottman, page 38, *Java Developer's Journal*, Volume 2 Issue 1, 1996.

Network Computers: Cost Control or Mind Control?, by Tom Halfhill, page 66, *Byte Magazine*, April, 1997.

Network Computers: Think Thin, by John Enck, page 128, *Windows NT Magazine*, October, 1996.

Programming an HTTP Web Server with Java, by Joseph DiBella, *Java Developer's Journal*, page 77, Volume 2 Issue 5, 1997.

Putting an embedded system on the Internet, by David Shear, page 37, *EDN Magazine*, Sept. 27, 1997.

Set-tops: For TV or PC? by Peter Varhol, page 18, *Computer Design*, March, 1997.

Set-top box reference designs attack cost concerns, by Peter Varhol, page 77, *Computer Design*, November, 1997.

Set-top box requires a specialized RTOS, by Ken Morse, page 94, *EETimes*, Oct. 27, 1997.

Software Design Issues for Digital Set-top box applications, by Stephen Christian, page 44, *Multimedia Systems Design*, Volume 1 Issue 1, 1998.

STB operating systems gear up for flood of data services, by Tom Williams, page 67, *Computer Design*, February, 1996.

Thin clients are heavyweight performers, by Mark Roberts, page 78, *EETimes*, Nov. 18, 1996.

Thin Is In, by Mark Smith, page 102, *WindowsNT Magazine*, September, 1997.

Time is coming for network computing, by David Weiss, page 106, *EETimes*, Nov. 18, 1996.

To NC or not to NC?, by Steven Vaughan-Nichols, page 29, *netWorker,* March/April, 1997.

Truth about NCs, by Gene Koprowski, page 22, *Client/Server Computing*, March, 1997.

TV can act as a looking glass for the Web, by Kevin Goldstein, page 128, *EETimes*, Nov. 18, 1996.

Web Server Technology for Windows CE, by Greg Corcoran, page 44, *Windows CE Tech Journal*, March, 1998.

2. Computer/ System Architecture

Architecture is key to execution in Java, by George Shaw, page 119, *EETimes*, June 16, 1997.

Challenges to combining general purpose and multimedia processors, by Thomas M. Conte, et. al., page 33, *IEEE Computer*, December, 1997.

Compiling C on a Multiple Stack Architecture, by Claus Assmann and Andreas Huth, page 60, *IEEE Micro*, October, 1996.

Compcon: 1994, 1995, 1996, 1997. (Association of Computing Machinery).

Computer Organization and Design, by David Patterson and John Hennessy (Morgan Kaufmann Publishers, 1994).

High Performance Computing, by David Loshin (AP Professional, 1994).

Hot Chips Symposium: 1996, 1996, 1997. (IEEE Computer Society).

How Multimedia workloads will change processor designs, by Keith Diefendorff and Pradeep Dubey, page 43, *IEEE Computer*, September, 1997.

Inside Sun's Java Chips, by Peter Wayner, page 79, *Byte Magazine*, November, 1996.

Microjava chip set emerges, bit, by bit, by Harlan McGhan, page 96, *EETimes*, Jan. 12, 1998.

Microprocessor Forum Proceedings: 1994., 1995., 1996., 1997. (Ziff Davis Publishing).

Stack based Java (chip). a back to the future step, by David Greve and Mathew Wilding, page 92, *EETimes*, Jan. 12, 1998.

Superscalar Microprocessor Design, by Mike Johnson (Prentice Hall, 1990).

Ultasparc I: A four issue processor supporting multimedia, by Marc Tremblay and J. Michael O'Connor, page 42, *IEEE Micro*, April, 1996.

Windows Hardware Engineering Conference (WinHEC).: 1995, 1996, 1997 (Microsoft Corp).

3. Computer Software/Operating Systems

Advanced Operating Systems: Special Report, page 117, *Byte Magazine*, January, 1994.

ActiveX Demystified, by David Chappell and David Linthicum, page 56, Byte Magazine, September, 1997.

Alternative Programming Languages (Special Issue), *Dr. Dobb's Sourcebook*, Winter, 1994.

C++ vs Java Software development, by Barry Boone, page 39, Java Report, April, 1997.

Choosing between CORBA and DCOM, by Mark Roy and Alan Ewald, page 24, *Object Magazine*, October, 1996.

Client Server Programming with Java and CORBA, by Robert Orfali and Dan Harkey (Wiley and Sons, 1997).

Component Architectures: OLE versus Java, by Allen Holub, page 62, *Dr. Dobbs Sourcebook*, Sept./Oct., 1996.

DCOM versus CORBA, by P.E. Chung, et.al., page 18, *C++ Report*, January, 1998.

Distributed Objects On the Web (Special Report), pages 30 - 46, *Web Apps Magazine*, March/April, 1997.

Distributed Objects Survival Guide, by Robert Orfali, Dan Harkey and Jeri Edwards (Wiley and Sons, 1996).

The Feel of Java, by James Gosling, page 53, *IEEE Computer*, June 1997.

Focus on Servelets: Special Report, by Ajit Segar, *Java Developer's Journal*, Volume 3 Issue I, 1997.

Graphical Applications with TCL and Tk, by Eric Johnson (MIS:Press, Inc. 1996).

Hot Topics In Operating Systems Workshop, 1994, 1995, 1996, 1997 (IEEE Computer Society Press).

Inside WindowsNT, by Helen Custer (Microsoft Press, 1993).

Internet Agents, by Fah-Chun Cheong (New Riders Publishing, 1996).

Internet Programming, by Kris Jamsa and Ken Cope (Jamsa Press, 1995).

Java and DSP Integration for Wireless Multimedia, by Jason Brewer and Marion Lindberry, page 20, *Communications Systems Design*, April, 1998.

Java as a programming language, by Arthur Van Hoff, page 51, IEEE *Internet Computing*, January, 1997.

Java brews better with Perc-olator on, by Kelvin Nilsen and Thomas Cheng, page 70, *EETimes*, Aug. 4, 1997.

Java for C/C++ Programmers, by Michael Daconta (Wiley and Sons, 1996).

Java's limitations in network communications, by Gerald Brose, Klaus Lohn and Andre Spiegel, page 50, *Object Magazine*, Dec. 1997.

Java Multimedia Environments, by Chris Behrens, page 23, *Java Developer's Journal*, Volume 2 Issue 9, 1997.

Java needs fixes to be portable language, by Carl Dichter, page 120, *EETimes*, June 16, 1997.

Java RMI breaks new ground in distributed computing, by Dough Sutherland, page 16, *Internet and Java Advisor*, December, 1966.

Java versus OLE, by Alan Holub, *Dr. Dobb's Sourcebook*, Jan./Feb., 1997.

Java Programming, by Neil Bartlett, Alex Leslie and Steve Simkin (Corolis Group Books, 1996).

Java Remote Method Invocation (Java Q&A), by Cliff Berg, page 101, *Dr. Dobbs Journal*, March, 1997.

Java, RMI and CORBA, by Dave Curtis, page 46, *Java Developer's Journal*, Volume 2, Issue 6, 1997.

New C++ spin targets embedded markets, by John Carbone, page 126, *EETimes*, June 16, 1997.

Objects and the Web, by Kevin Dick, page 34, *Object Magazine*, July, 1996.

Objects on the Web, by Ron Ben-Natan (McGraw Hill, 1997).

Objects Push a Paradigm Shift in Distributed Computing, by Michel Gien, page 125, *Electronic Design*, Oct. 23, 1997.

Operating systems issues for continuous media, by Henning Schulzrinne, page 268, *ACM's Multimedia Systems*, Volume 4 Number 5, 1996.

OS architects target information access, by Mark Moore, page 128, *EETimes*, June 16, 1997.

OS-9 Insights: An Advanced Programmer's Guide, by Peter Dibble (Microware Systems Corp., 1990).

QNX Neutrino Microkernel System Architecture Manual (QNX Systems Software Ltd.

Reinventing the Web: XML and DHTML, by Scott Mace, Udo Flohr, Rick Dobson and Tony Graham, page 58, *Byte Magazine*, March, 1998.

Unauthorized Windows 95, by Andrew Schulman (IDG Books Worldwide, 1994).

Unix versus NT, by Tom Halfhill, page 42, *Byte Magazine*, May 1996.

Web Programming Languages, by Steve Ford, David and Nancy Wells, page 24, *Web Apps*, Jan./Feb., 1997.

Windows NT: The next generation, by Len Feldman (SAMS Publishing, 1993).

Win98 Roadmap, by John Montgomery, page 66, *Byte Magazine*, June 1998.

XML: Door to automated Web Applications, by Rohit Khare and Adam Rifkin, page 78, *Internet Computing*, July/August, 1997.

4. Reliable Computing

ActiveX and Java: Virus Carriers?, by Eve Chen, page 38, *Computer Technology Review*, Spring, 1997.

Computer Immunology, by Stephanie Forrest, Steven Hofmeyr, and Anil Somayaji, page 88, *Communications of the ACM*, October, 1997.

Evolution of Fault Tolerant Computing, edited by A. Avizenis, et.al (Springer-Verlag, 1987).

Fatal Defect: Chasing Killer Computer Bugs, by Ivars Peterson (Random House, 1995).

Fault-Tolerance and Fault-Resilience, by Roger Addelson, *Network VAR*, May, 1996.

Fault-Tolerant Systems (Special Issue). , *IEEE Computer*, July, 1990.

Fighting Computer Viruses, by Jeffrey Kephart, Gregory Sorkin, David Chess and Steve White, page 88, *Scientific American*, November, 1997.

High Assurance Systems: Special Report, pages 67-94, *Communications of the ACM*, January, 1997.

How Microsoft Builds Software, by Michael Cusumano and Richard Sel, page 53, *Transactions of the ACM*, June, 1997.

How Secure?— Special Report, pages 39-60, *Java Report*, February, 1977.

How Software Doesn't Work, by Alan Joch, page 49, *Byte Magazine*, December, 1995.

Japan's Software Factories, by Michael Cusumano (Oxford University Press, 1991).

Java Security: Present and Near Future, by Li Gong, *IEEE Micro*, May/June, 1997.

Java Security, by Gary McGraw and Edward Felten (Wiley and Sons, 1997).

Protected mode boosts assurance in X86, by Paul Rosenfeld, *Electronic Engineering Times*, Dec. 16, 1996.

Safeware: System Safety and Computers, by Nancy Leveson (Addison-Wesley Publishing, 1995).

Security in Computing, by Charles Pfleeger (Prentice Hall PTR, 1997).

Software For Reliable Networks, by Kenneth P. Birman and Robbert Van Renesse, page 64, *Scientific American*, May, 1996.

Software Reliability and the Cleanroom Approach, by Michael Deck, page 218, 1998. *Proceedings of the Annual Reliablity and Maintainability Symposium.*

Writing Solid Code, by Steve Maguire (Microsoft Press, 1993.

5. Networks/Internet/Servers-Routers

Client/Server Survival Guide, by Robert Orfali, Dan Harkey and Jeri Edwards (Van Nostrand Reinhold, 1994).

Computer Networks, by Andrew Tanenbaum (Prentice Hall PTR, 1996).

Future Web servers will need 64-bit OSes, by Rajeev Bharadwaj and Robert Dominguez, page 108, *EETimes*, Nov. 18, 1996.

Hello, NUI, Good, by GUI, by Tom Halfhill, page 60, *Byte Magazine*, July, 1997.

Information Retrieval on the World Wide Web, by Venkat N. Gudivada et.al., page 58, *IEEE Internet Computing*, Spetember/October, 1997.

Internet Developers Forum: 1995., 1996., 1997. (Informant Communications Group).

International World Wide Web Conference: 1995., 1996., 1997. (World Wide Web Consortium.

Internet System Handbook, edited by Daniel Lynch and Marshall Rose (Addison-Wesley Publishing, 1993).

Internet: Bring Order From Chaos (Special Report), pages 49-83, *Scientific American*, March, 1997.

Internet Protocol Next Generation, edited by Scott Bradner and Allison Mankin (Addison-Wesley, 1996).

IP Switching, by Art Wittman, page 46, *Network Computing*, June 15, 1997.

Ipv6: The New Internet Protocol, by Christian Huitema (Prentice Hall PTR, 1996).

Mbone: Interactive Multimedia on the Internet, by Vinay Kumar (New Riders Publishing, 1996).

Making the most of Client/Server Technology, by Ellen Ullman, page 96, *Byte Magazine*, June, 1993.

Multimedia Networking: Special Issue, *ACM's Multimedia Systems*, Volume 4 Number 6 1996.

Network Evolution and Multimedia Communication, by Heintrich Stuttgen, page 42, *IEEE Multimedia*, Fall, 1995.

New Internet Navigator, by Paul Gilister (Wiley and Sons, 1995).

Pick a processor tailored for networking, by Peter Palm, page 114, *EETimes*, Nov. 18, 1996.

Real-time Internet Service, by George Lawton, page 14, *IEEE Computer,* November, 1997.

Routing in the Internet, by Christian Huitema (Prentice Hall PTR, 1995).

Scalability Seen as Key to the Web's Growth, by Robert Horst, page 94, *EETimes*, Nov. 18, 1996.

Server I/O Conference: 1995., 1996., 1997., 1998. (Strategic Research Corp).

Thin-client/Server Computing, by Joel Kanter (Microsoft Press, 1998).

Understanding Networked Multimedia, by Francois Fluckiger (Prentice Hall PTR, 1995).

Web Server Technology, by Nancy Yeager and Robert McGrath (Morgan Kaufmann Publishers, 1996).

6. Hypermedia and Multimedia

Digital Video: An Introduction to MPEG, by Barry Haskell, Atul Puri and Arun Netravali (Chapman and Hall, 1997).

Elements of Hypermedia Design, by Peter Gloor (Birkhauser Boston, 1997).

Enabling technology for distributed multimedia applications, by J.W. Wong, et. al., page 489, *IBM Systems Journal*, Volume 36 Number 4, 1997.

How Intel Built MMX Technology, by Alex Peleg, Sam Wilkie and Uri Weiser, page 24, *Communications of the ACM*, January, 1997.

Hypermedia/Hypertext Handbook, edited by Emily Berk and Joseph Devlin (McGraw Hill, 1991).

Hyper-G and Hyperwave: The Next Generation Web Solution, by Herman Maurer (Addison-Wesley, 1996).

Human Factors in Computing Systems Conference Proceedings:1991.-1995. (Association For Computing Machinery).

Industrial Strength Streaming Video, by George Avgerakis and Beck Waring, page 46, *New Media*, September 22, 1997.

Interactive Media: an Internet reality, by Richard Comerford, page 29, *IEEE Spectrum*, April;, 1996.

International Conference on Multimedia Proceedings: 1995., 1996., 1997. (Association For Computing Machinery Inc).

Java and VRML, by Jon Steiner, page 26, *Java Developer's Journal*, Volume 2 Issue 11, 1997.

MPEG 4 Special Report, *IEEE Transactions on Circuits and Technology for Video Technology,* February, 1997.

Multimedia: Computing, Communications and Applications, by Ralf Steinmetz and Klara Nahrstedt (Prentice Hall PTR, 1995).

Streaming Multimedia, by Craig Preston, page 38, *Web Techniques*, August, 1997.

Toward Support for Hypermedia on the World Wide Web, by Michael Bieber and Fabio Vitali, *IEEE Computer*, January, 1997.

Visual Information Management (Special Report). , *Communications of the ACM*, December, 1997.

VRML: Browsing and Building Cyperspace, by Mark Pesce (New Riders Publishing, 1995).

7. Embedded Hardware and Software Design

Architecture of Pipelined Computers, by Peter Kogge (McGraw Hill, 1985).

Debugging: Techniques and Tools for Software Repair, by Martin Stitt (Wiley and Sons, 1992).

The 80960 Microprocessor Architecture, by Glenford Myers and David Budde (Wiley and Sons, 1990).

Embedded Systems Conference: 1994., 1995., 1996., 1997., 1998. (Miller Freeman Publishing).

High Performance Computer Architectures, by Harold Stone (Addison-Wesley, 1990).

Java Perks up embedded systems, by Richard Quinnell, page 38, *EDN Magazine,* August 1, 1997.

Microprocessor-based Design, by Michael Slater (Mayfield Publishing, 1989).

Programming Embedded Systems, by Jack Ganssle (Harcourt Brace Jovanovitch, 1992).

Real time Java, by Kelvin Nielson, page 26, *Java Developer's Journal*, June, 1996.

RISC Architecture, by Daniel Tabak (Wiley and Sons, 1987).

RISC Architectures, by J.C. Heudin and C. Panetto (Chapman and Hall, 1992).

RISC/CISC Development and Test Support, by Marvin Hobbs (Prentice Hall, 1992).

32-bit Microprocessors, by H.J. Mitchell (McGraw Hill, 1990).

Appendix B

Websites for Further Information

I. NCs, NetPCs, Information Appliances And Settop Boxes

1. http://www.agranat.com/
   Agranat Systems, Inc.
2. http://www.allegrosoft.com/
   Allegro Software Development Corp.
3. http://www.apache.org/
   Apache Servers Project (public domain server)
4. http://www.bandai.co.jp/eng/enghome.htm
   Bandai Corp.
5. http://www.broadcom.com/
   Broadcom Corp.
6. http://www.diba.com/
   Diba, Inc. (now a division of Sun Microsystems Inc.)
7. http://www.divicom.com/
   DiviCom Inc Web Page
8. http://www.dmtf.org/
   DMTF, Desktop Management Task Force
9. http://www.gi.com/
   General Instrument Corp.
10. http://www.hds.com/
    HDS Network Systems Home Page
11. http://www.ima.org/
    IMA, Interactive Multimedia Association

12. http://connectedpc.com/sites/connectedpc/
    Intel's Connected PC Homepage
13. http://www.ptsc.com/PSC1000/
    Patriot's PSC1000 Microprocessor
14. http://www.lsilogic.com/
    LSI Logic Inc.
15. http://www.lucent.com/inferno/
    Lucent Technologies' Inferno
16. http://www.magmainfo.com/
    Magma Information Technologies Inc.
17. http://www.nc.ihost.com/
    NC, Network Computer Inc.
18. http://www.oracle.com/
    Oracle Corp.
19. http://www.quiotix.com/
    Quiotix Corp.
20. http://www.st.com/
    SGS Thomson Microelectronics, Inc.
21. http://www.spyglass.com/
    Spyglass Inc.
22. http://www.sunriver.com/
    SunRiversion Data Systems
23. http://www.transphone.co.uk/
    Transphone Ltd.
24. http://www.uplanet.com/
    Unwired Planet Inc.
25. http://www.webtv.net/pc/wtvnet.html
    WEBTV Corp. (now a division of Microsoft Corp.)
26. http://www.webdevices.com/
    Web Devices Inc. (formerly CNiT)

II. Embedded Hardware And Software

1. http://www.arm.com/
   Advanced RISC Machines
2. http://www.amd.com/
   Advanced Micro Devices Inc.
3. http://www.amc.com/internet.htm
   Applied Microsystems Inc.

4. http://www.chips.ibm.com/products/embedded/index.html
   IBM PowerPC Embedded Controllers Page
5. http://www.isi.com/
   Integrated Systems, Inc. (ISI)
6. http://www.intel.com
   Intel Corp.
7. http://www.inmet.com
   Intermetrics, Inc.
8. http://192.94.39.7/microtec/
   Mentor Graphics Microtec Division
9. http://www.chipanalyst.com
   MicroDesign Resources
10. http://www.microware.com/index.html
    Microware Systems Corp.
11. http://www.mot.com/
    Motorola Corp.
12. http://www.nsc.com/
    National Semiconductor Corp.
13. http://www.osicom.com/navbar.html
    Osicom Technology Inc.
14. http://www.ptsc.com/
    Patriot Scientific Corp.
15. http://www.versalogic.com
    Versalogic Corp. PC/104-plus products
16. http://www.pharlap.com/
    Phar Lap Software Inc.
17. http://www.semiconductors.philips.com/ps
    Philips Semiconductor
18. http://www.qnx.com/
    QNX Software Systems Ltd.
19. http://qedgate.qedinc.com/
    Quantum Effect Design Inc.
20. http://www.radisys.com/home.eet.html
    RadiSys Corp.
21. http://www.st.com
    SGS Thomson Microelectronics Inc..
22. http://www.sun.com
    Sun Microsystems Inc.
23. http://www.ti.com
    Texas Instruments, Inc.

24. http://www.wrs.com
    Wind River Systems Inc.

III. Desktop And Connected Multimedia PCs

1. http://www.apple.com/
   Apple Computer Inc.
2. http://apt.usa.globalnews.com/
   APT Client-Server Home Page
3. http://www.acm.org/
   Association of Computing Machinery
4. http://www.research.att.com/
   AT&T Bell Laboratories Research
5. http://www.compaq.com/
   Compaq Computer Corp.
6. http://www.emc.com/mkt/main/directory.htm
   EMC Corp.
7. http://www.hp.com/
   Hewlett-Packard Co.
8. http://www.hitachi.com
   Hitachi America Ltd.
9. http://www.ibm.com/
   IBM Home Page
10. http://www.intel.com/
    Intel Corp.
11. http://www.mips.com/
    MIPS Technologies Inc.
12. http://www.mpact.com/
    Mpact — Technical Info
13. http://www.sgi.fr/
    Silicon Graphics Inc.
14. http://www.sun.com/
    Sun Microsystems Inc.
15. http://www.tandem.com/
    Tandem Computers Inc.
16. http://www.toshiba.com/
    Toshiba America, Inc.
17. http://www.veritas.com/
    VERITAS Software Inc.
18. http://www.sco.com/

Santa Cruz Operations (SCO)
19. http://www.symbios.com/
Symbios Logic Inc.
20. http://www.zitel.com/
Zitel Corp.

IV. Internet/Web Technology

1. http://www.mcs.com/~lunde/web/aboutwww.html
About WWW and Internet Standards
2. http://www.bellcore.com/WWWCONF/
4th International WWW Conference - LIVE
3. http://www.citrix.com/
Citrix: Windows without Walls
4. http://inferno.bell-labs.com/inferno/index.html
Inferno Operating System
5. http://www.intel.com/iaweb/director/
Intel Internet Resource Directory
6. http://chatsubo.javasoft.com/current/
Java Distributed Systems
7. http://java.sun.com/
Java: Programming for the Internet
8. http://www.javasoft.com/
JavaSoft Home Page
9. http://www.att.com/lucent/
Lucent Technologies - Bell Labs Innovations
10. http://www.javasoft.com/products/personaljava/
PersonalJava Specifications
11. http://www.qdeck.com/
Quarterdeck Corp.
12. http://www.ietf.cnri.reston.va.us/proceedings/
Seattle 1994 IETF Proceedings
13. http://www6conf.slac.stanford.edu/
Sixth International World Wide Web Conf.
14. http://www.phoenix.ca/sie/ibj-home.html
Internet Business Journal
15. http://www.randomhouse.com/tid/
Internet Directory 2.0 Homepage
16. http://www.gnn.com/wic/
The Whole Internet Catalog

17. http://www.ul.ie/
    University of Limerick WWW Server Homepage
18. http://www.ai.sri.com/~heller/www-vrml/
    Distributed Computing and VRML

V. Server /Router/Switch Technology

1. http://www.3com.com/   3Com Corp.
2. http://pscc.dfw.ibm.com/aixtra/aixtra8/tech/64-2.htm
   64 bit Architectures: AIX and RS/6000
3. http://www.zenon.com/64bit_e.htm
   64-bit NT takes back seat to NT 4 anglais
4. http://www.as400.ibm.com/level2/C1.htm
   64-bit RISC Processing
5. http://www.cisco.com/univercd/data/doc/cintrnet/prod_cat/
   pc7500.htm
   Cisco 7500 Series
6. http://www.decus.org/S95sess-abs/vs023.html
   A Technical Introduction to 64-Bit Addressing
7. http://www.decus.org/S96sess-abs/vs026.html
   Taking Advantage of 64-bit Virtual Addressing
8. http://www.unix.digital.com/unix/64bit/lp64wht/P001.html
   Digital UNIX - 64 Bit White Paper: Page 1 of 2.
9. http://infopad.eecs.berkeley.edu/CIC/industry_hl.html
   Industry Web Servers
10. http://www.eecis.udel.edu/~saxe/857/857home.html
    Interconnection of Networks - Routers
11. http://www.knosof.co.uk/64bit.html
    Migrating from 32 to 64 bit systems
12. http://www.baynetworks.cz/
    Bay Networks Inc.
13. http://www.next-generation.com/news/120195d.html
    PCs to challenge 64-bit
14. http://qedgate.qedinc.com/
    Quantum Effect Design Inc.
15. http://sunsite.cs.msu.su/sunworldonline/swol-12-1995/swol-12-
    64bit.html
    SunWorld Online - December - News: 64-bit survey results
16. http://www.vivid.newbridge.com/
    Newbridge Inc., Vivid Server Subsidiary

VI. Networked Interactive Multimedia

1. http://www.8x8.com/
   8x8 Corp.
2. http://www.chromatic.com/
   Chromatic Research Inc.
3. http://www.at.infowin.org/
   Multimedia Services Domain
4. http://www.microunity.com/
   MicroUnity Systems Engineering Corp.
5. http://www.mpact.com/
   Mpact Technical Info
6. http://www.stud.ee.ethz.ch/~rggrandi/
   MPEG 4 Versus. other compression standards
7. http://wwwam.hhi.de/mpeg-video/
   MPEG Video Group Homepage
8. http://drogo.cselt.stet.it/mpeg/mpeg_7.htm
   MPEG-7 context and objectives version 3
9. http://www.ccir.ed.ac.uk/mpeg4/
   MPEG4 Forum Home Page
10. http://goblins.yonsei.ac.kr/~lsj/mpeg4.html
    MPEG4 Page
11. http://www.es.com/mpeg4-snhc/
    MPEG4-SNHC Web Home
12. http://www.stud.ee.ethz.ch/~rggrandi/mpeg4.html
    MPEG4 coding of A/V objects
13. http://www.ee.princeton.edu/~katto/m1274.htm
    NEC Proposal for MPEG4/SNHC
14. http://amalia.ist.utl.pt/~fp/artigo55.htm
    MPEG4: Evolution or Revolution?
15. http://www.vdo.net/text/info/
    VDOnet Corp.

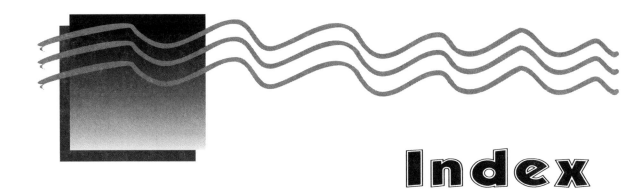

Index